Nanometer Technology Designs
High-Quality Delay Tests

Mohammad Tehranipoor • Nisar Ahmed

Nanometer Technology Designs
High-Quality Delay Tests

Springer

Mohammad Tehranipoor
University of Connecticut
Electrical and Computer Engineering
Storrs, CT
USA

Nisar Ahmed
Texas Instruments
Austin, TX
USA

Library of Congress Control Number: 2007938282

ISBN 978-0-387-76486-3 e-ISBN 978-0-387-75728-5

Printed on acid-free paper.

9 8 7 6 5 4 3 2 1

springer.com

To Maryam, Bahar and Parsa:
Thanks for your inspiration and support
MT

To my advisor, my parents and my wife:
Many thanks for all the encouragement and support
NA

Acknowledgement

We would like to thank Ken Butler, Vinay Jayaram and C.P. Ravikumar of Texas Instruments for their contribution in Chapters 3, 4, 7, and 8. Also, thanks to Mehrdad Nourani from the University of Texas at Dallas for his contribution in Chapters 10 and 12. Thanks to Jim Plusquellic of University of Maryland Baltimore and Abhishek Singh of Nvidia for their contribution in Chapter 1. Also, special thanks to Semiconductor Research Corporation (SRC), William Joyner and David Yeh for supporting our projects. Thanks to Jeremy Lee for his contribution in Chapters 6 and 11. Finally, we would like to thank Alex Greene of Springer for making this book possible.

Mohammad Tehranipoor
Nisar Ahmed

Preface

Modern electronic devices have become an inseparable part of the human life. To meet the market demand, next generation of technology appears with an ever increasing speed and performance driving manufacturing process to its limit. The demand for low power consumption in battery operated devices, higher frequency and higher functional density has introduced new challenges to design and test engineers. Reducing power supply will lower the total power consumption but increases the circuit sensitivity to noise since the voltage threshold is not scaling proportionally. Higher frequency and functional density will increase the power consumption producing more heat in the design and result in larger power supply noise. Integration of several cores for higher performance and throughput leads to longer interconnects thereby increasing coupling capacitance. As a result, performance verification has become one of the most challenging tasks for nanometer technology designs.

Delay test has gained popularity in industry over the past several years as a reliable method for post-silicon performance verification. Industry began using functional patterns first, but as the design size became larger, high cost of generation as such patterns usually are generated manually, and low fault coverage forced functional at-speed test as a supplement to structural test in many semiconductor companies' design-for-test (DFT) flow. Instead, scan-based delay fault test methods gained attention primarily due to the very high fault coverage and their simple procedure to generate patterns. Scan-based path delay fault test and transition delay fault test, together, can provide a high quality test. However, there are new challenges surfacing in nanometer technologies mainly due to the difference in the operating conditions during test mode and normal mode. For instance, power during test mode is 2-3X higher than normal mode resulting in higher power supply noise which in turn impacts circuit performance. Other important issues include implemention of scan-based methods using low-cost testers, improving fault coverage and reducing pattern count. Increasing population of small delay defects also need to be considered as they present quality and reliability issues.

This book contains twelve chapters that address these issues and present novel DFT architectures and pattern generation solutions to the aforementioned problems. In each chapter, we briefly describe the current state of knowledge and shortcomings and then present our method. Chapters 1 and 2 provide introduction to very large scale integration (VLSI) testing and a brief survey on future at-speed test challenges, respectively. The next four chapters, i.e. Chapters 3, 4, 5, and 6, present design-for-test methods to improve the quality of current delay test pattern generation and application methods. Chapters 7 and 8 deal with screening small delay defects, which is an important issue in nanometer technology designs. Chapters 9 and 10 address power supply noise issues during test mode. The last two chapters, i.e. Chapters 11 and 12, deal with pattern generation for crosstalk and signal integrity at the chip and system-chip level. In the following, these chapters will be introduced in more details.

Chapter 1 provides a brief introduction to VLSI testing. It covers various topics such as structural test, functional test, voltage and current based testing methods, fault models, stuck-at fault model, delay fault model, system-on-a-chip (SOC) and their testing, low-cost testers, etc. The nanotechnology issues are addressed in Chapter 2 which in particular focuses on performance verification and delay testing. It discusses issues such as using low-cost testers for at-speed testing, improving quality of tests by increasing fault coverage and reducing pattern count, dealing with process and environmental variations, generation of supply noise tolerant test patterns, dealing with crosstalk issues, and developing timing aware automatic test pattern generators (ATPGs).

Each of the remaining chapters focuses on the individual problems, provides in-depth analysis and practical solutions. Chapter 3 presents an interesting solution to the problem of implementing launch-off-shift (LOS) method using low-cost testers. The method generates a local at-speed scan enable signal using a cell called last transition generator. The cell can be inserted anywhere in a scan chain and can be tested using flush (aka chain test) patterns. The experimental results show that this technique can also reduce the overall scan enable routing area. Traditionally, LOC method offers lower fault coverage and higher pattern count when compared to LOS. A new solution referred to as enhanced LOC is presented in Chapter 4, to improve the quality of test patterns generated using launch-off-capture (LOC). The technique controls the scan chains to operate either in function mode or shift mode. This provides higher controllability and results in higher fault coverage and lower pattern count.

A hybrid method to further increase the transition fault coverage is shown in Chapter 5. The method intelligently selects a small subset of scan chains to be controlled by LOS and the rest are controlled by LOC. This significantly increases the fault coverage (even higher than LOS) and reduces the pattern count. The scan enable design effort will also be significantly reduced since only a small subset of scan chains will be timing closed. Identification and avoidance of functionally untestable faults are discussed in Chapter 6 and a novel method to avoid such faults is presented. The basic idea is to modify the netlist only for testing purpose by inserting some constraints to avoid functionally invalid states that may occur during test pattern generation and random don't-care filling.

Chapter 7 presents a timing-aware ATPG method using current commercial timing un-aware ATPGs to detect small delay defects on long paths. Small delay defects pose both quality and reliability challenges to nanometer technology designs. Chapter 8 illustrates a novel faster-than-at-speed test to detect small delay defects on short and intermediate paths. The method not only considers the positive slack of targeted paths but also the performance degradation caused by IR-drop to determine the higher than normal frequency. It ensures that chip failures during test do not occur due to high IR-drop, i.e. reduces yield loss.

Chapter 9 addresses the issue of high power supply noise during the fast launch-to-capture cycle in delay test. It presents a new power model that can be used during both test pattern generation and validation. The method can be easily adopted in current ATPG and compression flow. It identifies high-supply-noise patterns, excludes them from the pattern set and replaces them with new set of low-supply-noise patterns. Chapter 10 presents a pattern generation method to maximize power supply noise. Such patterns can assist in diagnosis and failure analysis.

Chapter 11 addresses the issue of escape in nanotechnology designs. In this chapter, a method is presented to maximize stress around critical paths by maximizing the crosstalk effects. The method intelligently identifies the location of nearby paths and the transition direction on each net of critical path and nearby paths. Finally, Chapter 12 presents a fault model for signal integrity and proposes a method to test signal integrity loss on SOC interconnects. It modifies the boundary scan architecture so that the test patterns can be generated on chip. Maximum aggressor and multiple transition fault models are used for pattern generation.

Although an attempt has been made to present the context and provide a brief introduction for each topic to an audience with little experience in IC design-for-test, however an experienced reader would have little trouble grasping the abstractions. We sincerely hope you enjoy reading this book.

Mohammad Tehranipoor

July 2007 *Nisar Ahmed*

Contents

1

Introduction

1.1 Introduction to VLSI Testing

At every stage of an integrated circuit (IC) life cycle the goal of a vendor is to ensure that the product meets its specifications and performs as predicted. This is achieved through product verification at each stage. ICs go through two main verification processes: 1) design verification and 2) manufacturing test. The goal of manufacturing test is to verify that the ICs were manufactured correctly, assuming that the design was correct. Due to the complex mechanical and chemical steps involved, a manufacturing process is inaccurate and induces imperfections that can cause a perfect design to violate its specifications. In a broad sense, the goal of testing is to identify parts (ICs) containing manufacturing defects or imperfections that can cause them to fail or violate specifications before the predicted life span. These include, (a) parts that fail or violate specifications due to defects (random or systematic) during production test, and (b) parts that pass or escape the production test but may violate specifications during their operational life. These are referred to as reliability failures. Devices that fail during the early phase of their operational life are called as infant mortalities and mostly can be identified during burn-in and stress tests.

Testing is done using testers also called as automatic test equipment (ATE). ATEs apply test patterns to a chip and collect the responses to each pattern. The measured result (in case of logic testing) at the output pins is then compared with an expected result obtained from simulation (called fault-free response). Digital ICs are subjected to both AC and DC parametric tests, in addition to logic testing. In these tests the AC or DC signals/responses of the IC are measured at the power (e.g. IDDQ and iDDT methods) or output pins (e.g. delay) and processed using a set of digital signal processing (DSP) operations to obtain parameters that are compared against certain threshold to determine the pass/fail status of an IC. Testers can measure analog parameters, such as input and output (or power) pin voltages (or currents), or propagation delay times. ATE testing is done at both wafer and package levels.

The fundamental purpose of VLSI testing is detection of defects. Defect detection does not correspond to finding the defect (type/cause) and its location within the IC. Identifying the physical location of defect is called as fault/defect diagnosis and the procedure to determine its failure mechanism is referred to as failure analysis (FA) or failure mode analysis (FMA). Commonly, the diagnosis procedures use logic based techniques to produce a prioritized list of nets in which a fault (generally stuck faults) may exist. While, other techniques, based on AC and DC current/voltage signal measurements, attempt to trace the possible physical location of the defect/fault in the IC. Failure analysis involves use of optical, mechanical, and chemical procedures to determine the potential cause and mechanism of the defect.

When the first roadmap for semiconductors was published, the common wisdom was that a new process technology would be put into production every three years. Due to tremendous advancements in fabrication techniques and the involved materials that the pace has accelerated and the semiconductor industry is driving new process technology every two years [1] [2]. This has resulted in severe pressure on the manufacturing sector for research and continuous development of test techniques to deal with new defects in the fabrication process. The manufacturing process eventually dictates the occurrence of a particular type of fault over the other. Logically speaking, interlayer issues create more bridge faults while intralayer issues create more opens. Therefore, depending on the weak spots in the manufacturing process (copper, low K dielectric, and resist defects), one or the other type is likely to dominate. This makes the development of test strategies a continuous challenge to semiconductor and electronic design automation (EDA) industry.

Test engineers need to introduce new design-for-test (DFT) techniques or improve existing DFT techniques to keep up with new process technology. The major challenges that the test community faces are: (a) dealing with nanotechnology challenges, (b) reduction of test time with designs getting more complex, (c) dealing with environmental and process variations, (d) testing faster circuits with slower and cheap testers, and (e) developing fault models that accurately reflect permanent (due to fabrication processes) as well as transient faults. Since 2000, many researchers in academia and industry began investigating these challenges and suggest novel methods to help reduce the production test cost.

Consider a simplified design cycle. The designer receives the circuit specifications and uses CAD tools to create and verify a new cell design and/or composes and verifies the system. When the design creation itself has been completed, the test program generation begins. The generated test patterns are then applied to test the device under test (DUT) at the wafer level (or pre-test) and/or when the device has been packaged. The earliest feedback of defects can theoretically be provided after test failure analysis. Critical feedback may also be provided by the customer who will identify shipped failures which is likely to be a serious issue in terms of cost and reputation of the supplier. If the product life is longer than one technology generation or requires several production batches spread over many years, improvements in the test program itself may solve the shortcomings and increase the test coverage. In the worst case, the cell or system may have to be partially modified. In general, the feedback is provided far too late, and test generation is not synchronized with design creation.

1.1.1 Defects

A defect can be defined as an unintended physical difference between the manufactured (fabricated) and design-intended implementation of a device. Different fabrication processes and process technologies (feature sizes) have different classes of defects. Defects can be induced at any step during manufacturing. Some of the common defects associated with silicon CMOS devices are: gate-oxide shorts, salicide opens, metal trace opens, metal trace bridges, open and plugged vias and shorts to Power (VDD) and Ground (VSS).

Defects can cause two types of failure in a device - catastrophic and parametric. Catastrophic failure occurs when the induced defect causes impaired functionality also called as functional or logic failures. These failures are generally caused by hard/complete/perfect opens or bridges. There are also defects that do not cause functional/logic failures (or catastrophic failures) and are called parametric failures. These failures are caused by defects such as weak/incomplete/resistive bridges and resistive opens (called parametric defects), and/or unfortunate process variations, that cause the device to fall out of its parameter specifications. For example, a circuit with weak/resistive bridge may perform correct logic operation, but exhibits high current leakage and/or degraded device performance.

Ideally, the violation of parameter specifications, caused by process variations and/or parametric/weak defects, should allow unambiguous detection of parametric failures. However, the adverse effects of technology scaling can make the detection of such defects extremely challenging. The increased background leakage due to constant field scaling can washout the effects of defect-elevated current. Also, delay or performance variations caused by process variations make it difficult to separate slow defect-free devices due to process variations from defective devices due to weak defects. Distinguishing defective devices from defect-free population, in midst of these adverse effects, is the biggest test challenge and can result in test escape of defective device or false detection of a defect-free device, hence yield loss. Some defective devices with weak defects may pass functional and parametric tests during production test due to aliasing effects of process variations. However, they are very likely to fail during their normal operation, much before their predicted operational life-span. Such failures are called reliability failures or infant mortality failures. On the other hand, some defect-free devices may also be falsely detected as defective. The former causes yield loss, while the latter causes field returns, both resulting in loss of revenue.

1.1.2 Fault Models

The sheer number of possible defects in a device can be unmanageable and exact behavior of the defect in the device can be extremely complex. These factors not only make defect analysis challenging, but also preclude generation of logic test patterns that can be used to detect the defects. In order to make the number and complexity of defects tractable and alleviate the test pattern generation complexity, one needs to model the actual defects that may occur in a chip with a fault model at higher levels of abstraction. Thus, a fault model is the representation of a defect behavior at a higher level of design abstraction. This process of fault modeling considerably reduces the burden of testing because it obviates the need for deriving tests for each possible defect. This is made possible by the fact that many physical defects may (theoretically) map to a single fault at the higher level. This in general also makes the fault model more independent of the technology. Fault models also enables the evaluation of test-coverage.

The goal of fault modeling is to model a high percentage of the physical defects that can occur in the device at the highest possible level of abstraction. The high level of abstraction reduces the number of individual defects that must be considered and lowers the complexity of the device description used in generating the test. The result is that test generation can occur earlier in the design cycle in less time with less expensive computing resources. Modeling of faults is closely related to the design modeling. Similar to design process, higher the level of model representation, (i.e. the more abstract) less technology-dependent it becomes. The levels of abstraction at which the design modeling is traditionally done are: behavioral, functional or RTL-level, structural or logic or gate-level, switch-level, and geometric or layout-level.

Fault Models at Different Abstraction Levels

The advantage of using a fault-model at the lowest level of abstraction is that it closely corresponds to the actual physical defects, and it is thus more accurate. However, the number of defects that may have to be modeled at such an abstract level would be overwhelming. For example, a chip consisting of 50 million transistors could have more than 500 million possible types of defects. Therefore, to reduce the number of fault types and hence the testing burden, one can go up in the design hierarchy, and develop fault models which are perhaps less accurate, but more practical. In fact, a good strategy may be to first derive tests for fault models at higher levels, and then determine the percentage of faults covered at the lower levels by these tests. Typically, the number of faults becomes smaller as we go up in the level of abstraction and the time taken for test generation becomes shorter. Also, since a fault model at higher level models the behavior of many faults at a lower level, such a test set should cover a high percentage of lower-level faults.

Although these advantages of fault modeling are attractive, in todays technology they seem far fetched. The rapidly decreasing feature size, and the need to accommodate increasingly large number of devices on the same silicon size demands increase in the number of metal layers required for I/O routing and power distribution. The increase in the device static leakage current not only increases the chips static power consumption but also makes it difficult to test. The available alternatives for limiting leakage current entails use of unconventional semiconductor materials and complex processing steps for device implementation. These not only increase the probability of occurrence of defects, but also gives rise to complex failure mechanisms and failure modes that are extremely difficult to model. The soaring power consumption accompanied by thermal dissipation, further aggravates the problem by increasing the possibility of reliability failures. The direct implication of these problems is on fault modeling. Most of the defects are not easy to model and a large number of them cannot be abstracted at logic level or higher. This renders the tests generated at higher levels (with conventional high level fault models), incapable of covering all physical defects. Such defects must be modeled and targeted at the lower level of design hierarchy, where the defect behavior can be captured more effectively and accurately.

In summary, fault modeling offers three fundamental advantages, (a) they make analysis of complex and innumerable defects and failure modes tractable, (b) they allow automation of test pattern generation, and (c) they enable test pattern grading in terms of fault coverage metric. Fault models have been developed at each level of abstraction as elaborated below.

- **Behavioral Fault Models:** These fault models are associated with high-level hardware descriptions of digital systems. They are related to failure modes of behavioral constructs in hardware description languages as System C, Verilog, VHDL etc. The importance of these models comes from the increasing desire among designers to start the product modeling from a behavioral level.

- **Structural/Gate/Logic Level Fault Models:** At this level the design model is represented in terms of netlist of gates. The single stuck-at fault (SSF) model developed at this level is the most popular fault model in digital testing. Other fault models at this level are delay faults and bridging faults. These fault models are further detailed in later sections. Generally SSF model is classified as DC (timing-independent) and delay fault models, such as gate and path delay, are classified as AC (timing-dependent) fault models.

- **Transistor/Switch/Component Level Fault Models:** Transistor level fault modeling deals with faults in transistors at a transistor-level description of a circuit. The fault model has mostly been used with MOS technologies, specifically CMOS technology, thus making these faults technology-dependent. The two main members in this category are the stuck-open and stuck-on fault models.

• **Geometric Level Fault Models:** Geometric fault models are derived directly from the layout of the circuit. Such models exploit the knowledge of line-widths, inter-line and inter-component distances, and device geometries to determine what defects are most likely to occur. Mostly, opens and shorts are considered at this level. With shrinking geometries, the percentage of bridging defects can be expected to increase even more. Bridging faults can be modeled at various levels of abstraction. However, at geometric level, such a fault model is the most accurate representation of a bridging defect. Fault modeling at this level requires detailed device level simulators such as SPICE. For non CMOS technologies, bridging fault between two lines is assumed to result in an AND or OR function being realized on the two lines. These kind of faults are referred to as wired-AND and wired-OR, respectively. In case of CMOS technology, whether the shorting of two nodes creates a logic 0 or 1 depends on the relative impedances of the PMOS and NMOS network of the gates driving the shorted nodes. These impedances vary as a function of input vector applied to the driving gates. As a result the voltage at the shorted nodes may be 0 or 1 or an intermediate value. Such faults cannot be modeled using logic based models. They must be modeled at layout level and the analog circuit parameters of the defect must be captured in the fault model. These faults, can then be detected using parametric techniques such as those based on IDDQ or iDDT measurement.

1.1.3 Types of Defects

Failures of electronic devices, in general, can be catastrophic or noncatastrophic. Catastrophic failures render the device totally nonfunctional, while noncatastrophic failures result in an electrically operating device that shows parametric degradation and limited performance. The catastrophic failures are also referred to as functional failures and are caused by two types of defects namely, bridging defects and open defects. The noncatastrophic failures are referred to as parametric failures and are caused by either weak bridging or open defects (called as parametric defects), or variations in environmental or circuit parameters in defect-free devices, collectively referred to as parametric failure modes. Hawkins et. al. [3] classified the cause of device failures into three general classes based on their electrical behavior, (a) bridging defects, (b) open defects, and (c) parametric failure modes. Within these general classes there can be several defect (or failure mode) classes.

• **Bridging Defects:** A bridge or shorting defect is an unintended electrical connection between two or more circuit nodes. Bridges can induce abnormal electrical behaviors that depend on certain circuit parameters and the resulting circuit topology. The behavior of a bridge defect depends on several variables such as:

1. Transfer characteristics of bridging defect e.g. ohmic or nonlinear.
2. Hierarchical location of defect e.g. between I/O nodes of separate logic gates (inter-gate), between transistor nodes of the same logic gate (intra-gate), between power and ground rails.
3. Resulting circuit topology caused by the defect e.g. combinational or sequential circuit.
4. Type of interconnect material e.g. metal, polysilicon, diffusion region etc.
5. Critical resistance of the bridging defect as a function of transistor drive strength and W/L ratios.

• **Open Defects:** Breaks are a common type of defects that occur during IC manufacturing process. Breaks in a digital CMOS circuit fall into different categories depending on their location. A break can occur inside a CMOS cell affecting transistor drain and source connections, disconnect a single transistor gate from its driver, or disconnect a set of logic-gate inputs from their drivers; thus causing these inputs to electrically float. In order for a break to disconnect a set of logic-gate inputs from their drivers, the break must occur in the interconnect wiring. In today's CMOS ICs with several metal layers, interconnect wiring is probably the most likely place for a break to occur. Also, vias are especially susceptible to breaks and the number of vias exceeds the number of transistors in large designs [17].

Critical Resistance

Critical resistance is an important parameter associated with bridge defects, that relates the defect resistance to the electrical properties of the surrounding circuitry and its induced behavior. The critical resistance is defined as the value of defect resistance above which the circuit functions correctly. In other words, the defect resistance below which the circuit fails to function corresponds to its critical resistance. The critical resistance is a function of the contending transistor current drive strengths and therefore varies with the circuit design, logic input levels to contending logic gates, and process variations. Hawkins et. al. [3] reported that the critical resistance decreases as the transistor size shrinks.

Based on their location in a circuit, bridging defects are classified under (a) signal-node to power/ground short, (b) signal-node to signal-node short (inter-gate), and (c) intra-gate short.

1.2 Types of Testing

The goal of manufacturing test is to detect any defect that occurs during the manufacturing process and a variety of test methods can be found in the literature. There is no optimal (low defect level) test strategy that can be quantified and acceptable to both customer and supplier. Most companies follow different test philosophies to best suite their designs and achieve their test goals. Testing can be categorized in two different ways: 1) Classification based on paradigm of testing, and 2) Classification based on nature of measurement.

1.3 Classification Based of Paradigm of Testing

Under this classification, the testing techniques are listed in terms of their fundamental philosophies. The two most prevalent testing philosophies include functional and structural. The disciples of these philosophies claim to achieve the above mentioned goal of testing in two, entirely different ways. Nevertheless, a typical test setup involves a mix of both of these philosophies. Functional testing verifies the correct fabrication of the IC by assuming that the defect occurred during the manufacturing process would cause the IC to fail to meet its functional specifications. While, structural testing classifies physical defects into faults based on their effect on the modeled features of the IC at various abstraction levels. If these erroneous effects are observed in the IC, during testing, it is classified as defective.

Functional Testing

The goal of design verification is to verify the correctness of design. Functional testing is just a physical extension to design verification. In functional testing, the test vectors generated during/for design verification are applied to and the response is captured from the DUT by the functional tester. The tester (or ATE) interfaces to the DUT through probe-card or interface-board (DUT-board or load-board). For a microprocessor, a functional test may involve booting the operating system. The most common way to run functional patterns is to first load them in an on-chip RAM and direct the CPU to jump to that address and start executing code. While CPU is running, the tester will wait in a loop observing output pins, and detect fails during code execution or on pass/fail indications at the end. The interface used to deliver code to an on-chip RAM, i.e. the front side bus (FSB), needs to operate at the operational speed. For example, the FSB speed in case of Pentium 4 processor is 400MHz (100MHz quad pumped i.e. data is transferred twice in each clock cycle, on the rising and falling edge, and also two words of data are transferred at a time giving a multiplier of 4). The higher the interface bandwidth and the frequency, the smaller is the overall test time. Before using on-chip RAMs to store code, they should be thoroughly tested using appropriate memory test techniques.

Functional testing is based on the fundamental assumption that if a defect occurred in the DUT, during the manufacturing process, it would manifest as functional failure. Conversely, if the DUT passes all the functional tests then it can be assumed that the DUT is free of manufacturing defects. However, this fundamental assumption is not entirely true, as, not all defects manifest themselves as functional failures. Another established belief about functional testing is that, when the tests applied to a DUT invoke functions that it is designed for, then in doing so, a significant portion of DUTs logic gets exercised. Moreover, functional tests exercise the logic inside the DUT, in the way similar to its normal mode of operation thus, providing them significant defect detection capabilities. Since, maximum clock speed or $Fmax$ is the most common functional specification in practical designs, the functional test patterns are generally applied at-speed. This further boosts the defect detection capabilities of functional testing given that, a larger number of defects manifest as increased delay in the logic transition, discussed later in this chapter.

It must be emphasized that functional tests are not written with the purpose of catching any defects that occur during manufacturing. The test engineer writes the tests with a particular functional scenario in mind and not a physical defect. Therefore, functional tests do not target manufacturing defects. However, during the process of verifying the design functionality at the physical level, a significant number of defects are detected. As a result, functional test patterns are believed to possess high collateral coverage.

Structural Testing

Unlike functional testing, structural testing takes a more focused approach towards detecting manufacturing defects by considering, how a defect manifests at the logic level. Structural testing does not care about the functionality of the DUT, instead it targets each node in the DUT (logic level i.e. gate netlist) and tests them for presence of a fault (or faults e.g. stuck-at 0 and stuck-at 1 at each node). Since, fault represents a possible defect manifestation at logic level, structural testing serves as a explicit means of detecting manufacturing defects, in comparison to functional testing. The term structural is believed to indicate the fact that it tests the structure of the DUT for presence of faults. The difference between functional and structural testing arise from the purpose of test vector generation. In case of functional testing test patterns are generated to cause various operations or sequence of operations in the modeled design (listed in the product specification). While, in case of structural testing, the test generation program does not care whether the test vectors cause an operation or sequence of operations that falls under its specification-behavior. The test vector is generated to simply excite a targeted fault (at a specific node) and propagate its induced error to specific DUT output or outputs.

Fault models became the vehicle of the 1970-1980 era for automatic generation of test patterns and evaluation of fault coverage. Single stuck-at-fault (SSF) coverage evolved as the de facto test metric in 1980s. SSF model is the earliest fault model and still the most common. In the SSF model, a single node, somewhere in the circuit, is assumed to have taken a fixed logic value (and is thus stuck-at either 0 or 1) independent of anything else. Thus, intuitively, a stuck-at fault represents defects that cause a direct connection between a node and power (for stuck-at 1) or ground (for stuck-at 0). There are many different SSF models. However, the most commonly used one requires two conditions:

1. The circuit is modeled as interconnected elementary gates - AND, OR, NAND, NOR, INVERTER (no EXOR or EXNOR gates)

2. Two single stuck faults - a stuck-at-0 and a stuck-at-1 fault corresponding to each gate input and each gate output

The Boolean form of SSF model enabled straightforward (albeit oversimplistic) test generation and fault grading with limited knowledge of circuit to be tested. Only the logic level netlist was needed for SSF test pattern generation. This property was significant in driving proliferation of SSF, beginning with bipolar ICs and extending to CMOS technologies. The main advantage of structural testing based on SSF model is that the test patterns generated for SSF model cause many patterns to be applied to each individual gate and the resulting gate output to be propagated to the primary circuit output. This is good because it is believed that defects are localized and thus only one gate should be faulty. Thus, unlike functional testing where, only few paths in the DUT are tested, structural testing, based on SSF model, allows a uniform testing of the entire DUTs gate netlist by applying near gate-exhaustive tests to each gate (3 patterns out of 4 for a two-input gate). The simple Boolean representation of SSF model makes it the basis for almost every ATPG algorithm that targets other faults such as pseudo-stuck-at, bridging, delay, stuck-open etc.. Every ATPG tool uses SSF fault model in one way or another.

The reasons for popularity of stuck-at fault model are as follows:

1. **Simplicity:** There are exactly two faults for every circuit signal node. The SSF model is usually applied at the gate level and each logic gate has two faults for each of its inputs and two for its output. For example, a n-input gate requires 2(n+1) SSFSSF faults, considering stuck-at-0 and stuck-at-1 at each input and output node.

2. **Availability:** Most commercially-available test software (ATPG, fault simulation, etc.) supports only the SSF fault model.

3. **Logical behavior:** Each fault is confined to a single location in the circuit and its logic nature readily allows its effect to be modeled as a Boolean equation. As a result, equivalence relationships can be developed, for example a stuck-at 0 fault on the output of an AND gate is equivalent to a stuck-at 0 fault on any of its inputs. This greatly reduces the size of fault list and provides test patterns that can detect multiple faults, thus decreasing the number of test vectors.

4. **Tractability:** The number of faults is directly proportional to the size of the circuit, so very large circuits can be analyzed. Stuck-at fault based test generation and simulation can be applied to designs containing millions of logic gates in a single pass, provided that suitable design-for-test circuitry is present (e.g. full or near-full scan).

5. **Measurability:** Because the number of faults can be counted and fault behavior is so precise (logic 0 versus 1 or vice versa), it is possible to precisely determine whether or not a given set of circuit inputs detects a fault. By collecting these measurements, it is possible to quantify the total fault coverage of a test set.

1.4 Classification Based on Measurement Parameters

Digital IC testing mainly uses voltage-based and current-based measurements. Voltage-based testing measures the potential at IC output pins. It examines Boolean correctness by applying logic voltages to the input pins and measuring the voltage levels at the output pins of the IC. The expected high and low logic output voltages are computed from expensive logic simulations and stored in computer memory for comparison with measured values during test. Voltage-based testing, in practice, is often done at slow application rates, using fault models, or at high clock rates that evaluate subsets of IC functionality over a range of temperatures and power supply voltages (VDD).

Current-based testing analyzes the rich information content of the IC power supply current in either the DC quiescent logic state (IDDQ) or in the transient current (iDDT) pulse. The quiescent current test, called the IDDQ test, measures the power supply current when all logic states are settled at their stable, steady-state values. The IDDQ test is by far the most sensitive of the CMOS digital IC tests [15]-[18]. The transient current test takes several forms, depending upon when in the transient time period the current is measured, or may involve statistical techniques on the whole waveform utilizing its charge (time-domain), magnitude or phase spectrum (frequency-domain) information. Various forms of voltage- and current-based techniques are described next.

1.4.1 Voltage-Based Testing

The major voltage based test techniques are (1) *functional testing*, (2) *stuck-at fault testing*, and (3) *delay fault testing*. These three sometimes overlap, but understanding their individual strengths and weaknesses is necessary for defect-based test practice.

Functional Testing

As described earlier, functional testing literally means to test the IC for its intended use by the customer. The main problem with functional testing is that it is not exhaustive (except for small designs, where exhaustive testing is feasible). Even relatively simple ICs are numerically intractable to the task of replicating customer function. Microprocessors, microcontrollers, and DSP (digital signal processing) ICs are even worse. In spite of this profound weakness, functional testing plays a pragmatic role in testing. When large number of functional test vectors are cycled through an IC at fast clock rates (at-speed), the defect detection is significant enough to warrant its use even though we cannot predict what defects are actually detected. Also, functional testing is the method of choice for testing at high clock rates to separate the product into various speed performance bins (speed binning). These test patterns are usually derived from design verification simulations. The main drawbacks associated with functional testing are the very high time to generate patterns and low fault coverage.

Single Stuck-at-Fault (SSF) Testing

Test engineers in early days of digital IC production quickly realized that the time to generate test vectors became prohibitive as die complexity rose. A wonderfully simple fault model is the SSF model whose hypothesis is that the IC failed its Boolean logic exercise due to one of its node is stuck to 0 or 1. The SSF test does not concern itself with bridge defects and their critical resistance, or even open defects. It just tests for Boolean upset at a signal node. SSF behavior arises when input or output node of a logic gate is clamped at zero Ohms to the strong power supply (VDD) or ground (GND). Another SSF behavior arises when an open defect causes a floating gate that takes one of the rail voltages because of its strong coupling. If floated to VDD, then the nMOS transistor would be permanently on, and the pMOS transistor should be permanently off, causing an output stuck-at-0.

Numerous ATPG algorithms have been developed to generate test vectors using SSF model. Most of these algorithms assume that the circuit is combinational (with or without DFT). One of the earliest ATPG algorithm for SSF model is D-algorithm [4] [5]. It generates a primitive D-cube for each logic gate and uses four major steps, D-propagation, line justification, consistency operation, and node implication. However, the D-algorithm has been shown to take exponential time to generate tests in the worst case. Other algorithms that use testability measures to come up with heuristics have been developed that improve upon the D-algorithm. Some of the frequently used ones are PODEM [6], FAN [7] and SOCRATES [8]. Once the SSF test generation is completed, the percentage of detected SSFs can be calculated. Faults models provide a metric for evaluating test effectiveness in terms of fault coverage in contrast to functional testing.

Speed Testing

Speed testing refers to any voltage-based test that examines the IC at its user-specified clock frequency or clock period. This test is also used to classify and separate parts depending on their operating frequency. This procedure is called speed binning. SSF testing is done at slower clock rates, but ICs must be tested at their rated speeds and needs other forms of testing.

• **Fmax Test:** The Fmax test applies several hundreds to thousands of functional test vectors to the IC at a certain clock period. If the part passes, then the clock period is shortened until a failure occurs. Fmax is equal to the reciprocal of the minimum period of functionality. It measures how fast an individual IC can be clocked. Typically, the Fmax test is done on parts that passed all other tests, so that Fmax is often considered a speed binning procedure on good parts rather than a test. However, the Fmax measurement has significant role in testing resistive vias and contacts. Fmax test is expensive in terms of tester time since it replicates the test pattern set at multiple frequencies until a failure occurs.

• **Delay Fault Test:** Delay fault testing seeks to quantify speed test coverage of the IC. Delay fault tests evaluate individual timing paths or the signal nodes of each logic gate, testing for their rise- and fall-time capability. The delay fault test uses a two-vector sequence. The first vector sets the initial conditions of the test, and the second vector activates a signal that typically tests for transition time of a gate or a path propagation delay. The resulting signal at the designated logic gate must be carefully steered through a sensitive path in the logic between that gate and a primary output where the tester makes a pass-fail timing measurement. The number of delay paths in most ICs is usually numerically intractable and therefore targeting gate node transitions is more practical. However, targeting only gate delays may result in a bad delay test for many parts since gate level test usually target shorter timing paths, leaving long ones untested. The delay fault test presently cannot resolve timing defects other than at a gross level. Die-to-die statistical variation and inability to test propagation delay in different signal paths forces delay fault testing to gross test limits. Typically, the period of a test is set at the nominal clock period of the product plus a guard band for tester noise and product variances, and adjustment for yield considerations. This degrades the ability to detect finer-resolution defects that may cause customer failures. Delay fault test methods and their challenges in nanotechnology era will be discussed in detail later in this chapter and also in Chapter 2.

1.4.2 Current-Based Testing

Data taken during late 1980s and early 1990s showed that many CMOS IC defects caused intermediate voltage at internal nodes (i.e. voltage values out of the established ranges of logic 0 and 1). This caused circuit behaviors that could not be described using voltage-based approaches since internal node intermediate voltages are restored by successive gates, and non-intermediate voltages are generated at the circuit outputs. These defect-induced behaviors were better described by circuit parameters other than the logic response. Current-based testing techniques were shown to be effective parametric test methods for CMOS ICs. Technology scaling pushed current-based testing methods from single-threshold IDDQ to more elaborate quiescent current techniques such as current-signatures, delta-IDDQ, current-ratios, nearest-neighbor statistics.

IDDQ Test

The use of quiescent power supply current as a means of testing CMOS ICs started in early 1970s and 1980s [9] [10], and later became known as IDDQ test [11]. It was observed that since CMOS circuits had virtually zero supply current in the quiescent state. Therefore, an elevation in this current would indicate the presence of a defect, design error, or variation in processing parameters. IDDQ is measured when the circuits reaches steady state after the application of a test vector and all transient currents have settled.

Transient Current (iDDT) Testing

As discussed above, some defects such as, certain opens, do not induce intermediate voltages into the circuit, and therefore cannot be detected with the IDDQ test method. In addition, IDDQ testing has several limitations when applied to deep-submicron ICs. To solve some of these limitations, a number of techniques based on analyzing the transient portions of the supply current have been studied [12]-[14].

The transient portion of the supply current tracks the charge transfer that takes place during the switching activity in the circuit. Existence of current-flow paths between power, ground and gate output nodes, through the nMOS and pMOS transistors during the logic transition, enables the condition of internal circuitry to be reflected in the amplitude and shape characteristics of the transient signals. In addition, the start and end of the charge transfer phenomena relates to the propagation delay of the gate. Thus, the width of the transient current pulse captures the delay characteristics of the circuit, which can be used to detect the presence of defects. In summary, transient current signatures incorporates information on not only the current transfer characteristics, but also the propagation delay of the switching circuit. These properties of transient current signatures suggests that iDDT testing can be used to exploit the advantages of both current and delay based parametric testing.

1.5 System-on-Chip (SoC)

SoC is defined as a complex IC that integrates the major functional elements of a complete end-product into a single chip or chipset. In general, SoC design incorporates a programmable processor/DSP, on-chip memory, and accelerating function units implemented in hardware. It also interfaces to peripheral devices and/or the real world. SoC designs encompass both hardware and software components. Because SoC's can interface to the real world, they often incorporate analog components, and can, in future, also include opto/microelectronic mechanical system components.

1.5.1 SoC Testing

As electronic systems become more complex, so do the chips that go into these systems. Higher chip complexity results in more complex and expensive manufacturing test techniques. Unfortunately, the test cost of a chip is rising faster than the revenue it generates, i.e., test cost is becoming a larger percentage of total chip cost.

SoC testing refers to the testing of multiple independent cores assembled on a single chip. Typically, a SoC design is implemented by integrating multiple cores also known as intellectual property (IP) and efficient testing is best done block by block. A test access mechanism (TAM) is designed to provide full controllability and observability of each core in the SoC. Today, designers can install a specialized, pre-designed, configurable embedded system to test and debug each block. Using such an embedded system, a designer can specify the test speed, fault coverage, diagnostic options, and test length for testing any random logic block.

Structurally, today's SoCs are not much different from those developed several years ago. The big different is that they are much faster with extremely high density of smaller transistors. This, of course, creates new challenges for SoC testing such as at-speed testing, large test data volume, increased test application time and high power consumption.

1.5.2 SoC Test is Expensive

SoC technology has found its way into every facet of our lives, particularly in the consumer and communications market. The complexity and speed of SoCs both of which continue to rise rapidly has taxed the ability to efficiently and effectively test these chips prior to their insertion into target systems. Among the SoC trends that increase test cost are:

- High pattern volume due to increasing number of logic gates and transistors with relatively small number of device pins hindering ATE access to silicon cores.
- An increasing number of device pins requiring expensive ATEs.
- Higher device speed, both on the chip and through high-speed I/Os.
- More on-chip analog components.
- An increase in resistive and speed-related defects, adding test complexity to the traditional stuck-at fault detection.
- Longer test times to account for more complex system conditions.
- Higher power consumption during test.

The cost of manufacturing test can be defined as:
Cost of Test = (Fixed Costs + Recurring Costs)/(Yield x Utilization x Throughput)
where:

- Fixed Costs = Cost of chip tester and handler depreciated over lifetime plus cost of floor space
- Recurring Costs = Utilities, labor, supervision, maintenance, service contracts, engineering support, consumables, and others
- Yield = Number of good devices
- Utilization = Time the tester is used
- Throughput = Number of devices tested per unit time

The test cost can be reduced by decreasing the cost of the test system, increasing product yield (Note that, the cost of test is really the cost of testing for good chips only), increasing tester utilization (test multiple chips at same time), or testing devices faster using adaptive test techniques. The test equipment cost is controlled by ATE manufacturers but its throughput and utilization is affected by EDA vendor design-for-test (DFT) tools. Device yield is a complex parameter, dependent on the silicon process and specific chip-design details.

1.5.3 SoC Tester: An Example

There are many companies developing low cost ATE's. Agilent [20] is a leader in this area and has developed a series of SoC testers. The 93000 SoC DFT series promises users a one-cent-per-second cost of test. As such, this version of the Agilent Technologies 93000 SoC platform would become the most cost-effective SoC test solution available.

The 93000 DFT series includes a test processor per pin architecture (TPPA), which allows users to set up cores for independent operation and concurrent test. Ultimately, this is expected to reduce test time by 30% to 50%. The TPPA also provides enhanced capabilities for SoC DFT diagnostics, such as selective BIST (built-in self-test) capture, to help diagnose failures detected during the test process. Also, the series includes SmartTest PG CTL Browser. This test-program generation environment directly supports the proposed IEEE P1450.6 Core Test Language (CTL) and other standards. CTL provides a standard interface between EDA and automatic-test-equipment environments, enabling faster turn-on and debug cycles for SoCs reducing time-to-market.

1.6 Design For Testability (DFT)

The product-development cycle for today's complex ICs is constantly shrinking. This trend necessitates a smooth transition from the design to the testing phase of the product life cycle. It also demands a concurrent, rather than serial, design-process mindset. Consider, for example, the increased test cost for million-gate ICs. The cost to test per transistor is almost greater than the cost to manufacture it. One way to address this cost imbalance is to make sure that test is considered a design requirement, especially manufacturing test. This is the basic idea behind the popular DFT approach to IC development.

DFT is the set of rules and methods to be applied during the design implementation, to add physical structures that will enable high quality production test. The physical structure(s) implemented will not add any additional functionality to the defined functional specification but will make it possible to reduce the test time in production and time means money. The production tests are normally performed as off line tests. The testing of a digital system, or a single integrated circuit, is extremely vital to the ultimate goal of achieving high reliability, availability, safety, and maintainability, at as low a cost as possible. DFT is the only method to guarantee that a system functions correctly when first placed into operation. Detecting defects in manufacturing process increases the cost to repair/remove the faulty items, the costs increase by a factor of 5-10 for each step later in the manufacturing process when a fault is discovered.

DFT which envisions testing capabilities in the early part of the design process reduces the testing complexity. DFT approaches cut the cost of test by reducing the overall number of test patterns needed to verify the "goodness" of a chip. The automatic test pattern generator (ATPG) generates test vectors for the ATE. Increase in chip complexity corresponds to rise in the number of test patterns required to ensure an equivalent level of testing. This, of course, means more memory - an expensive upgrade commodity for ATE systems. Eliminating the need for additional ATE memory when testing more complex chips is the aim behind using test data compression or BIST techniques.

Another aspect that complicates the testing of fully integrated system chips is the various types of IP utilized in them. Each IP typically has different native testing requirements. This means that every IP must be tested in a serial fashion that, in turn, results in longer test time and higher manufacturing costs.

Due to shrinkage of feature sizes, in the future system-on-chip (SOC) will contain various types of circuits such as digital, analog, RF and mixed signal in a single chip. The DFT for such designs are very limited and requires considerable research to avoid high cost test equipments for testing such complex designs. DFT based test approaches require new techniques and methodologies to improve the defect coverage and quality of test. DFT also reduces cost by reducing the I/O data rate requirements, low pin count for testing and less expensive test equipments. Research has to be done on faults due to dynamic circuit behavior and in high speed circuits which are common in advance applications like vector applications.

Existing testing methods for stuck-open and bridging fault, like IDDQ testing are becoming infeasible or not suited for high speed designs. The test data required is increasing at a faster rate than that of I/O data rate; this makes techniques like test data compression, built-in self-test (BIST) and at-speed scan-based testing very essential. The amount of test data directs much effort for new approaches in DFT for mixed signal designs.

New DFT techniques are also required for high-speed I/O, signal integrity, analog and mixed signal [1] [2]. Note that traditional ATEs cannot support the latest ICs frequency. One idea is to move functionality from the automatic test equipment (ATE) to infrastructure intellectual property (IIP) on chip in order to reduce the requirements and costs of the ATE. The requirements include speed, resolution, and bandwidth. For example, testing signal integrity loss on SoC interconnects can be implemented using boundary scan and on-chip pattern generator. This provides the ability to perform at-speed test and it reduces test application time. Briefly, test costs have to be reduced, test time needs to be reduced and testability and test quality need to be improved. Demands on high speed expensive external tester has to be reduced, the use of digital only testers for testing mixed signal functions needs to be promoted.

1.7 DFT Techniques

1.7.1 Built-In Self-Test (BIST)

BIST has emerged as a promising solution to the VLSI testing problems. BIST is a DFT methodology that aims at detecting faulty components in a system by incorporating the test logic on chip. In BIST, a test pattern generator (TPG) generates test patterns and a signature analyzer (SA) compacts test responses. Figure 1.1 shows BIST architecture. The entire process is controlled by BIST controller. BIST is well known for its numerous advantages such as:

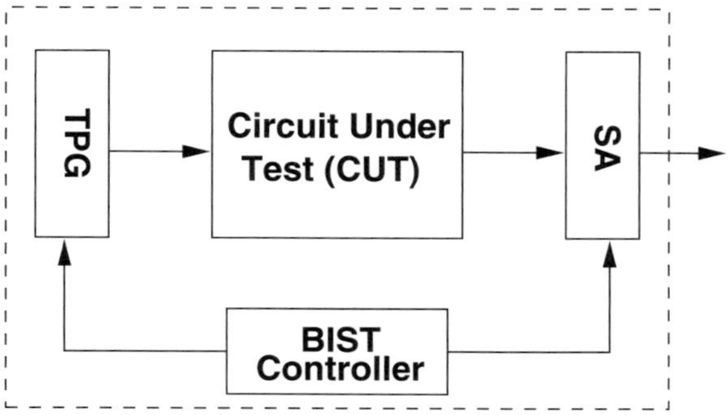

Fig. 1.1. Built-in self-test architecture.

- No expensive test equipment is needed.
- Field test during operation and maintenance also possible.
- Supports concurrent testing.
- Can be used for different test levels.
- At-speed testing.
- Improved testability.

The disadvantages of using BIST are listed below:

- Additional BIST hardware overhead.
- Performance degradation, timing issues.
- Excessive power consumption.
- Inability to achieve full fault coverage.

1.7.2 Scan or Full Scan

The main idea in *scan design* is to obtain controllability and observability for flip-flops [27], which eventually increases the overall test coverage. This is done by adding a test mode to the circuit such that when the circuit is in this mode, all flip-flops functionally form one or more shift registers. The input and output of these shift registers (also known as scan registers) are made into primary inputs and primary outputs. Thus, using the test mode, all flip-flops can be set to any desired states by shifting those logic states into the shift register. Similarly, the states of flip-flops are observed by shifting out the content of the scan register. All flip-flops can be set or observed in a time (in terms of clock periods) that equals the number of flip-flops in the longest scan register. Today's large circuits contain a long scan chain sometimes more than tens of thousands of flip-flops.

The foundation for any structured DFT methodology is scan architecture. Scan lets the large sequential functions implemented in a design be partitioned into small combinational blocks during test. Scan is the standard DFT infrastructure for delivering test data to internal nodes of the circuit and for

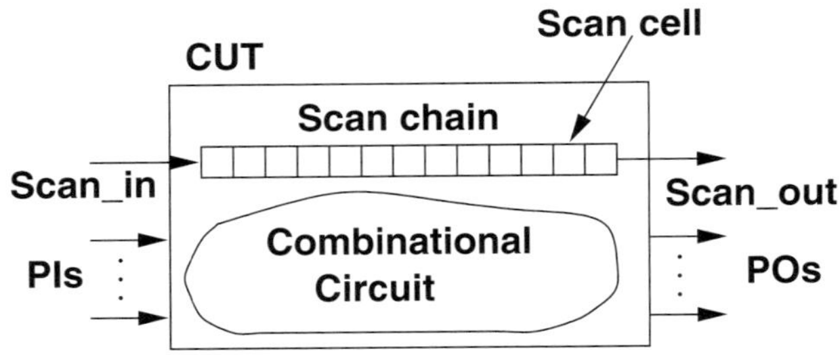

Fig. 1.2. Scan architecture.

observing their responses. With today's multimillion-gate SoC designs, scan is required to ensure efficient generation of high-quality manufacturing tests. Indeed, scan is considered the basic building block for automating the entire test-generation process.

Scan and automatic test pattern generation (ATPG) are the solutions of choice for ensuring the highest-quality test during manufacturing. Functional test strategies are losing popularity across the industry due to their high development cost. It has also become difficult, if not impossible, to grade the effectiveness of functional tests for large multimillion-gate designs. The simplicity and effectiveness of ATPG and scan-based test patterns directly address the problems of functional test patterns and offer several advantages.

On the other hand, using scan-based test patterns provides access to internal nodes of the design and simplifies the problem into much smaller blocks of logic. Additionally, scan and ATPG enable the entire process of generating the test patterns to be fully automated. This ensures very high coverage in test patterns, as well as a predictable and repeatable process. Today, ATPG tools can generate very high-coverage test patterns for multimillion-gate designs in a matter of hours.

Scan-Based BIST is a new DFT that allows test designer to design and add a TPG, i.e. LFSR, to the scan architecture to generate and apply random patterns serially. This reduces the need for an expensive external tester while it ables to perform at-speed testing and applying test patterns with the normal operating frequency of the circuit under test. A signature analyzer is used to generate a signature of the test responses. This technique is becoming increasingly acceptable in industries.

The main concerns about scan architectures are the long length of scan chains and huge number of test patterns, considering limited ATE's bandwidth, both significantly increase the test application time. Power consumption during test is another important issue for scan architectures. In this book, we address these issues and present solutions to alleviate them. Scan architectures, e.g. full scan, scan-based BIST, and boundary scan, are the main focus of this work.

1.7.3 Boundary Scan (BS)

Most designs today implement some sort of boundary scan so that chip I/O can be accessed and interconnect can be tested at the board level. The boundary scan controller has also emerged as the standard mechanism on SoC designs for initiating and controlling the multiple internal memory BIST controllers. Boundary scan is now a well-known and documented IEEE standard, and some test software vendors offer automated solutions.

IEEE 1149.1, also known as JTAG or boundary scan, was introduced in 1990 [19]. This standard endeavors to solve test and diagnostic problems arising from loss of physical access caused by the increasing use of high pin count and BGA devices, multi-layer PCBs, and densely packed circuit board assemblies. The standard outlines predefined protocols for testing and diagnosing manufacturing faults. It also provides a means for on-board programming of non-volatile memory devices such as Flash, or in-system programming of devices like PLDs and CPLDs. The boundary scan architecture is shown in Figure 1.3.

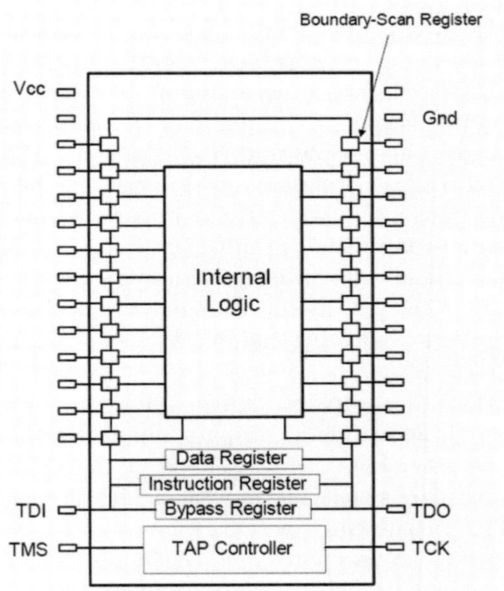

Fig. 1.3. Boundary scan architecture [19].

An RTL designer can design his/her own IEEE-compliant boundary scan chain and associated controller. But to improve efficiency and time-to-market, an automated boundary scan tool can be used to let the register transfer level

(RTL) designers focus on critical areas of the functional design. Automating boundary scan can save weeks in initial implementation, and even more in all subsequent revisions that affect I/O and device-pinout assignments.

The ultimate goal of manufacturing test in SoC designs is to screen out bad devices from good devices. Eventually, it comes down to quality. The better the manufacturing tests, the less likely it is a defective part will escape the test process and make it to the end customer. As explained, the most common test methodologies including scan and ATPG, memory BIST, and boundary scan are available today to fully automate the creation and insertion of test logic, and the creation of the final manufacturing test patterns. High-quality manufacturing tests and improved time-to-market don't have to be at odds.

As an standard, JTAG was not intended to be used for high speed testing because of its serial nature of scan. In this book, an extended version of JTAG is presented to generate and apply test pattern high speed and to be used for interconnect testing in SoCs.

1.8 Delay Fault Testing

As technology scales, test engineers are facing new challenges. Over the years, test methodologies for digital circuits have evolved along with technology. The deep-submicron (DSM) effects are becoming more prominent with shrinking technology, thereby increasing the probability of timing-related defects [21] [22]. For DSM designs, the stuck-at and I_{DDQ} tests alone cannot ensure high quality level of chips because of increasingly growing number of timing-related defects in a chip. In the past, functional patterns were used for at-speed test. However, functional testing is not a viable solution because of the difficulty and time to generate these tests for complex designs with very high gate density. Therefore, more robust at-speed techniques are necessary as the number of timing-related defects is growing and effectiveness of functional and I_{DDQ} testing is reducing [23] [24].

Figure 1.4 show the concept of timing failure in a circuit. The inputs can be flip-flops in the circuit and primary inputs (PIs). The outputs can include primary outputs (POs) and flip-flops in the circuit. It is assumed that the input changes are synchronized with the input clock. All outputs are expected to stabilize in one clock period. Figure 1.4(b) shows two different output waveforms to an input change, i.e. O_{j1} and O_{j2}. O_{j1} is the fault-free response to an input pattern where the I_i input changes its state while O_{j2} is the faulty response to the same input. As seen, the O_{j2} output stabilizes eventually but with a delay. This delay could have been caused either by a gate or interconnect in the circuit. It could have also been caused by multiple gates or interconnects or both.

As seen, in order to examine the timing operation of a circuit we must examine signal transition; i.e. patterns must be applied such that they create signal transition at the circuit inputs. It is also seen in Figure 1.4(b) that the output signal may contain several glitches but the final correct value must be present at the next clock edge to be read either by tester or latched by a flip-flop. The location of these glitches in the circuit output depends on the circuit topology and the length of combinational paths. There are several paths with varying lengths from every net in the circuit to observation points (flip-flops or POs). This could result in glitches in the circuit. The longest path in a design is called *critical path.* The delay of critical paths determines the smallest clock period at which the circuit can function correctly [27].

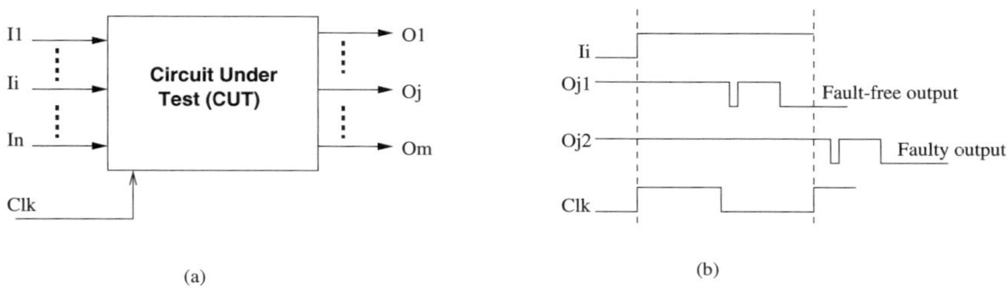

Fig. 1.4. Delay fault problem: (a) Circuit under test and (b) Input and output signal waveforms.

The known fault models that represent or generate timing-sensitive voltage measurements are the *path-delay* and the *transition fault* models. The path-delay fault model takes into account the sum of all delays along a path while the transition fault model accounts for localized faults (delays) at the inputs and outputs of each gate. In the following, these two models are described in details.

1.8.1 Path-Delay Faults

The path delay fault is an important fault model used in delay testing. It is commonly used for performance verification during production test. The delay defect in the circuit is assumed to cause the cumulative delay of a combinational path to exceed the clock period. The combinational path begins at a primary input or a flip-flop and contains a chain of gates. The path ends at a primary output or a flip-flop. The propagation delay is the time that a signal transition takes to travel through the path. The propagation delay includes both switching delay of gates and transport delays of interconnects on the path. For a fault-free response, the propagation time must be less than the clock period. Each path must be tested for both $0 \rightarrow 1$ and $1 \rightarrow 0$ transitions generated at the input of the path [27].

The total number of paths in a large circuit can be extremely high. In general, any combination of paths in the circuit can be faulty and must be targeted during test. In practice both signal path delay and multiple path delay faults can exist.

Non-robust Path Delay Test

A test that guarantees to detect a path delay fault, when no other path delay fault is present, is called a non-robust test for that path.

Robust Path Delay Test

A robust path delay test guarantees to generate an incorrect value at the output of a path under test if the path delay is longer the clock cycle in presence of other faulty paths in the circuit.

1.8.2 Transition Delay Faults

A transition delay fault on a line makes the signal change on that line slow. The two possible faults are slow-to-fall (stf) and slow-to-rise (str) types. If any of such faults on a gate or interconnect is present then it may slow down a path such that the propagation delay of the path becomes longer than the clock period causing a timing failure.

In a design containing n lines, there can be a maximum $2n$ transition faults (a str and a stf fault on each line), but there potentially can be more than n^2 (or exponential) path delay faults. Since all the paths cannot be tested, the path delay model requires identification and analysis of critical paths in the design. This makes it more complicated to use on large designs. Consequently, the transition delay fault model has been accepted in industry as a good method to test for delay faults.

Scan-based structural tests generated by an automatic test pattern generator (ATPG) are increasingly used as a cost-effective alternative to the at-speed functional pattern approach by providing high controllability and observability [24]. The transition fault and path delay fault testing together provide a relatively good coverage for delay-induced defects [25] [26]. Path delay model targets the cumulative delay through the entire list of gates in a pre-defined path while the transition fault model targets each gate output in the design for a slow-to-rise and slow-to-fall delay fault [27]. The transition fault model is more widely used than path delay because it tests for at-speed failures at all nets in the design and the total fault list is equal to twice the number of nets. On the other hand, there are billions of paths in a modern design to be tested for path delay fault leading to high analysis effort; this makes path delay fault model very cost intensive compared to transition fault model.

A transition fault test requires a pattern pair $(V1, V2)$ to be applied to the circuit-under-test (CUT). Pattern $V1$ is termed as the initialization pattern and $V2$ as the launch pattern. The response of the CUT to the pattern $V2$ is captured at the operational functional speed. The entire operation can be divided into three cycles: 1) Initialization Cycle (IC), where the CUT is initialized to a particular state ($V1$ is applied), 2) Launch Cycle (LC), where a transition is launched at the target gate terminal ($V2$ is applied) and 3)

Capture Cycle (CC), where the transition is propagated and captured at an observation point.

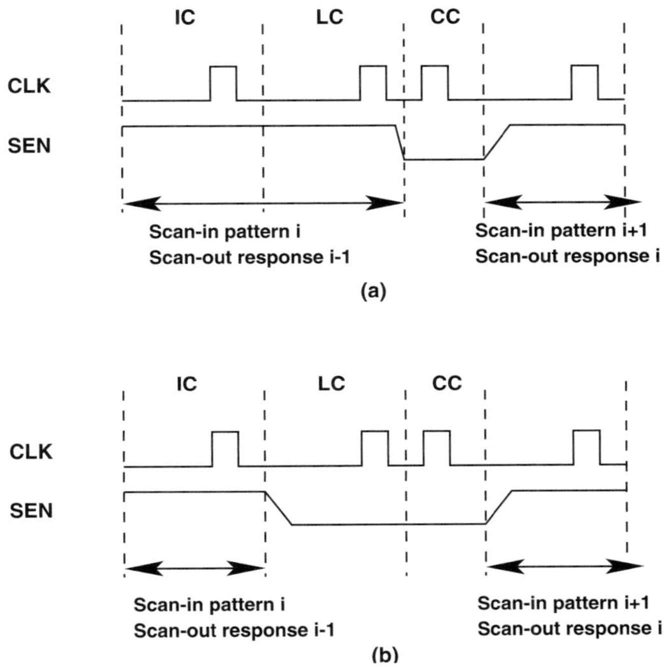

Fig. 1.5. Transition delay fault pattern generation methods: (a) Launch-off-shift and (b) Launch-off-capture [33].

Depending on how the transition is launched and captured, there are three transition fault pattern generation methods. In the first method, referred to as *launch-off-shift* (LOS) or *skewed-load* [28], the transition at the gate output is launched in the last shift cycle during the shift operation. Figure 1.5(a) shows the launch-off-shift method waveform for a multiplexed-DFF design; similar approach can be applied to an LSSD. The LC is a part of the shift operation and is immediately followed by a fast capture pulse. The scan enable (SEN) is high during the last shift and must go low to enable response capture at the CC clock edge. The time period for SEN to make this $1 \rightarrow 0$ transition corresponds to the functional frequency. Hence, LOS requires the SEN signal to be timing critical. Skewing the clock (CLK) creates a higher launch-to-capture clock frequency than standard shift clock frequency. Saxena et al. [31] list more launch and capture waveforms used by launch-off-shift approaches. In [31], implementation of scan-based transition fault testing and its low cost test challenges are discussed.

Figure 1.5(b) shows the waveforms of the second approach, referred to as *launch-off-capture* (LOC) or *broadside* method [29]. In this method, the launch cycle is separated from the shift operation. At the end of scan-in (shift mode), pattern $V1$ is applied and CUT is set to an initialized state. A pair of at-speed clock pulses are applied to launch and capture the transition at the target gate terminal. This relaxes the at-speed constraint on the SEN signal and dead cycles are added after the last shift to provide enough time for the SEN signal to settle low.

The third technique, known as *Enhanced Scan* [35] requires that two vectors $V1$ and $V2$ are shifted into the scan flip-flops simultaneously. The drawback on enhanced scan is that it needs hold-scan flip-flops and is area-intensive [35], making it unattractive for ASIC designs.

The LOS method is more preferable based on ATPG complexity and pattern count compared to LOC method. The LOC technique is based on a sequential ATPG algorithm, while the LOS method uses a combinational ATPG algorithm. This will increase the test pattern generation time in case of LOC and also a high fault coverage cannot be guaranteed due to the correlation between the two patterns, $V1$ and $V2$; note that $V2$ is the functional response of pattern V1. Due to its dependence on the functional capture path for launching transitions, LOC ATPG can be quite slow, and on large designs, can consume several days of ATPG run-time. The main concern about LOS is its requirement to at-speed scan enable (SEN) signal.

References

1. International Technology Roadmap for Semiconductors 2001 edition - Executive summary.
2. International Technology Roadmap for Semiconductors 2001 edition - Test and test equipment.
3. C. F. Hawkins, J. M. Soden, A. W. Righter, and F. J. Ferguson, “Fault Classes - An Overdue Paradigm for CMOS IC Testing,” in Proc. *Int. Test Conf. (ITC’94)*, pp. 413-425, 1994.
4. J. P. Roth, “Diagnosis of Automata Failures: A Calculus and a Method,” IBM Journal, vol.10, pp. 278-291, 1966.
5. J. P. Roth, W. Bouricius, and P. Schneider, “Programmed Algorithms to Compute Tests to Detect and Distinguish Between Failures in Logic Circuits,” IEEE Transactions on Electronic Computers, EC-16(5), pp. 567-580, 1967.
6. P. Goel, “An Implicit Enumeration Algorithm to Generate Tests for Combinational Logic Circuits,” IEEE Transactions on computers, vol. 30, no. 3, pp. 215-222, 1981.
7. H. Fujiwara and T. Shimono, “On the Acceleration of Test Generation Algorithms, IEEE Transactions on Computers, vol. 32, no. 12, pp. 1137-1144, 1983.
8. M. H. Schulz, E. Trischler, and T. M. Sarfert, “Socrates: A Highly Efficient Automatic Test Pattern Generation System,” IEEE Transaction on CAD, pp. 126-137, 1988.

9. E. E. King, and G. P. Nelson, “Radiation testing of an 8-bit CMOS microprocessor,” IEEE Transactions of Nuclear Science, NS-22, 5, pp. 2120-2123, 1975.
10. M. Levi, “CMOS is MOST Testable,” IEEE Proceedings of International Test Conference, pp. 217-220, 1981.
11. J. Soden, and C. F. Hawkins, “IDDQ Testing: A Review,” Journal of Electronic Testing: Theory and Applications (JETTA), vol. 3, no. 4, pp 291-304, 1992.
12. J. F. Frenzel and P. N. Marinos, “Power Supply Current Signature (PSCS) Analysis: A New Approach to System Testing,” IEEE proceedings of International Test Conference (ITC), pp. 125135, 1987.
13. M. Hashizume, et al., “Fault Detection of Combinatorial Circuits based on Supply Current,” IEEE proceedings of International Test Conference (ITC), pp. 374380, 1988.
14. J. S. Beasley, H. Ramamurthy, J. Ramirez-Angulo, and M. DeYong, “iDD Pulse Response Testing of Analog and Digital CMOS Circuits,” IEEE proceedings of International Test Conference (ITC), pp. 626634, 1993.
15. S. Sengupta, S. Kundu, S. Chakravarty, P. Parvathala, R. Galivanche, G. Kosonocky, M. Rodgers, TM Mak, “Defect-Based Test: A Key Enabler for Successful Migration to Structural Test, Intel Technology Journal, Q1, 1999 (http://developer.intel.com/technology/itj/q11999/articles/art_6.htm).
16. P. Wiscombe, “A Comparison of Stuck-at Fault Coverage and IDDQ Testing on Defect Levels, in Proc. *International Test Conference*, pp. 293-299, 1993.
17. H. Konuk, “Voltage- and current-based fault simulation for interconnect open-defects,” IEEE Trans. on Computer-Aided Design of Integrated Circuits and Systems, vol. 18, no. 12, pp. 1768-1779, 1999.
18. E. J. McCluskey, and C.W. Tseng, “Stuck-faults vs. Actual Defects, in Proc. *International Test Conference*, pp. 336-343, 2000.
19. IEEE Standard 1149.1-2001, *“Standard Test Access Port and Boundary-Scan Architecture”*, IEEE Standards Board, 2001.
20. www.agilent.com
21. S. Natarajan, M.A. Breuer, S.K. Gupta, “Process Variations and Their Impact on Circuit Operation,” in Proc. *IEEE Int. Symp. on Defect and Fault Tolerance in VLSI Systems*, pp. 73-81, 1998.
22. R. Wilson, “Delay-Fault Testing Mandatory, Author Claims,” *EE Design*, Dec. 2002.
23. G. Aldrich and B. Cory, “Improving Test Quality and Reducing Escapes,” in Proc. *Fabless Forum, Fabless Semiconductor Assoc.*, pp. 34-35, 2003.
24. X. Lin, R. Press, J. Rajski, P. Reuter, T. Rinderknecht, B. Swanson and N. Tamarapalli, “High-Frequency, At-Speed Scan Testing,” *IEEE Design & Test of Computers*, pp. 17-25, Sep-Oct 2003.
25. K. Cheng, “Transition Fault Testing for Sequential Circuits,” *IEEE Transactions on Computer-Aided Design of Integrated Circuits and Systems*, vol. 12, no. 12, pp. 1971-1983, Dec. 1993.
26. T. M. Mak, A. Krstic, K. Cheng, L. Wang, “New challenges in delay testing of nanometer, multigigahertz designs,” *IEEE Design & Test of Computers*, pp. 241-248, May-Jun 2004.
27. M. Bushnell, V. Agrawal, *Essentials of Electronics Testing*, Kluwer Publishers, 2000.
28. J. Savir, “Skewed-Load Transition Test: Part I, Calculus,” in Proc. *Int. Test Conf. (ITC’92)*, pp. 705-713, 1992.

29. J. Savir and S. Patil, "On Broad-Side Delay Test," in Proc. *VLSI Test Symp. (VTS'94)*, pp. 284-290, 1994.
30. B. Dervisoglu and G. Stong, "Design for Testability: Using Scanpath Techniques for Path-Delay Test and Measurement," in Proc. *Int. Test Conf. (ITC'91)*, pp. 365-374, 1991.
31. J. Saxena, K. M. Butler, J. Gatt, R. Raghuraman, S. P. Kumar, S. Basu, D. J. Campbell, J. Berech, "Scan-Based Transition Fault Testing-Implementation and Low Cost Test Challenges," in Proc. *International Test Conference (ITC'02)*, pp. 1120-1129, Oct. 2002.
32. E. Park and R. Mercer, "An efficient delay test generation system for combinational logic circuits," IEEE Trans. on Computer-Aided Design of Integrated Circuits and Systems, vol. 11, no. 7,1992, pp. 926938.
33. N. Ahmed, C.P. Ravikumar, M. Tehranipoor and J. Plusquellic, At-Speed Transition Fault Testing With Low Speed Scan Enable, in proc. *IEEE VLSI Test Symposium (VTS'05)*, pp. 42-47, 2005.
34. X. Liu, M. S. Hsiao, S. Chakravarty and P. Thadikaran, "Techniques to Reduce Data Volume and Application Time for Transition Test," in Proc. *Intl Test Conf. (ITC'02)*, pp. 983-992, 2002.
35. B. Dervisoglu and G. Stong, "Design for Testability: Using Scanpath Techniques for Path-Delay Test and Measurement," in Proc. *Intl Test Conf. (ITC'91)*, pp.365-374, 1991.

2

At-speed Test Challenges for Nanometer Technology Designs

As the electronic design automation (EDA) industry focuses on design-for-manufacturability (DFM), the older problem of design-for-test has almost been forgotten. But ICs built at 90 nanometers and below pose new and complex challenges for design-for-testability (DFT) tools and techniques. At those geometries, small delay defects become a major contributor to chip failures, but they can't be detected by conventional automatic test pattern generation (ATPG) tools since they are timing unaware. Low-power ICs, which will include most chips at 65nm technology node, demand new approaches to low power scan design and pattern generation. Test data run over many dice and wafers can provide valuable diagnostic information that helps foundries and designers ramp up their yields. In this sense, DFT meets DFM and becomes a critical element in the attempt to mitigate process variability.

2.1 Technology Scaling Effects

Technology scaling continues to drive the increase in chip frequency and density. Ensuring that designs meet performance specifications using transition delay test has become a very challenging task for test engineers. Traditional transition fault testing methods do not guarantee accurate performance verification in nanometer scale era unless all the nanotechnology issues are very well addressed. Therefore, path delay test for a number of selected critical paths will be a necessity. Traditionally, nominal or worst-case static timing analysis is done to find the critical paths. However, this may not be quiet accurate for nanometer technology designs due to the increase in delay variations. Process variations, manufacturing defects, and noise can appear as delay variations and are considered to be the major factors affecting the quality of nanotechnology designs.

As functional density and operating frequency increase, the number of interconnects and length of interconnects are expected to increase as well. Over the years, the number of metal layers has incrementally increased from a single layer and currently using six to eight metal layers in industry is a common practice. Increasing number of metal layers in turn increases the number of vias which are proven to be the main sources of defects. The situation will grow worse since the number of metal layers will further increase rising up to 10 and 12 in the next few years.

The material of the layers used in fabrication processes has also undergone a major change from aluminum to copper. Using copper provides a better scalability compared to aluminum. As technology scales and more transistors are integrated on a chip, the interconnects become longer. For high-speed nanometer technology designs the interconnect delay dominates the gate delay [1]. It is predicted that in the near future the longest path in the design will be the critical one. In nanometer technology era, crosstalk will be a major contributor to interconnect delay. To keep the resistance of the wires low as technology scales, the interconnects are becoming narrower by width and taller by height. This results in large cross coupling capacitances which are now dominating substrate capacitances.

To reduce the power and minimize the negative impact of hot career, which causes reliability issues overtime, the power supply is reduced [2]. However, the transistor voltage threshold is not scaling proportionally which results in increase in the circuit sensitivity and reduction in noise margin. The scaling also increases the leakage current. In 65nm technology, the static power consumption contributes to 50% of total power consumption while it is expected to further increase in 45nm and 22nm technologies. Another negative impact of increased leakage current is rendering IDDQ and ΔIDDQ methods less effective as these two rely on the static currents drawn from power grid after applying patterns in quiescent mode. Reduced effectiveness of IDDQ and stuck-at fault models in detecting timing-related defects along with increasing number of such defects in nanotechnology requires new fault models and novel DFT methods.

Transition and path delay fault models have become very popular in the past decade as they significantly increase the production test quality and reduce the defective parts per million (DPPM) levels. The path delay test targets delay defects on critical paths in a design to verify the target performance. Due to various delay variation sources, selecting critical paths using static timing analysis tool may not seem to be the best option. Process variations, supply noise and crosstalk can have significant impact on the delay of paths and need to be considered during critical path selection procedure. Below, we briefly describe some of the delay variations in nanometer technology.

Crosstalk noise can slow down the signal traversing a path causing delay faults. Supply noise can also cause delay of a path to increase due to reduced voltage reaching the devices on the target path. Heat induced during the continuous operation of the circuits can potentially degrade the performance (up to 20% reported in [2]). Historically, a small number of critical paths have been selected for path delay testing. However, considering the above mentioned delay-alteration factors, a larger number of paths will be critical to design's performance and must be tested. On the other hand, due to process variations and pattern-delay-dependency of the designs, inter-die critical path selection must be performed as one path might be critical on one chip but not on the other. Critical path selection, therefore, will be a challenging problem that test engineers face. Statistical timing analysis and simulation will be one possible solution to take all variables into account. The delay of logic elements and interconnects must be modeled as correlated random variables.

2.1.1 Crosstalk Effects

Signal integrity can be significantly affected by crosstalk in nanometer technology designs. An aggressor line will induce a current upon a victim line based on the voltage change in the aggressor and the coupled capacitance between the two lines. The aggressor voltage rate of change is dependent on the rise and fall time. The coupling capacitance is proportional to the density of interconnects around the critical path. However, two additional factors affect the signal integrity of a path, namely transition direction of aggressors and timing of transitions on aggressors. Each of these components have a direct effect on coupling and propagation delay. In this book, we focus on transition direction due to the difficulty in correctly predicting actual circuit timing without extensive simulation. Identifying the timing of transitions on aggressor and victime lines requires an extensive statistical analysis since various sources of delay, such as process variations, supply noise, etc, can have significant impact on the signal arrival time on each net.

As technologies continue to shrink beyond the ultra-deep submicron level, the parasitic coupling capacitance will continue to grow worse and will play a greater role not only in the design of a chip but also testing. Figure 2.1 shows sidewall capacitance between nearby wires in addition to interconnect C model (R is not shown for the sake of simplicity). Since the height of the wires remains almost the same as technology shrinks and the fact that the length of interconnects are becoming longer, the cross coupling effects are becoming increasingly dominant when compared to substrate capacitances. Due to reduced wire spacing, it is expected that the coupling effects will increase. The switching activity on the aggressor lines may either speed-up or slow-down a transition signal on the victim line. If the aggressor lines switch in the same direction as victim, they cause a speed-up on the signal traveling through victim line. The slow down on victim line occurs if the aggressor lines switch in opposite direction of victim line. The effect of aggressor lines transitions on a quiescent victim line appears as glitches which can cause logic failure [6] [7] [8] [9].

Design and test engineers must assess, analyze and deal with these problems before signing off on a tapeout and after fabrication during manufacturing test (mainly during delay testing). Note that since the direction of transitions and transition arrival time on aggressor lines directly impact the

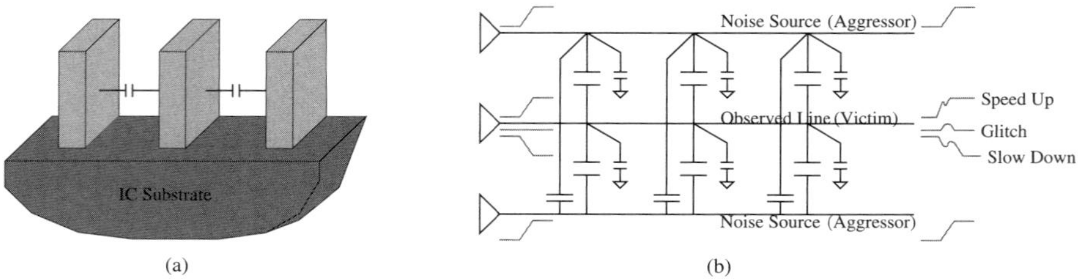

Fig. 2.1. (a) Sidewall capacitance effects increase with shrinking feature sizes and (b) Interconnect C only model (for simplicity R is not shown).

signal delay on victim line then effective pattern generation procedures are required to maximize such effects. In other words, the total number of switchings around critical paths do not ensure maximum crosstalk. What is important is to intelligently generate patterns such that they magnify such effects on critical paths.

2.1.2 Power Supply Noise Effects

Power supply noise (PSN) due to switching current has become an important factor for nanometer technology designs. As the number of interconnect layers and gate density increases, the switching activity increases leading to an increase in current density and voltage drop along the power supply net. Increasing the frequency and decreasing the rise/fall transition time in today's designs causes more simultaneous switching activity within a small time interval and increases the instantaneous currents. The power supply noise can increase the signal delay and result in signal integrity loss and performance degradation [10]. It may also cause logic errors, degradation in switching speed and hence timing failures. To deal with large supply noise, designers perform wire sizing and decap (decoupling capacitance) insertion to compensate the large instantaneous currents drawn from power rails.

In general, PSN includes two components: inductive ΔI noise and power net IR voltage drop and is given by PSN = Ldi/dt + IR. The inductive ΔI noise (Ldi/dt) depends on the rate of change of the instantaneous current; the inductance is mainly due to package leads and wire/substrate parasitics. Simultaneous switching of a large number of gates often induces a very large current spike on the power/ground lines in a short time interval. With low-k copper (Cu) interconnects being used in nanotechnology designs, the resistance of the wires is drastically reduced. This will generate considerable inductive noise Ldi/dt even though the inductance L can be relatively small. The simulation results in literature, e.g. [11], shows that inductive noise can dominate the resistive noise in sub-100nm technology.

For the worst case analysis, the idea is to generate the steepest maximum switching current spike and observe its impact on path delay. In order to create maximum switching noise, it is important to analyze the characteristics of the switching current waveform [11] [10]. Note that the circuit tolerance to voltage fluctuation and noise is reducing and both IR-drop and Ldi/dt will have significant impact on chip performance. In this book, we will be focusing on IR-drop and Ldi/dt effects will be considered in the future research.

The circuit model, shown in Figure 2.2, will be used for power/ground pin and power/ground network. Each V_{dd} and V_{ss} pin is modeled as an RLC circuit. The pin parasitics are R_p, L_p and C_p for V_{dd} pin and R_s, L_s and C_s for V_{ss} pin. The power/ground network is essentially modeled as a lumped RLC network.

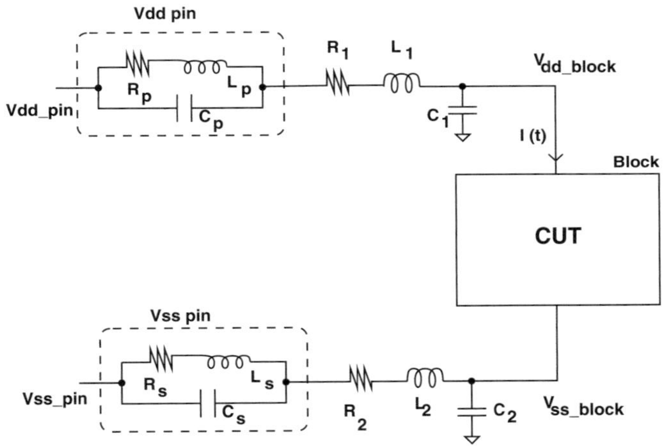

Fig. 2.2. The circuit model and power supply noise measurement [10].

Whether we use traditional at-speed test or faster-than-speed test to target small delay defects, the supply noise will be a major issue to deal. Compared to the scan shift where lower frequency is used, functional frequency is used during launch-to-capture cycle of at-speed test. In general, the supply noise is much larger during at-speed delay test compared to normal circuit operation since larger number of transitions occur within a short time frame. Existing commercial ATPG tools do not consider the excessive supply noise that might occur in the design during test pattern generation.

On the other hand, increasing the frequency during faster-than-at-speed test impacts the performance of the chip due to adverse IR-drop effects [13]. Faster-than-at-speed test method detects small delay defects by reducing the positive slack of the path under test. This book will present the practical issues during at-speed and faster-than-at-speed delay tests. A case-study of a design is performed and the increase in both peak and average IR-drop effects due to at-speed and faster-than-at-speed pattern application are illustrated. Increase in IR-drop directly relates to performance degradation due to effective voltage reduction reaching the gates in the circuit.

New methods are needed to generate IR-drop tolerant transition delay fault test pattern for at-speed test. Also, new techniques are needed for the application of transition fault patterns generated for faster-than-at-speed test while considering IR-drop effects. The physical design implementation, power/ground distribution network, and pattern-delay analysis must be carefully studied. Detailed IR-drop analysis and associated performance degradation due to effective voltage reduction for transition test patterns application at rated functional speed and faster-than-at-speed to detect small delay defects must be performed.

Figures 2.3 and 2.4 show the average IR-drop plots on the ground (VSS) network for two transition delay fault patterns, *P1* and *P2*. The design used for this experiment was ITC'99 benchmark *b19*. The IR-drop plots were obtained from the Cadence SOC Encounter tool [25] measured across the respective switching time frame for each pattern. Note that, for pattern P2, the IR-drop in a large portion of the chip increases which results in reduced effective voltage difference between the VDD and VSS ports observed by each gate in that region. This might result in higher performance degradation or functional failure of the circuit due to excessive noise.

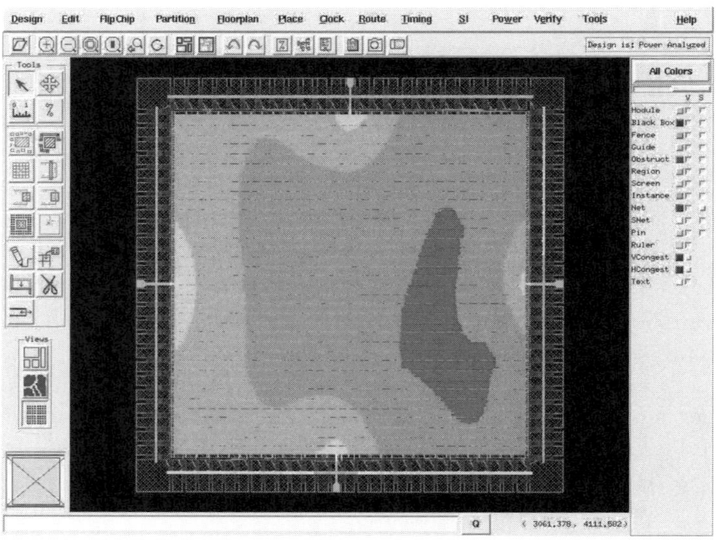

Fig. 2.3. IR-drop plot in VSS net for pattern *P1*.

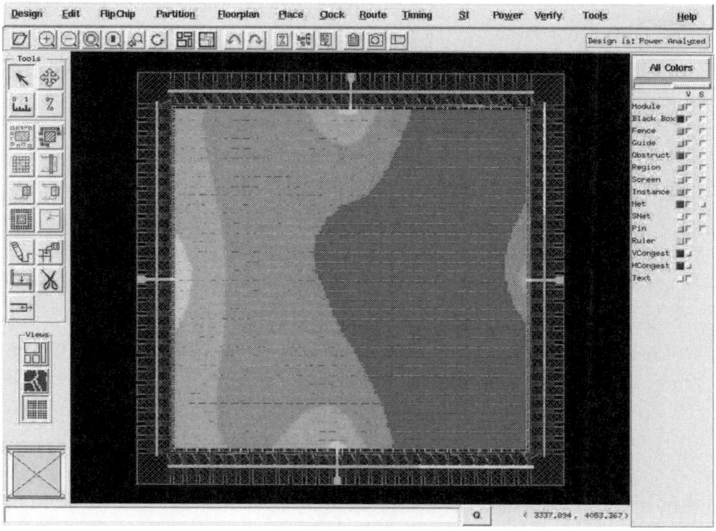

Fig. 2.4. IR-drop plot in VSS net for pattern *P2*.

2.1.3 Process Variations Effects

Process variations due to imperfect fabrication process exist across the die and wafer, from one die to another, from one wafer to another, and from lot to lot. These variations are all independent of one another and in most cases they are systematic. Many variations will not get worse as technology scales but many others will be significantly affected by it.

Usually designers develop process technology files (mostly worst-case conditions) in order to deal with the variations in their designs. They then simulate their design using these process files to ensure that their subchips are functional and that the specific timing behaviors are met for static timing analysis.

There are several constraints arisen in the photolithography area. For instance, the wavelength of the light used for imaging the geometries is longer than the geometry desired for printing. For example, a designer uses an almost 200nm light source for a 130nm gate length. Since light will inadvertently reflect off neighboring geometries, the printed images will differ from their intended shape. This results in inaccurate prints and designers must deal with these difficult issues. Usually the lithographic engineers generate shaping rules that add or subtract geometries to the mask so that the final printed shape is what the designers intended. Although, the process was successful for 180nm and 130nm technologies, process variations were still affecting the designs and were strong enough to affect the performance. However, the situation is growing worse for sub-100nm technologies. Another, more-crucial effect relates to the printing of the polysilicon layer that defines the transistors gate length, which, in turn, defines the transistors performance.

Not only designers must fully comprehend process variations and take it into account during design but also test engineers must be able to sort out if, for example, all the paths (mostly critical ones) meet a particular frequency specification. The more variability in the process, it is more likely that more paths become timing sensitive and will require testing. The traditional model of testing a few selected critical paths (path delay fault testing) based on their nominal or worst-case delays will fall apart.

Process variations also affect interconnect characteristics [19] [22]. Chemical mechanical polishing technology helps planarize the metal layers or the interlayer dielectrics for subsequent layer deposition. As an example, overetching, together with the dishing of interlayer dielectrics, can then produce higher-resistance wires that have a higher coupling capacitance with the geometries above them, hence affecting the interconnect delays which might have impact on high-speed designs' performance. As the industry has moved toward using copper instead of aluminum unfortunately process variation effects have become significant. Note that, due to such variations, the resistance and capacitance of interconnects within a die can be different. Same holds true from one die to another.

Even though all these variations are to a certain extent predictable because of the extensive research done in each semiconductor industry or foundry and possibly avoidable during design because of several avilable EDA tools and in-house tools in industry, avoiding them in all circumstances would be difficult. Aberrations could still affect interconnect and gate delays, hence design performance.

2.1.4 Thermal Effects

In general, due to the current surge by millions of active transistors during computation, high-performance circuits tend to run hot. In high-performance designs, the average number of transistors operating in each clock cycle is higher compared to other designs and that generates more power and heat in the chip. The increased leakage currents in nanometer technology designs increases the power dissipation as well.

The heat generation depends on the switching activity in a design, clock frequency, and circuit supply voltage. The large instantaneous switching activity can produce local hot spots in the design. This local hot spot can be several tens of degrees hotter than the rest of the die, even after the application of the best package cooling techniques. It is estimated that a temperature difference of 40 Celsius could result in a 20% performance variation [2]. Usually designers consider 125 Celsius as the temperature corner condition when they simulate their designs under worst-case performance conditions.

The situation may grow worse during test as larger number of switchings may occur both in scan shift and fast launch-to-capture cycle during at-speed test. Scan-based at-speed test or at-speed built-in self-test (BIST) would produce more power than normal functional operation. All this makes it difficult to pinpoint what temperature to set and what patterns to apply for testing. Setting the temperature too high might overstress the noncritical paths and such overly conservative design may result in a chip that is too slow or too big. However, setting it too low might result in undertesting of the critical paths which may pass the test but fail in the field causing reliability problem.

2.1.5 Statistical Analysis

Because of cost and complexity issues, engineers usually perform path delay testing for a very limited number of critical paths. Identifying a small set of paths out of millions of paths in a large design is a complex task. For higher technology nodes, designers have based critical-path selection on nominal or worst-case timing analysis. With these discrete timing models, a critical-path selection step either selects all paths whose delays fall into a predefined percentage of the delay time range, or simply chooses a fixed number of the longest paths (i.e. near critical paths) [16]-[24].

Practically, the worst-case conditions must be set based on process variations, supply noise and temperature which are the main delay variation sources. These can significantly alter the timing assumed in the discrete models and increase the number of paths with delays close to critical. Considering all these conditions may result in a large number of process corners during simulation. The more number of process corners the more conservative the design process is. On the other hand, due to these variations in the process, the set of critical paths can differ from chip to chip. Therefore, the traditional approaches for selecting the optimal set of paths might be ineffective for delay testing of sub-100nm designs. Also note that establishing very conservative design rules may result in chips that are slow.

The effects of noise and process variations on delays can be modeled as a distribution function. Such timing behavior strongly suggests that statistical analysis and simulation should play a role in the selection and testing of critical paths. However, developing such models requires significant effort, tools, and algorithms. The researchers in industry and academia have focused on various aspects of this problem and novel and practical models will soon be developed. In future, statistical timing and noise analysis will appear in the industry tool flows for design sign-off.

2.2 High Quality Test Patterns

As explained in the previous sections, delay test pattern generation faces a number of challenges for successful and reliable screening of defective chips. Traditional ATPG algorithms based on simple fault models and zero delay gate models no longer suffice for nanometer technologies. Advanced fault models are required to handle crosstalk, bridging defects and small delay defects. There is an growing concern from the semiconductor industry for timing-aware ATPG tools pushing EDA vendors for competitive solutions. Industry had adopted various DFT alternatives such as clock-gating and divide-and-conquer test adding more area and routing overhead to generate IR-drop tolerant pattern generation rather than improving the ATPG tools.

Pattern generation techniques must become physical aware and consider technology dependent parameters to stress appropriate parts of the chip better, which are more susceptible and prone to defects. Constant feedback is required from foundries especially during the early stages of adopting new technologies to target special physical locations with certain characteristics more comprehensively during pattern generation.

2.3 Small Delay Defects

Transition fault model is widely practiced in industry to test delay-induced defects and it is a cost-effective alternative to functional pattern generation [14]. Traditionally, transition fault tests were generated assuming a fixed cycle time for each clock domain (generally delay tests are generated/applied one clock domain at a time). Under the above assumption, a delay defect will be detected only when it causes a transition to reach an observe point (primary output or scan flip-flop) by more than the positive slack of the affected path. Slack of a path is a measure of how close a transition on the respective path meets the timing to an observable point, relative to the test cycle time.

A delay defect with defect size not large enough to cause a timing failure under the fixed cycle time notion is referred to as a small delay defect. A small delay defect might escape during test if is tested through a short path. While the same defect might be activated on a longer path during functional operation and it may cause a timing failure. Also, small delay defects might become a *reliability* issue as the defect might magnify during subsequent aging in the field and cause a failure of the device. Hence, it is important to detect such defects during manufacturing test using efficient techniques [15].

Encounter True-Time Delay Test Tool TM [25] uses actual design timing (Standard delay format (SDF)) information for ATPG. The tool uses efficient ATPG algorithms and pseudorandom data to achieve high coverage in fewer patterns. It also uses back-annotated timing information (SDF) to apply them at faster-than-at-speed. It sets the transition test capture frequency based on the slack of the paths exercised and also includes the ability to test non-critical paths faster-than-at-speed for small delay defects. The possible limitations of such techniques are: 1) the SDF must be calibrated with very high accuracy to correspond with the tester operating conditions and 2) the timing information must also take into account the process variation effects and dynamic effects, such as the IR-drop and crosstalk, 3) the compression will be low due to masking longer paths when generating patterns for shorter paths.

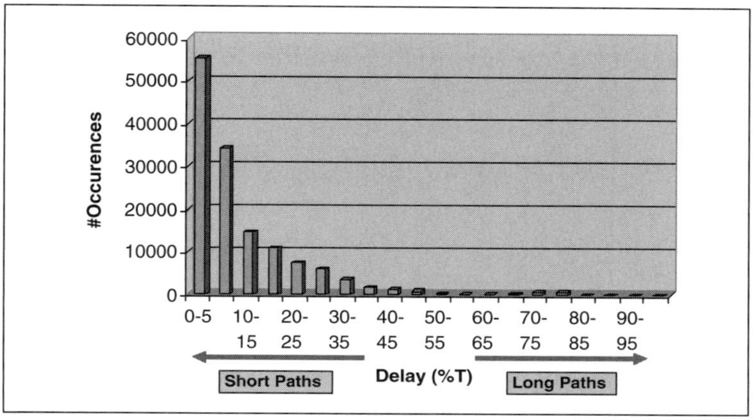

Fig. 2.5. Path delay distribution for no-timing ATPG transition fault pattern set (benchmark s38584) [12].

Figure 2.5 shows the delay distribution of a transition fault pattern set generated using a timing unaware ATPG tool for an ISCAS'89 benchmark (s38584). The patterns were generated using launch-off-capture technique (*broadside*) for the total transition fault list (52874 faults) and the gross delay fault coverage and pattern count were 76.92% and 372 respectively. It can be noticed that majority of the paths exercised for delay fault detection are short paths. In this particular example, most of the paths affected are less than 30% of cycle time. A delay defect size of at least 70% cycle time is required for the faults to be detected. Therefore, more robust at-speed techniques are required to improve the effectiveness of transition fault testing to affect more longer paths and screen the small delay defects better.

2.4 Using Low-Cost Testers to Reduce Capital Test Cost

The cost of testing is becoming a concern in the nanometer era due to larger number of faults, more fault types, and the resultant increase in pattern count. There is a concern that test cost may well exceed the cost of design and manufacture in the near future. Various companies began using low-cost testers to reduce the capital test cost.

The low-cost testers are normally optimized for use with scannable and BISTed designs. Such testers are available in a 256 or 512 pin configuration. In order to keep the cost contained, it is constructed largely from off the shelf components, and the timing accuracy requirements are relaxed when compared to a high performance VLSI tester. Such testers usually contain a single high frequency source, capable of moderate length clock bursts of up to several hundred megahertz. Due to the limitations posed by low-cost testers, design-for-test (DFT) engineers must devise new test and pattern generation methods to achieve high quality test. Two of the implementation challenges are described below:

2.4.1 Local At-Speed Scan Enable Generation

When implementing launch-off-shift method, an at-speed scan enable signal is required to switch the design from test mode to functional mode. Since the low-cost testers, may not necessarily be able to provide such at-speed signal, new methods are required to generate it. On-chip scan enable signal generation is one possible solution to the problem. The signal must be synchronized with clock and easy to generate (please see Chapter 3 for more details).

2.4.2 At-Speed I/O Testing

Due to limited number of pins with the ability to send and receive test data at-speed, chip I/O pins and paths between I/O pins and internal flip-flops cannot be tested. As a result, the fault coverage reported from ATPG with *no input changes* and *no output measures* is low. Thus, this inability will result in yield loss which can be significant for chips with large number of I/Os. A strategy can be used to place all I/Os in a form of scan wrapper. Such I/O wrapper will provide test engineers with the ability to test paths between internal flip-flops and I/O wrapper scan flip-flops. However, the paths between I/O pins and wrappers flip-flops are still not tested. Another issue is that todays large designs normally have multiple frequency domains; this complicates the task of wrapper insertion considering the low-cost testers limitations.

References

1. International Technology Roadmap for Semiconductors 2006 edition.
2. T. M. Mak, A. Krstic, K. T. Cheng and L. C. Wang, "New challenges in delay testing of nanometer, multigigahertz designs," IEEE Design & Test of Computers, vol. 21, no. 3, May-June 2004, pp. 241 - 248.
3. K. Roy, K. T. Cheng and T.M. Mak, "Test Consideration for Nanometer-Scale CMOS Circuits," *IEEE Design & Test of Computers*, pp. 128-136, 2006.
4. K. Kim, S. Mitra and P.G. Ryan, "Delay Defect Characteristics and Testing Strategies," IEEE Design & Test of Computers, vol. 20, no. 5, Sept.-Oct. 2003 pp. 816
5. R. Wilson, "Delay-Fault Testing Mandatory, Author Claims," *EE Design*, Dec. 2002.

6. M. Tehranipour, N. Ahmed and M. Nourani, "Testing SoC Interconnects for Signal Integrity Using Extended JTAG Architecture," *IEEE Transactions on CAD*, vol. 23, issue 5, pp. 800-811, May 2004.
7. M. Cuviello, S. Dey, X. Bai, and Y. Zhao, "Fault Modeling and Simulation for Crosstalk in System-on-Chip Interconnects," in Proc. *Int. Conf. Computer-Aided Design (ICCD'99)*, pp. 297-303, 1999.
8. A. Sinha, S. K. Gupta, and M. A. Breuer, "Validation and Test Issues Related to Noise Induced Byparasitic Inductances of VLSI Interconnects," *IEEE Trans. Adv. Packaging*, pp. 329339, 2002.
9. W. Chen, S. Gupta, and M. Breuer, "Test Generation for Crosstalk-Induced Delay in Integrated Circuits," in Proc. *Int. Test Conf. (ITC'99)*, pp. 191200, 1999.
10. M. Nourani, M. Tehranipoor and N. Ahmed, "Pattern Generation and Estimation for Power Supply Noise Analysis," in proc. *IEEE VLSI Test Symposium (VTS'05)*, pp. 439-444, 2005
11. S. Zhao and K. Roy, "Estimation of Switching Noise on Power Supply Lines in Deep Sub-micron CMOS Circuits," in Proc. *Thirteenth Int. Conf. on VLSI Design,*, pp. 168-173, 2000.
12. N. Ahmed, M. Tehranipoor and V. Jayaram, "Timing-Based Delay Test for Screening Small Delay Defects," in Proc. *Design Automation Conference (DAC06)*, pp. 320-325, 2006
13. N. Ahmed, M. Tehranipoor and V. Jayaram, "A Novel Framework for Faster-than-at-Speed Delay Test Considering IR-Drop Effects," in Proc. *Int. Conf. on Computer-Aided Design (ICCAD'06)*, 2006.
14. X. Lin, R. Press, J. Rajski, P. Reuter, T. Rinderknecht, B. Swanson and N. Tamarapalli, "High-Frequency, At-Speed Scan Testing," *IEEE Design & Test of Computers*, pp. 17-25, Sep-Oct 2003.
15. B. Kruseman, A. K. Majhi, G. Gronthoud and S. Eichenberger, "On hazard-free patterns for fine-delay fault testing," in Proc. *Int. Test Conf. (ITC'04)*, pp. 213-222, 2004.
16. S. Borkar, "Designing Reliable Systems from Unreliable Components: The Challenges of Transistor variability and Degradation," IEEE Micro, vol. 25, no. 6, Nov.-Dec. 2005, pp. 10-16.
17. A. Agarwal, D. Blaauw, V. Zolotov "Statistical Timing Analysis for Intra-Die Process Variations with Spatial Correlations," in Proc. *Int. Conf. on Computer-Aided Design*, pp. 900-907, 2003.
18. D. Blaauw and S. Nassif, "Static Performance Analysis under Process and Environment Variations," http://domino.research.ibm.com/acas.
19. K. Agarwal, D. Sylvester, D. Blaauw, Frank Liu, S. Nassif, and S. Vrudhula, "Variational Delay Metrics for Interconnect Timing Analysis," in Proc. Design Automation Conference (DAC'04), pp. 381-384, 2004.
20. J. Le, X. Li, and L. Pileggi, "VSTAC: Statistical Timing Analysis with Correlation," in Proc. *Design Automation Conference (DAC'04)*, pp. 343-348, 2004.
21. B. F. Romanescu, M. E. Bauer, S. Ozev and D. J. Sorin, "VariaSim: Simulating Circuits and Systems in the Presence of Process Variability," Technical Report, DUKEECE, no. 2007-3, June, 2007,
22. Z, Lin, C. Spanos, L. Milor, Y. Lin, "Circuit Sensitivity to Interconnect Variation," IEEE Trans. on Semiconductor Manufacturing, vol. 11, no. 4, 557-568, 1998.

23. A. Agarwal, D. Blaauw, V. Zolotov "Statistical Clock Skew Analysis Considering Intra-Die Process Variations," in Proc. *Int. Conf. on Computer-Aided Design*, pp. 914-921, 2003.
24. V. Mehrotra and D. Boning, "Technology Scaling Impact of Variation on Clock Skew and Interconnect Delay," in Proc. *Intl. Interconnect Tech. Conf.*, pp. 122-124, 2001.
25. Cadence Inc., "http://www.cadence.com,", 2005.

Further Reading

26. L. Wang, P. Bastani, M. S. Abadir, "Design-Silicon Timing Correlation A Data Mining Perspective," in Proc. *IEEE Design Automation Conference*, pp. 384-389, 2007.
27. B. N. Lee, L. C. Wang, M. S. Abadir, "Refined statistical static timing analysis through learning spatial delay correlations," in Proc. *Design Automation Conference*, 2006.
28. L. C. Wang; T. M. Mak, K. T. Cheng, M. S. Abadir, "On path-based learning and its applications in delay test and diagnosis," in Proc. *Design Automation Conference*, 2004.
29. L. C. Wang, "Regression simulation: applying path-based learning in delay test and post-silicon validation," in Proc. *Design, Automation and Test in Europe Conference*, vol. 1, pp. 692-693, 2004.
30. R. Putman, R. Gawde, "Enhanced timing-based transition delay testing for small delay defects," in Proc. *VLSI Test Symposium*, 2006.
31. Y. Sato, S. Hamada, T. Maeda, A. Takatori, Y. Nozuyama, S. Kajihara, "Invisible delay quality - SDQM model lights up what could not be seen," in Proc. *Int. Test Conference*, 2005.
32. M. Kumar and S. Tragoudas, "High-Quality Transition Fault ATPG for Small Delay Defects," IEEE Trans. on Computer-Aided Design of Integrated Circuits and Systems, vol. 26,no. 5, pp. 983-9.
33. X. Lin; K. Tsai; M. Kassab, J. Rajski, T. Kobayashi, R. Klingenberg, R. Klingenberg; Y. Sato, S. Hamada, T. Aikyo, "Timing-Aware ATPG for High Quality At-speed Testing of Small Delay Defects," in Proc. *Asian Test Symposium*, pp. 139-146, 2006.
34. V. Kumar, S. Tragoudas, "Quality transition fault tests suitable for small delay defects," in Proc. *Int. Conf. on Computer Design*, pp. 468-470, 2005.
35. S. Hamada, T. Maeda, A. Takatori, Y. Noduyama, Y. Sato, "Recognition of Sensitized Longest Paths in Transition Delay Test," in Proc. *Int. Test Conference*, pp. 1-6, 2006.
36. H. Yan and A. Singh, "Evaluating the effectiveness of detecting delay defects in the slack interval: a simulation study," in Proc. *Int. Test Conference*, pp. 242-251, 2004.
37. B. Kruseman, A. Majhi, G. Gronthoud, S. Eichenberger, "On hazard-free patterns for fine-delay fault testing," in Proc. *Int. Test Conference*, pp. 213-222, 2004.
38. M. Favalli and C. Metra, "Pulse propagation for the detection of small delay defects," in Proc. *Design, Automation & Test in Europe Conference & Exhibition*, pp. 1-6, 2007.

39. R. Tayade, S. Sundereswaran and J. Abraham, "Small-Delay Defect Detection in the Presence of Process Variations," in Proc. *Int. Symp. on Quality Electronic Design*, pp. 711-716, 2007.
40. K. T. Cheng and H. C. Chen, "Generation of High Quality Non-Robust Tests for Path Delay Faults," in Proc. *Design Automation Conference*, pp. 365-369, 1994.
41. H. Yan and A. Singh, "A delay test to differentiate resistive interconnect faults from weak transistor defects," in Proc. *Int. Conf. on VLSI Design*, pp. 47-52, 2005.
42. K. L. Shepard and V. Narayanan, "Noise in Deep Submicron Digital Design," in Proc. IEEE Int. Conf. on Computer Aided Design (ICCAD), pp. 524-531, 1996.
43. N. Ahmed, M. Tehranipoor and V. Jayaram, "Supply Voltage Noise Aware ATPG for Transition Delay Faults," IEEE VLSI Test Symposium (VTS), 2007.
44. N. Ahmed, M. Tehranipoor and V. Jayaram, "Transition Delay Fault Test Pattern Generation Considering Supply Voltage Noise in a SOC Design," Design Automation Conf. (DAC), 2007.
45. S. Pant, D. Blaauw, V. Zolotov, S. Sundareswaran and R. Panda, "Vectorless Analysis of Supply Noise Induced Delay Variation," in Proc. IEEE Int. Conf. on Computer Aided Design (ICCAD), pp.184-191, 2003.
46. J. Wang, Z. Yue, X. Lu, W. Qiu, W. Shi, and D. Walker, "A vector-based approach for power supply noise analysis in test compaction," in Proc. Int. Test Conf. (ITC), 2005.
47. J. Wang, X. Lu, W. Qiu, Z. Yue, S. Fancler, W. Shi, and D. Walker, "Static compaction of delay tests considering power supply noise," in Proc. IEEE VLSI Test Symp. (VTS), pp. 235240, 2005.
48. R. Ahmadi and F. N. Najm, "Timing Analysis in Presence of Power Supply and Ground Voltage Variations," in IEEE/ACM International Conference on Computer Aided Design, San Jose, CA, Nov. 2003, pp. 176-183.
49. A. Muhtaroglu, G. Taylor, and T. Rahal-Arabi, "On-die droop detector for analog sensing of power supply noise," Jour. of Solid-State Circuits, vol. 39, no. 4, pp. 651660, 2004.
50. Z. Abuhamdeh, B. Hannagan, J. Remmers and A. Crouch, "Consideration For a Production IR-Drop Screen On A Chip," IEEE Design & Test of Computers, May-June 2007.
51. P. Ghanta and S. Vrudhula, "Analysis of Power Supply Noise in the Presence of Process Variations," IEEE Design & Test of Computers, May-June 2007.
52. I. Pomeranz and S. M. Reddy, "An Efficient Nonenumerative Method to Estimate the Path Delay Fault Coverage in Combinational Circuits," IEEE Transactions on Computer-Aided Design, vol. 13, no. 2, Feb. 1994, pp. 240-250.

3

Local At-Speed Scan Enable Generation Using Low-Cost Testers

At-speed testing is becoming crucial for modern VLSI systems which operate at clock speeds of hundreds of megahertz to several gigahertz. In a scan-based test methodology, it is common to use transition delay fault model for at-speed testing. The test procedure is to create a transition at a node using scan chains for controllability, capture the results after a time period equal to one system clock cycle, and observe the contents of the scan chain through serial shift operation. The launching of the transition can be done either in the last cycle of scan shift (called launch-off-shift), or in a functional launch cycle that follows the scan shift and precedes the fast capture (called launch-off-capture). When comparing these two, the launch-off-shift technique offers significant advantages over the launch-off-capture in terms of coverage and pattern count, but since it requires the scan enable signal to change state in the time period of one functional clock cycle, considerable engineering resources and design efforts are required to close the timing on the scan enable signal. Usually, due to high-speed pin limitation, low-cost testers may not be able to provide the at-speed scan enable signal as required by launch-off-shift technique.

In this chapter, an efficient scheme is presented to practice the launch-off-shift methodology using low cost/speed testers as well as reducing the scan enable routing area. The scan enable control information for the launch and capture cycle is embedded in the test data itself. A new scan cell, called the last transition generator (LTG), generates the local fast scan enable signals used to control scan cells in a scan chain. The LTG cell has the flexibility to be inserted anywhere in the scan chain and the hardware area overhead is comparable to the pipeline scan enable approach. This allows the physical synthesis tool to only focus on minimizing the routing congestion caused by scan cells rather than the location of LTG cells.

In general, DFT insertion is performed with arbitrary selection of scan flip-flops across different scan chains controlled by the local scan enable signal which results in higher scan enable routing overhead. Unlike pipeline approach, the LTG cells drive a pre-determined set of scan flip-flops in a localized area stitched in a scan chain. This will result in a much lower signal enable routing area. This method poses no additional constraints for the place and route (PNR) tool and scan flip-flops selection method provides a better mechanism to reorder the scan cells to meet the timing closure of the local scan enable signals. Since the LTG cell becomes part of scan chain, it can easily be tested using flush patterns therefore no extra effort is required to test LTG cells.

3.1 Introduction

Semiconductor industry is adopting new fabrication processes to reduce the cost per chip and meet the area, power and performance requirements. As a result, modern ICs are growing more complex in terms of gate count and operating frequency. Traditionally, the defects seen in higher technology nodes (≥ 130 nm) were mostly static in nature, which could be targeted and detected by traditional stuck-at tests. In addition to stuck-at tests, I_{DDQ} testing became an effective way to increase test quality. This was because quiescent leakage current in CMOS was low enough for an unusually high leakage to be able to detect many of the non-static defects [1] [18] [16] [17] [19].

A transition fault test requires a pattern pair $(V1, V2)$ to be applied to the circuit-under-test (CUT). Pattern $V1$ is termed as the initialization pattern and $V2$ as the launch pattern. The response of the CUT to the pattern $V2$ is captured at the operational functional speed. The entire operation can be divided into three cycles: 1) Initialization Cycle (IC), where the CUT is initialized to a particular state ($V1$ is applied), 2) Launch Cycle (LC), where a transition is launched at the target gate terminal ($V2$ is applied) and 3) Capture Cycle (CC), where the transition is propagated and captured at an observation point.

Depending on how the transition is launched and captured, in general, there are two widely used transition fault pattern generation methods. In the first method, referred to as *launch-off-shift* (LOS) or *skewed-load* [10], the transition at the gate output is launched in the last shift cycle during the shift operation. Figure 3.1(a) shows the launch-off-shift method waveform for a multiplexed-DFF design. Note that a similar approach can also be applied to an level sensitive scan design (LSSD). The LC is a part of the shift operation and is immediately followed by a fast capture pulse. The scan enable (SEN) is high during the last shift and must go low to enable response capture at the CC clock edge. The time period for SEN to make this $1 \rightarrow 0$ transition corresponds to the functional frequency. Hence, LOS requires the SEN signal to be timing critical. Skewing the clock (CLK) creates a higher launch-to-capture clock frequency than standard shift clock frequency. Saxena et al. [13] list more launch and capture waveforms used by launch-off-shift approaches. In [13], implementation of scan-based transition fault testing and its low cost test challenges are discussed.

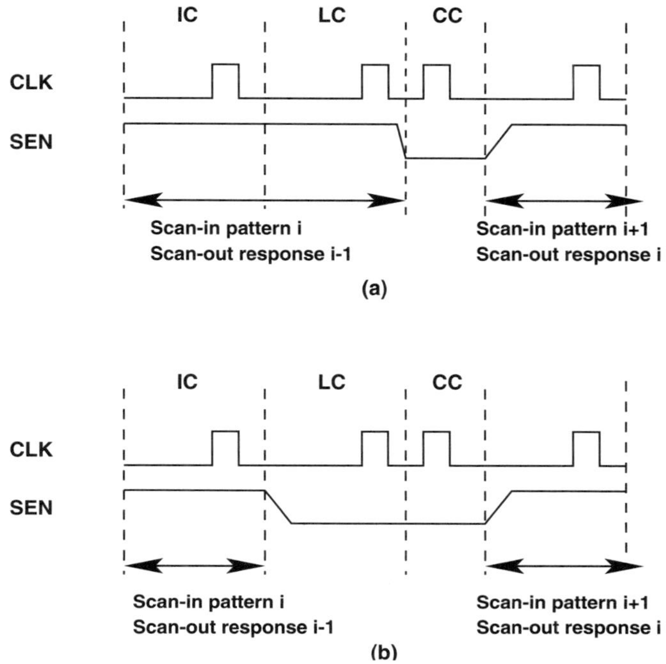

Fig. 3.1. Transition delay fault pattern generation methods: (a) Launch-off-shift and (b) Launch-off-capture [14].

Figure 3.1(b) shows the waveforms of the second approach, referred to as *launch-off-capture* (LOC) or *broadside* method [11]. In this method, the launch cycle is separated from the shift operation. At the end of scan-in (shift mode), pattern $V1$ is applied and CUT is set to an initialized state. A pair of at-speed clock pulses are applied to launch and capture the transition at the target gate terminal. Pattern $V2$ is the functional response of pattern $V1$. This relaxes the at-speed constraint on the SEN signal and dead cycles are added after the last shift to provide enough time for the SEN signal to settle low.

The LOS method is more preferable based on ATPG complexity and pattern count compared to LOC method [6]. The LOC method is based on a sequential ATPG algorithm, while the LOS method uses a combinational ATPG algorithm. This will increase the test pattern generation time in case of LOC and also a high fault coverage cannot be guaranteed due to the correlation between the two patterns, $V1$ and $V2$; note that $V2$ is the functional response of pattern V1. Due to its dependence on the functional capture path for launching transitions, LOC ATPG can be quite slow, and on large designs, can consume several days of ATPG run-time. The main concern about LOS however is its requirement to at-speed scan enable (SEN) signal.

As design sizes grow due to higher integration, the number of flip-flops in a design is increasing, and it is not uncommon to see designs with several hundred thousand flip-flops today. The physical design of the scan enable signal for LOS is therefore as challenging as that of the clock signal. An approach similar to clock tree synthesis is not used to address this issue due to the associated design cost [20]. The problem can be alleviated by increasing the number of scan enable ports thereby reducing the fanout of scan enable signal, but this is not practical when low-cost testers are employed [13].

In [21], a hybrid architecture is proposed which controls a small subset of selected scan cells by LOS and the rest are controlled by LOC approach. A fast scan enable signal generator is designed which drives the LOS-controlled scan flip-flops. Its design is such that the fast scan enable signal makes a transition only on the negative edge of the controlling clock. Therefore, all positive edge scan flip-flops loose half cycle for the scan enable signal. Moreover, the selection criteria of the LOS-controlled scan flip-flops and the order in which these flip-flops are stitched determines the effectiveness of the method. In some cases the number of patterns generated by the hybrid method exceeds the launch-off-capture pattern count [21].

A widely used method in industrial practice is to pipeline the scan enable signal [22]. In a multi-stage pipeline scan enable, the designer must carefully select the group of scan cells controlled by the respective scan enable signal. In order to meet timing closure of the pipeline scan enable signals, this selection criteria manifests into design iterations and additional design effort for the place and route (PNR) tool especially for scan reordering during routing step. A scan insertion tool stitches the scan chains based on alpha-numeric naming convention hierarchically with no knowledge of the physical location of the cells in the design. If the scan chain is allowed to be routed in an un-ordered fashion, the resultant routing occupies more available routing resources. A physical design tool reorders the scan chain starting from the scan input pin such that the Manhattan distance from the output port *(Q)* of a scan cell to the scan-in port *(SD)* of the following cell is minimum. Reordering the scan chain based on the physical placement information will optimize the amount of routing resources needed for scan chain routing but it might increase the routing area of the pipeline scan enable. This can be a cause of routing congestion in the design, and could force the layout designers to enlarge the given placement to accommodate this inefficient scan enable routing. A not careful reordering may cause a mess in terms of routing of scan enable signals in a chip.

Saxena et al. [13] discuss scan-based implementation of transition fault testing methods and the challenges posed by low-cost test equipments. A careful planning and execution is required when using low cost testers to test transition faults. Authors of [13] describe implementation of LOC method on modern designs which use a low-speed scan enable signal.

Several other researchers have also focused on other issues related to transition fault testing, such as fault coverage, ATPG, yield loss, functionally testable and untestable faults, etc. Authors of [23] proposed an enhanced scan-based delay fault testing which reduces the area overhead when compared to conventional enhanced scan [12]. However, the proposed technique offers high area and delay overhead compared to LOC and LOS methods. Instead of using an extra latch, a supply gating was proposed to be used at the first level of logic gates of a combinational circuit.

A transition fault automatic test pattern generator methodology for scan-based designs using broadside (launch-off-capture) test format is proposed in [24]. A replicate and reduce circuit transform maps the two time frame processing of transition fault ATPG to a single time frame processing on duplicated iterative blocks with reduced connections. The authors of [24] report high transition fault coverage using this ATPG method.

A path oriented test generation procedure called POTENT is proposed in [25] to generate high quality tests for transition faults. Conventional ATPG tools are unaware of the circuit states exercised during the functional operation of the circuit. In [26], the authors presented a concept of testing only functionally testable transition faults in broadside transition testing via a constrained ATPG. The method generates a test set for functionally testable transition faults and minimizes detection of functionally untestable transition faults by the assumption that ATPG knows illegal states.

A SAT-based ATPG for path-oriented transition faults is proposed in [27]. A transition fault is detected through the longest path. A false path pruning technique is utilized to identify the longest path through each fault site, which results in faster test pattern generation. Author in [28] argues that scan-based transition fault testing is shown to allow tests to be applied which are unrealizable in normal operation. In other words, scan-based fault testing methods have negative impact on yield and designer productivity due to over testing for transition faults.

3.1.1 A Big Picture of Low-cost Testers

The cost of testing is becoming a big concern in the nanometer era due to larger number of faults, more fault types, and the resultant increase in pattern count. There is a concern that test cost may well exceed the cost of manufacture in the near future. Currently semiconductor industry is taking a serious look at using low-cost testers instead of very expensive testers to reduce the overall capital test cost. However, to be able to apply high quality tests to design under test some functionality of modern testers must be moved into the design, e.g. built-in self-test for transition delay fault test, using PLL to generate on-chip test clocks, etc.

A low-cost tester is usually optimized for use with scannable and BISTed designs but may vary from one company to another. Such testers are available with a limited pin configuration (e.g. 256 or 512). In order to keep the cost contained, it is constructed largely from off the shelf (OTS) components, and the timing accuracy requirements are relaxed when compared to a high performance VLSI testers. The top speed attainable for functional pattern operation on such testers is in the range of few tens of megahertz. For most designs running in production today, such testers are incapable of delivering at-speed functional patterns primarily due to the low maximum speed of the pin electronics. However, most of such testers do contain a single high frequency source, capable of moderate length clock bursts of up to few hundred megahertz to one gigahertz with edge placement. The operation of its high speed internal oscillator is completely asynchronous with respect to the data on the low-speed tester channels. Also, the high frequency source is not dedicated to any particular tester channel, but may be mapped to any device pin as required.

Such low-cost testers are capable of executing LOC method since no at-speed SEN signal is required and SEN can be relaxed and be fed by a low speed external scan enable signal. Note that inserting dead cycles between the last shift and the launch cycle is a very common practice in industry. However, new techniques are required to implement and practice LOS using these kind of testers.

3.2 Background and Motivation

The main concern in implementing launch-off-shift (LOS) method is that it requires the scan enable (SEN) signal to switch at functional speed. Here, we describe the pipelined scan enable methodology that is able to overcome this problem. Figure 3.2 shows the basic implementation and operation of a single stage pipeline scan enable for launch-off-shift [22]. The scan enable port (SEN_port) is de-asserted asynchronously in the initialization cycle. The pipeline scan enable ($SEN\_pipeline$) goes low synchronously at the active edge of launch clock cycle. The SEN_port is asserted high after the capture cycle to re-enter the shift mode. The $SEN\_pipeline$ signal is generated synchronously and has one full clock cycle available for transition, assuming the pipeline scan enable signal is generated separately for positive and negative edge flip-flops. Also, the tester skew at the scan enable port is decoupled from the internal scan enable net.

Figure 3.3 shows a three stage pipeline scan enable architecture. A leaf pipeline stage is defined as the last stage of the scan enable pipeline tree which directly drives the scan enable of a subset of scan flip-flops in the respective logic cone. There are mainly two concerns in the pipeline scan enable insertion, 1) determine the scan flip-flops to be controlled by each leaf stage and 2) determine the level (module or top) at which the leaf stages are inserted. Earlier, the test engineers used to arbitrarily select the controlled scan flip-flops. In arbitrary selection, the leaf stage pipeline will control scan flip-flops across different scan chains. The physical design tool is challenged in two dimensions, 1) re-order scan flip-flops in respective scan chains to reduce each scan chain routing area and 2) placement of leaf pipeline scan enable

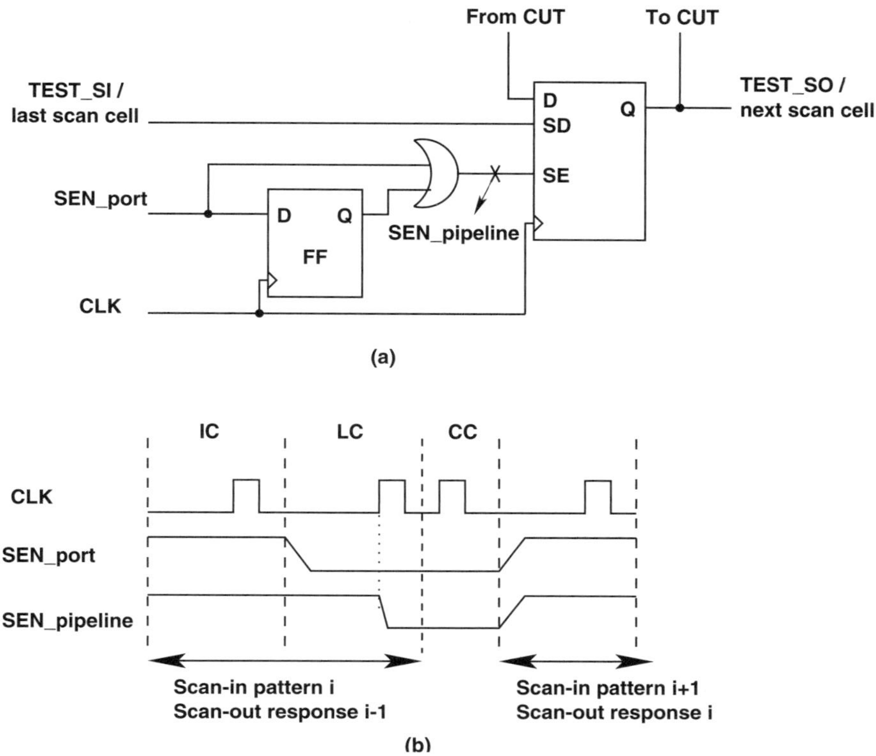

Fig. 3.2. Pipeline scan enable [15].

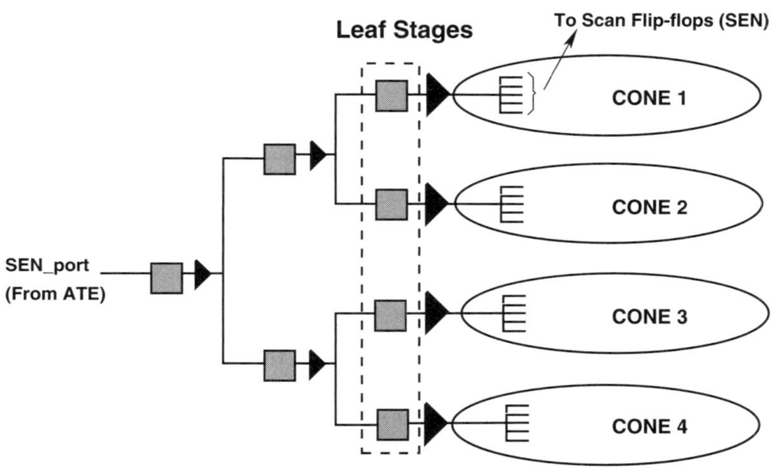

Fig. 3.3. Pipeline scan enable architecture [15].

stage to reduce area (repeater + routing) across chains (arbitrary selection might select flip-flops across scan chains). Selection of flip-flops in the same scan chain will reduce this two fold problem into one and the re-order of scan cells can be used in a positive sense to reduce the scan enable routing also.

Typically in very large designs, scan insertion is performed at module level and later the module level scan chains are stitched at the top level. Customarily, one pipeline stage is added per module and later at the top-level these are connected in the form of a tree network. Figure 3.4 shows this scan enable pipeline insertion methodology. In this example, there are four blocks A, B, C and D and each block is inserted with a pipeline stage which forms the leaf stage of the scan enable tree. The leaf stages are connected with additional stages at the top-level depending on the timing requirements. Since, each module contains different number of flip-flops, the number of scan cells driven by the leaf stages will be different, resulting in an imbalanced scan enable tree. Also, different clock domains have different clock insertion delays, the unbalanced tree may result in additional effort for physical designers to adjust the drive strengths of buffers in different branches of the tree.

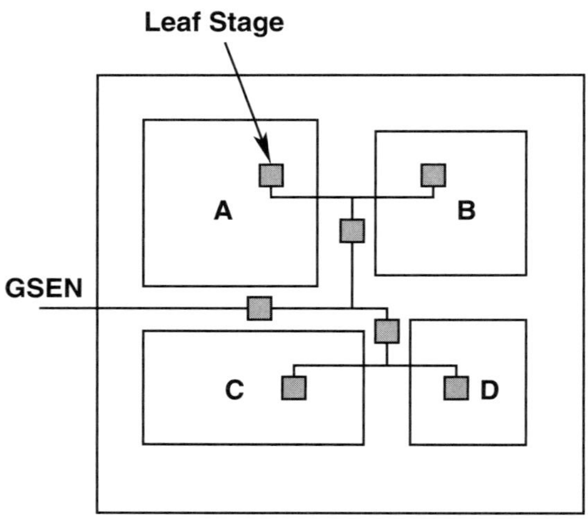

Fig. 3.4. One possible pipeline scan enable insertion technique [15].

In modern-day VLSI design flows, the physical design flow reorders the scan flip-flops in scan chains to reduce the routing congestion, after placement step based on the proximity of the scan flip-flops. Since, the leaf stages control scan flip-flops across different scan chains, the re-ordering of scan flip-flops may negatively impact the routing of the scan enable signal. The problem however is to determine the set of scan flip-flops controlled by each pipelined scan enable stage so that proximity of flip-flops to the leaf stage is taken into account.

This chapter presents generation of fast scan enable signals through the use of some special sequential cells called last transition generator (LTG) cells; these LTG cells are similar to scan flip-flops and one or more LTG cells are inserted into a scan chain. The LTG cell generates the scan enable signals for all or a subset of cells in *the same scan chain.* A physically-aware scan insertion tool can treat the LTG cells similar to scan flip-flops and can hence ensure proximity of the LTG cell to all the flip-flops that it controls. This is straight-forward when there is a single LTG cell per scan chain. Sometimes, when the scan chain has a large number of flip-flops, more than one LTG cell may become necessary. In this case, the scan insertion tool can still be used to first create smaller scan segments, each of which contains one LTG cell, and then stitch the scan segments together. In modern SoC designs that employ test data compression, it is common to use a large number of short scan chains [33]. The LTG-based method will be a natural fit in such situations. Moreover, in LTG-based method, the leaf stages are inserted in the scan path which increases the testability of the design and inserted cells. Note that the leaf stages in a pipelined scan enable cannot be easily tested unlike LTG cells inserted in scan chain.

The problem can be defined as to determine a set of scan flip-flops, controlled by the leaf pipeline scan enable stage, which are physically close to each other. Currently, the scan insertion tools are unaware of the physical location of the scan cells during scan insertion. The only information available to the test engineer are the number of scan cells and the number of scan chains to be stitched. In the LTG-based method, each leaf stage controls the scan cells of one particular scan chain. One can add either one/two/three leaf stages per scan chain. The fanout of the leaf node is used as a criteria to find the number of leaf stages. In this case, even if the scan cells are reordered, the leaf stage will be driving the same set of flip-flops, reordered to reduce the routing area. The leaf stages are inserted at the top-level during scan chain stitching. The scan enable routing will be more modular.

In the LTG-based method, since the number of scan chains and scan cells/chain is known ahead, the number of scan flip-flops driven by the leaf pipeline node can be a criteria to find the number of (LTG/leaf pipeline stage) cells. Either one/two/three cells per scan chain can be added depending the fanout cone of the leaf node. Moreover, the advantage of LTG cell is that it can be added anywhere which will be reordered later by the floor-planning tool. They can be added during top-level stitching between module scan chains or anywhere. In addition, the LTG cells can be fed either by the external tester scan enable directly or through additional pipeline stages similar to a pipeline scheme.

The pipeline scan enable method requires routing awareness before insertion. The number of pipelined flip-flops added in the scan enable path depends on the worst case delay. Multiple pipeline scan enable stages can be added depending on the distance of the last stage driving the scan enable network. However, multiple stage pipelines have increased susceptibility to noise glitches from tester or due to crosstalk. An assertion glitch on the first pipeline stage would force scan enable active for multiple clock cycles. Note that, the pipeline scan enable scheme is an industry practiced technique and the LTG-based technique discussed in the next section (Section 3.3) provides more robustness and can be practiced along with the pipeline scan enable scheme.

3.3 Local Scan Enable Signal Generation

Practicing the LOS method for at-speed testing requires high-speed testers. The tester provides an at-speed scan enable signal to launch patterns and capture the responses. As mentioned earlier, the main goal in this chapter is to test transition faults using LOS method implemented with low-cost testers. The pipelined scan enable methodology divides the fanout of the scan enable signal without using multiple external pins and eliminates the external scan enable port tester skew. It is unaware of the floorplanning step and does not provide much flexibility to the PNR tool in terms of selecting the scan cell group driven by the leaf pipeline scan enable stage, which, in some cases may result in scan enable routing overhead. A technique is presented that inherits the advantages of pipelined scan enable methodology and yet makes physical design easier.

In order to improve testability and provide more flexibility with all the advantages of the pipeline scan enable, the local scan enable generation method is presented such that scan enable generator cells are inserted *within* the scan chains. Therefore, the control information can be passed as part of the test data. Before describing the local scan enable generator cell, first, it is explained how the transition data can be passed through scan chain as part of the test data without inserting an LTG cell in the scan chain. Figure 3.5 shows a small example of generating the local scan enable signal from the test pattern data for LOS. The external scan enable signal from the tester is referred to as the global scan enable (GSEN). The internally generated scan enable signal is termed as local scan enable (LSEN). The main objective is to de-assert GSEN in the initialization cycle without the at-speed constraint and then generate the at-speed LSEN signal during the launch and capture cycle synchronously from the test data. There are eight scan flip-flops in the scan chain and the test pattern shifted is $b_8b_7b_6b_5b_4b_3b_2b_1$=10001110 assuming b_1 is the first bit shifted into the scan chain. The values shifted into the scan flip-flops during the various shift cycles are shown under each flip-flop. GSEN is de-asserted during the $(n-1)$th shift cycle (IC), where n=8.

For proper shift operation, the LSEN signal should be logic 1 in the $(n-1)$th cycle of the shift operation (IC) and logic 0 in the last shift cycle (LC) to enable capture in the next clock cycle. In other words, the LSEN signal must make a $1 \rightarrow 0$ transition at the launch edge. For this particular example, the pattern during the shift operation generates the required $1 \rightarrow 0$ transition at

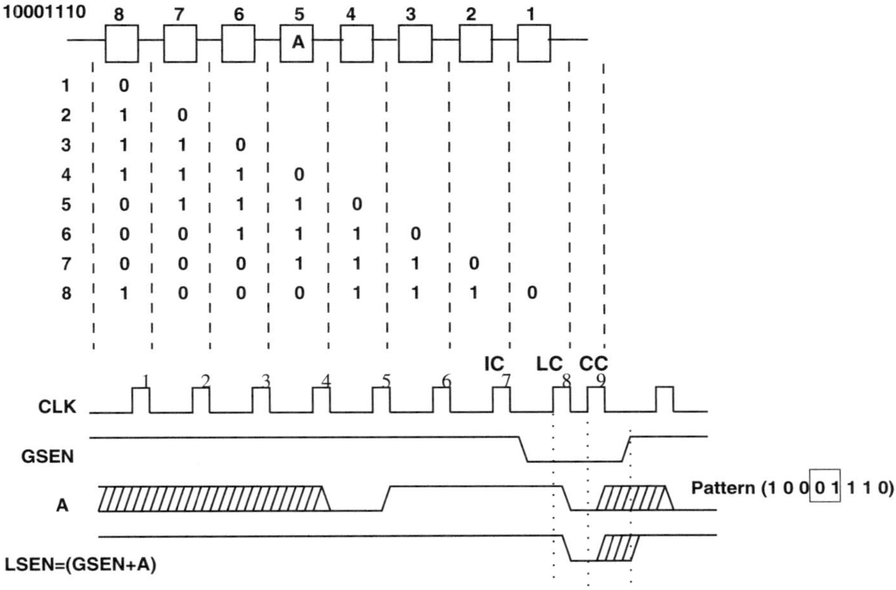

Fig. 3.5. Generation of local scan enable signal [14].

the output of scan flip-flop A. The output of scan flip-flop A is ORed with GSEN to generate the LSEN signal. Therefore, the final value of scan flip-flop (A) and its following scan flip-flop (b_5 and b_4) at the end of shift operation must be 0 and 1, respectively, so that A is loaded with logic 1 in IC and logic 0 in LC. A full functional clock cycle is available for LSEN to make the transition. After the capture cycle, the LSEN signal is asynchronously set to 1 by GSEN for scanning out the response.

3.3.1 Last Transition Generator (LTG)

As explained earlier, during LOS pattern generation, to generate the scan enable transition $1 \rightarrow 0$ at the launch edge, the scan flip-flop A should be 0 and the following scan flip-flop must be 1. This is very unlikely to occur in every pattern generated during ATPG. It can also be seen in Figure 3.5 that there is an unknown value loaded in A during the capture edge, which can cause a problem if the method is to be used for clock sequential pattern generation with multiple capture cycles. For such patterns, the LSEN signal must be de-asserted and multiple capture clocks must be applied. Applying capture signal may also change the content of scan flip-flop A since it is connected to the CUT. As a result, LSEN may change depending upon what value (0 or 1) is received as response bit into the scan flip-flop A. Hence, A

must be constrained to 0 during ATPG during pattern generation and scan flip-flop A must be selected such that the response coming back into A be 0. This constraint is necessary but not sufficient for proper operation as the value loaded in A from the functional path after applying the first system clock is not known. Therefore, after going to capture control state (0), LSEN must remain in this state as long as it is asynchronously set to 1 by GSEN. This requires additional atpg constraints for a conventional scan flip-flop architecture to control the functional path (D) to logic 0. This might lead to fault coverage reduction.

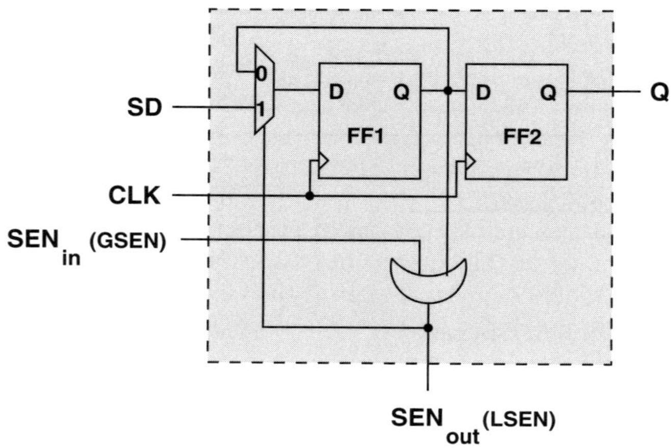

Fig. 3.6. Last transition generator (LTG) cell [14].

In order to avoid loss of coverage without significant change in the architecture, a new scan cell called *last transition generator* (LTG) is designed such that when the scan cell is loaded with logic 0 during pattern shift and the scan enable control is made logic 0, i.e. capture state, the cell will remain in that state until it is shift controlled by GSEN. This cell is inserted into scan chain to generate the $1 \rightarrow 0$ transition at the launch edge through the scan path.

Figure 3.6 shows the LTG cell architecture. It consists of two flip-flops which are used to load the control information required for the launch and capture cycles. The port definition is similar to a scan cell and the output of FF1 (Q) is fed back to the functional input port of the scan cell. The input D of LTG cell does not exist because it does not connect to CUT. The LTG cell consists of an input scan enable (SEN_{in}) pin which takes GSEN signal as input. An additional output scan enable (SEN_{out}) pin (GSEN+Q) represents the LSEN signal. FF2 does not allow the output of FF1 to be shifted into the next normal scan cell in scan chain in the last shift process. The LTG cell can be inserted anywhere in the scan chain and it is not connected to the CUT.

Therefore, any atpg constraint on the LTG cell does not affect the CUT fault coverage.

Theorem: The local scan enable signal generated by the LTG cell switches at-speed during the capture cycle.

Proof: SEN_{out} refers to the local scan enable signal in the LTG cell of Figure 3.6. The clock input to the LTG cell for launch-off-shift transition delay ATPG is of the form shown in Figure 3.1(a). It is assumed that the clock tree synthesis tool is capable of routing the clock signal so that the local clock signal at the input of the LTG cell switches at functional speed during the LC and CC cycles. During the scan shift cycle (IC), a "1" is scanned into the flip-flop FF1 of LTG cell at low frequency. During the last cycle of shift, denoted by LC, the clock switches at functional speed and the output of FF1 also switches to 0 state at the functional speed, since the number of flip-flops driven by SEN_{out} is an order of magnitude smaller than the total number of flip-flops in the design, thereby reducing the capacitive load on the local scan enable signal. The global scan enable signal switches to 0 during the beginning of the LC cycle. Let A refer to the output of FF1. SEN_{out} is the logical OR of the signal A and the global scan enable signal; therefore, SEN_{out} also switches at the speed of signal A, except for the small delay in the OR gate.

3.3.2 Operation of LTG Cell

Figure 3.7(a) shows the previous example with the LTG cell inserted in the scan chain. Note that, the LTG cell can be placed anywhere in the scan chain and it is not connected to the CUT. The values of the two flip-flops of the LTG cell in the test pattern are shown as X (1000[XX]1110). These flip-flops are constrained during atpg to assign specific values to the Xs. Figure 3.7(b) shows the pattern and the timing waveform for LOS. During the shift operation, at the last shift, the scan enable must make a $1 \rightarrow 0$ transition. Thus, FF1 of LTG cell should be loaded with 1 in IC, followed by 0 in the next cycle (LC). The SEN_{in} (GSEN) signal is asynchronously de-asserted in IC. The SEN_{out} (LSEN) signal is generated by the Boolean equation $SEN_{out} = A(Q_{FF1}) + SEN_{in}$. After the capture cycle, the LSEN is asserted to 1 asynchronously by GSEN.

For LOC, the GSEN signal is high during the entire shift operation, after which, GSEN is asynchronously de-asserted and LSEN must change to logic 0. For LOC, since LSEN must remain 0, the value in FF2 is a don't-care. The value of FF1 of LTG cell must be constrained to 0 during atpg. Figure 3.7(c) shows the pattern and the timing waveform for LOC. It can be noticed that SEN_{out} switches at the speed of GSEN (not at-speed).

Compare the LOS timing waveforms in Figures 1(a) and 6(b). In Figure 1(a), the scan enable signal may feed both positive edge and negative edge triggered flip-flops. In the worst case, therefore, the scan enable signal must

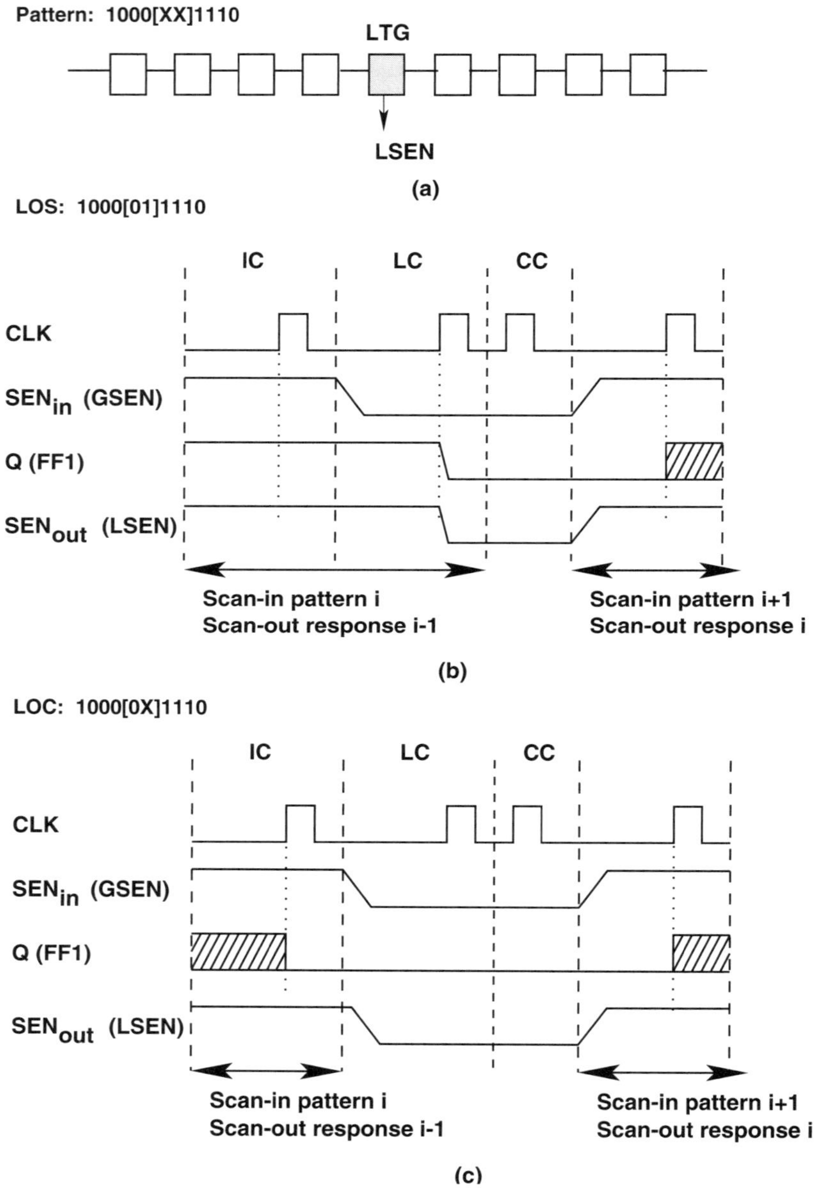

Fig. 3.7. Operation of LTG cell, (a) Example scan chain, (b) LOS and (c) LOC [14].

make a transition at the negative edge of the LC clock. This cuts down the time for scan enable transition to a half of the functional clock cycle, making timing closure harder. In the timing waveform of Figure 6(b), it can be seen that this problem can be eliminated. It is customary to place positive edge and negative edge flip-flops on separate scan chains. Therefore, a separate LTG cell can be used to control posedge and negedge flip-flops. As a result, every local scan enable signal has one complete functional clock cycle to transition.

3.4 DFT Architecture

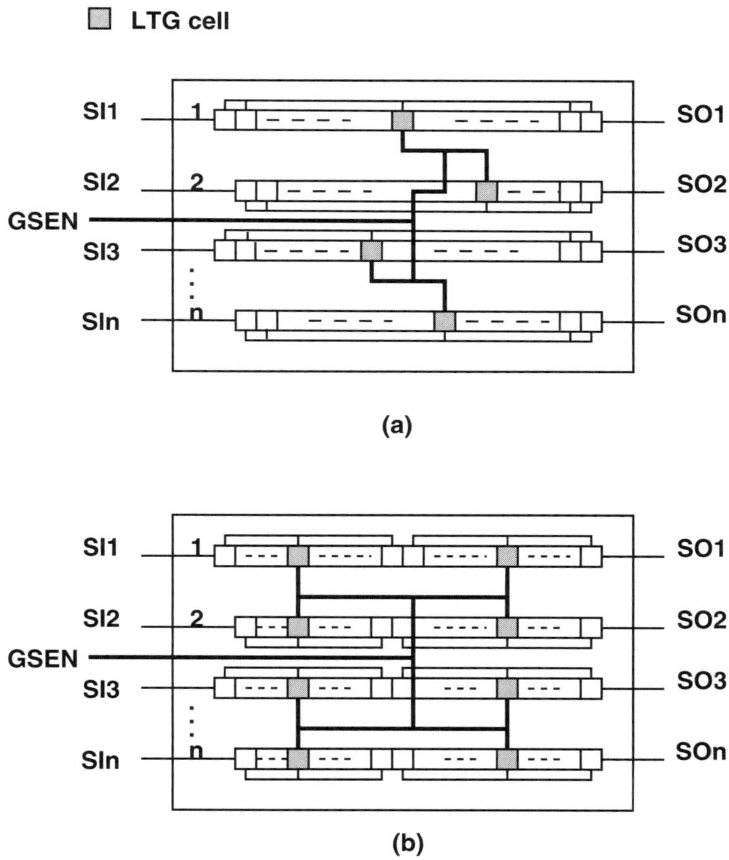

Fig. 3.8. Test Architecture (a) one LTG cell per scan chain and (b) two LTG cells per scan chain [15].

In conventional scan architecture the global scan enable (GSEN) is connected to all scan cells in every scan chain. While in the new architecture, the local at-speed scan enable (LSEN), generated by LTG cell using GSEN and pattern data, is connected to each and every scan cell in the respective scan chain. The GSEN signal is only connected to LTG cells. In general, there can be multiple scan chains in a design to reduce the test application time. Figure 3.8(a) shows a multiple scan chain architecture with n scan chains and each scan chain contains m scan cells. As shown, each scan chain i, where $1 \leq i \leq n$, consists of one LTG cell which generates the fast scan enable signal LSEN_i for m scan cells.

The area overhead of an LTG cell (equivalent to the sum of two flip-flops and some glue logic) is a few extra gates, which is negligible in modern designs. Note that, if scan enable timing is not met then multiple LTG cells can be inserted to generate multiple LSEN signals to control different segments of the same scan chain as shown in Figure 3.8(b). In this case, two smaller trees for local enable signal are designed.

The main area overhead of a tree design is the routing area and buffers inserted on the tree. The fanout load on the GSEN signal is reduced and the fanout load driven by a local scan enable signal is used as a constraint to find the number of LTG cells inserted. For example, for k total number of flip-flops in a design and l being the maximum number of flip-flops that can be allowed for the local scan enable to be timing closed for a particular operating frequency, the number of LTG cells are estimated by k/l.

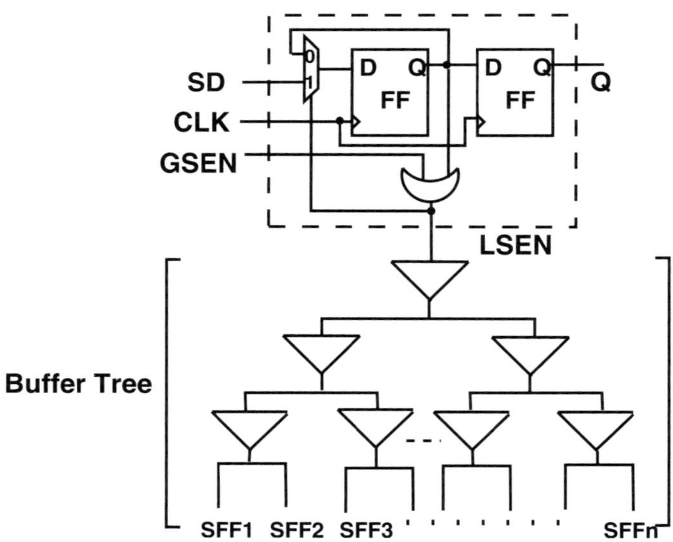

Fig. 3.9. LTG cell with buffer tree driving scan enable signal [15].

Although, the area of the LTG cell alone is negligible, the size of the buffer tree driving the local scan enable varies depending on the number of controlled scan flip-flops. Note that, in general, the buffer tree is required for all cases when using, 1) GSEN as an at-speed SEN signal generated using ATE, 2) pipeline method and 3) LTG-based method. For example, when using an at-speed GSEN drived from tester, clock tree is required to connect GSEN to all the scan flip-flops in the design. The size of the tree depends on the drive strength of the signal and the number of scan flip-flops.

Figure 3.9 shows the architecture of LTG cell and the buffer tree driving the local scan enable signal. The buffer size and the number of stages required are mutually exclusive. In the experiment, a minimum size buffer is used with a drive strength of two. Figure 3.10 shows the number of repeaters in the buffer tree required to drive different number of scan flip-flops. The operating frequency selected was 250MHz and a maximum number of 2000 scan flip-flop's scan enable, meeting the timing requirements, can be driven by a single LTG cell. As shown in Figure 3.10, the number of repeaters required in the buffer tree reduces drastically with decrease in the number of controlled scan flip-flops.

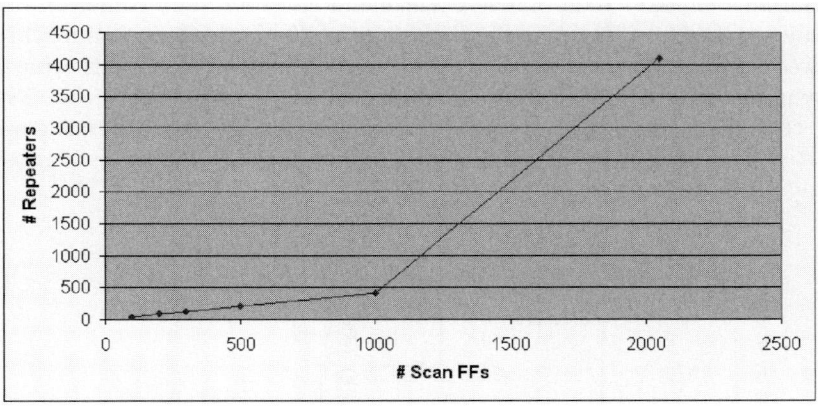

Fig. 3.10. Number of repeaters required to drive scan flip-flops by a single LTG cell [15].

Figure 3.11 shows the relationship between number of repeaters required and multiple LTG cell insertion when using 2000 scan flip-flops per scan chain with an operating frequency of 250MHz. The buffer tree size decreases using multiple LTG cells. The percentage reduction is not significant when inserting 2, 3 and 4 LTG cells. Note that, further increasing the number of LTG cells may not significantly reduce the buffer tree size while the area overhead of LTG cells will increase. Based on these results, when scan chain length=2000 and frequency=250MHz, it can be recommended to insert at least two to four

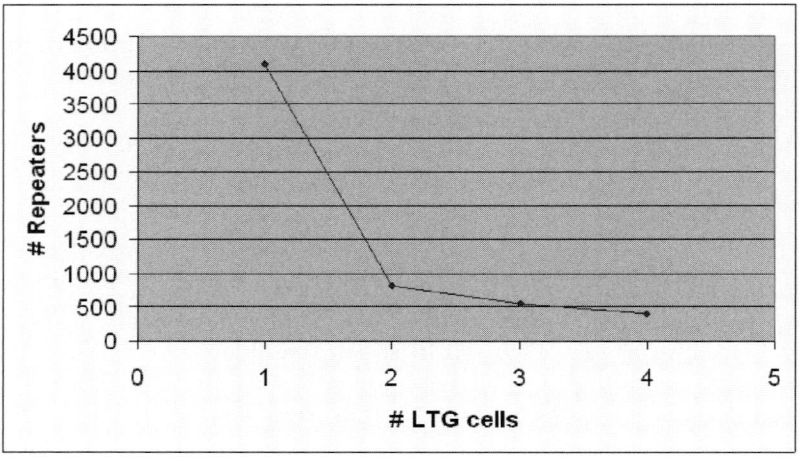

Fig. 3.11. Number of repeaters required for multiple LTG cells inserted per scan chain (scan chain length = 2000) [15].

LTG cells per scan chain. These experiments give only an estimation of the number of LTG cells required and the size of the buffer tree required. Note that, in general, the area of the scan enable tree is similar to the pipeline scan enable scheme. Also notice that due to increasingly using commercial test compression tools, in practice, the length of scan chains are considerably shorter than 2000. Thus, the size of clcok tree to drive the SEN signals will be significantly smaller.

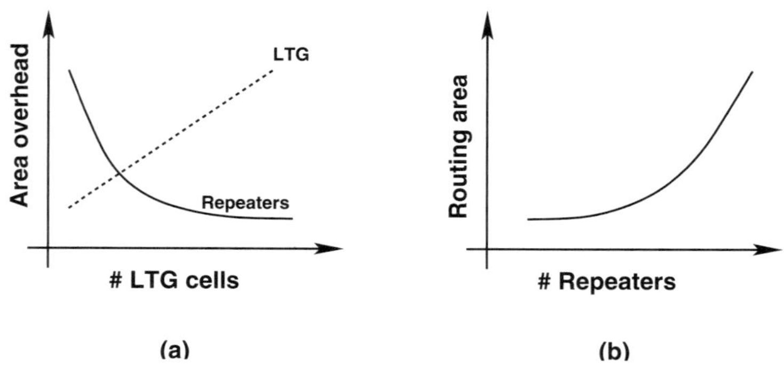

Fig. 3.12. (a)LTG cell and repeater area overhead (b)Routing area overhead versus number of repeaters [15].

The area overhead can be classified into three different categories: 1) LTG cell, 2) repeaters and 3) routing. Figure 3.12 shows the trend of each of these categories. It can be seen (Figure 3.12(a)) that with multiple LTG cell insertion, although the LTG cell area increases (negligible) but there is significant decrease in the repeater area required to drive the local scan enable tree. Also the routing area increases with increase in the number of repeaters (Figure 3.12(b)). Therefore, an efficient tradeoff would be possible to perform with carefully designing the buffer tree for LSEN and calculating the overall area overhead based on the scan enable tree area (number and size of repeaters), routing area and LTG cells.

3.4.1 Multiple Clock Domain Analysis

In general for a multi-clock domain design, the scan chains are clock mixed to balance the scan chains. To avoid setup and hold violations due to inter-clock domain skews, lock-up latches are inserted between flip-flops of different clock domains. Figure 3.13 shows an example with two clock domains $CLK1$ and $CLK2$ respectively. The LTG cell in the scan chain is connected to $CLK2$. A negative level sensitive lock-up latch is added between the scan flip-flop of $CLK1$ domain and the LTG cell. This will give an additional half cycle for the other clock domain to avoid skew problems. There are two possible cases, $Case1$: $CLK1$ edge occurs before $CLK2$ (Figure 3.13(b)) and $Case2$: $CLK1$ edge occurs after $CLK2$ (Figure 3.13(c)). In $Case1$, the LTG cell data shifts and the local scan enable is generated at $CLK2$ edge. In $Case2$, the local scan enable is generated at $CLK2$ edge, which occurs before $CLK1$. This might result in hold violations in $CLK1$ domain due to scan enable transition. Hence, to avoid hold violations, the LTG cell must be connected to the late arriving clock. The timing constraint for the LSEN signal is that it must be timing closed for the fastest clock domain in the scan chain. It can be represented as:

$$T_{SEN} = T_{period} - T_{skew} - T_{setup}$$

The transition fault patterns for LOS are generated per clock domain. If the pattern is shifted at a slow speed followed by a fast capture, the time from the launch edge of LC to the capture edge of CC is not really at-speed. Figure 3.14(a) shows the limitation of the clock timing waveform. The functional operating frequency is 125MHz. The launch edge in the last shift occurs at 45ns and the capture edge occurs at 2ns in the capture cycle of 8ns time period. The time from the launch edge to the capture edge is (55+2)=57ns. Figure 3.14(b) shows the modified at-speed clock timing waveform used for LOS. The last shift is done at-speed corresponding to the clock domain being tested and the capture clock is applied only for that clock domain. A dead cycle (DC) is added after the initialization cycle for the scan chain to settle down.

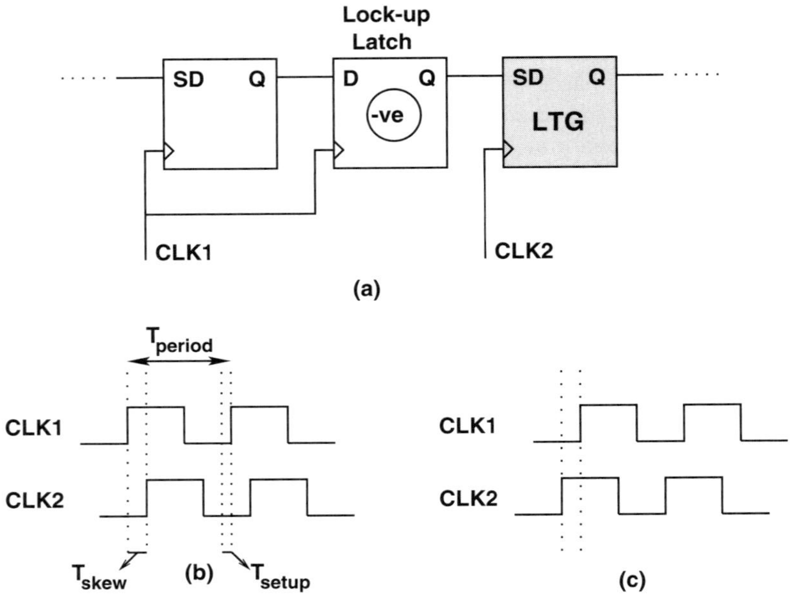

Fig. 3.13. Multiple Clock Domain Analysis [14].

3.4.2 LTG Insertion Flow

After determining the number of LTG cells to be inserted based on the optimization analysis explained in the early part of this section, scan insertion is performed. There are two issues relating to insertion of multiple LTG cells, 1) the number of LTG cells, 2) insertion flow of LTG cells (module level or top-level). There are two widely used scan insertion flows, 1) Top-level scan insertion flow and 2) Bottoms-Up scan insertion flow. For smaller designs, the preferred flow is top-level scan insertion and for large designs, the bottoms-up scan insertion flow is followed. Figure 3.8 shows the top-level scan insertion with a single LTG cell per scan chain.

The DFT insertion tool [32] allows a single internal signal to control the scan enable of all the scan flip-flops in a single chain. For a design with N scan chains and M LTG cells to be inserted per scan chain, the scan insertion tool is directed to insert $N \times M$ scan chains with one LTG cell per scan chain. In the next step, the scan chains are re-stitched into N scan chains. For example, if $N = 4$ and $M = 2$, there are 8 scan chains inserted with one LTG cell per scan chain and later these chains are re-stitched into 4 chains at top-level.

As the designs presently are large and often reuse IPs, the scan insertion is done at module level and then these chains are connected at the chip level. It is also referred to as bottoms-up approach. Figure 3.15 shows the flow 1 where each module is scan inserted and each scan chain having a individual scan enable control signal. The LTG cells are inserted manually between the

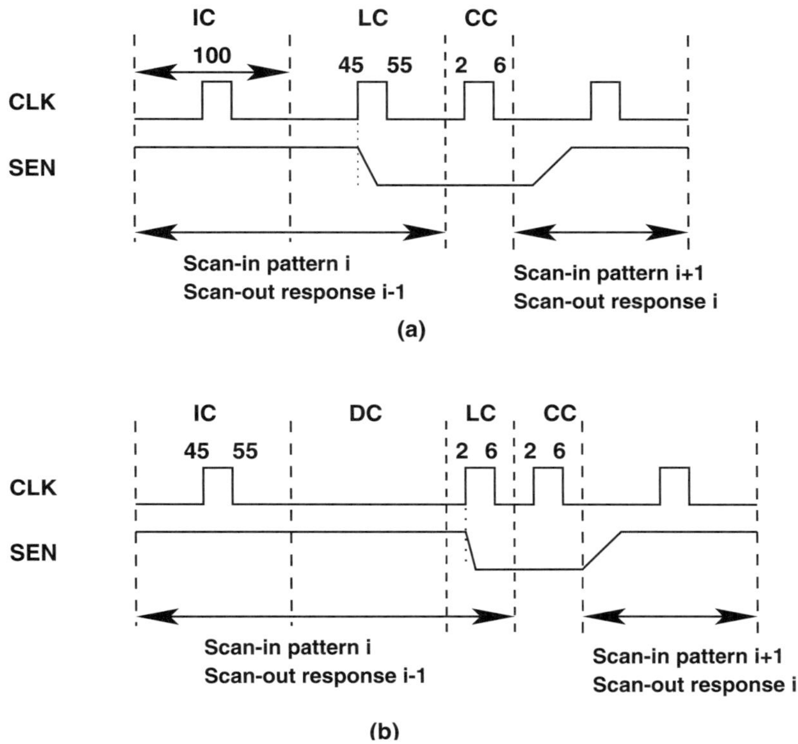

Fig. 3.14. LOS clock timing waveform [14].

modules and the scan chains are stitched at the top-level. In flow 2, each modules scan chains are inserted with LTG cells and the scan chains are stitched at the top-level (Figure 3.16). This might be more preferable as the test engineer will only have to hookup the scan chains appropriately and need not worry about LTG cell insertion.

3.4.3 ATPG

There is no fundamental difference in the ATPG methodology when the LTG-based solution is used. The scan enable signal for the flip-flops now comes from an internally generated signal. An ATPG tool must be able to place the scan enable signal in the active mode during scan shift operation. Notice that the OR gate in the LTG cell generates the local scan enable signal through a logical OR of the global scan enable and the Q output of the flip-flop FF1. The $1 \rightarrow 0$ transition of the local scan enable is generated by the pattern shift. During "design rule check (drc)" phase, the ATPG tool performs a single sensitization shift where it assumes a don't-care in each individual scan flip-flop and shifts a known value to observe it at the output of each individual scan flip-flop.

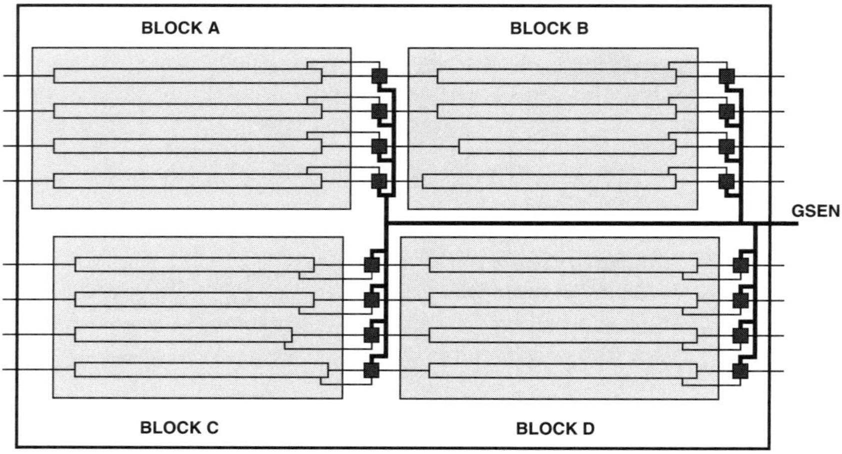

Fig. 3.15. Bottoms-Up scan insertion: Flow 1 [15].

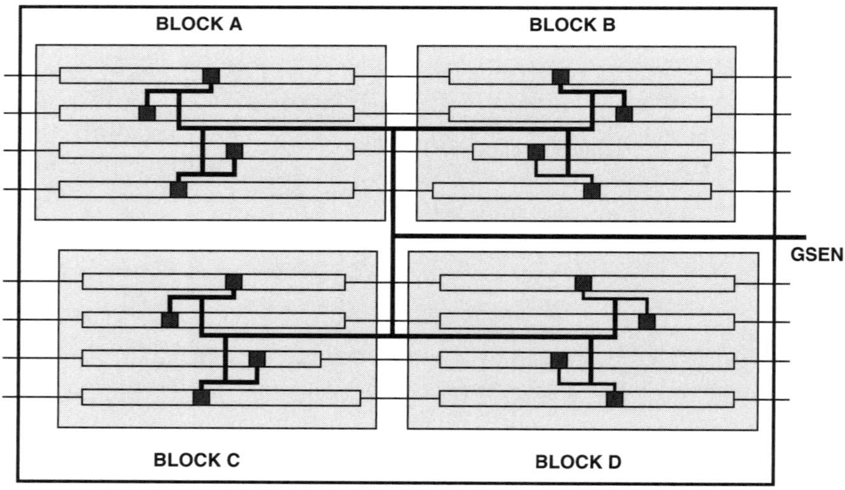

Fig. 3.16. Bottoms-Up scan insertion: Flow 2 [15].

The tool starts from the scan output and performs this process till it reaches the scan in pin and checks whether the scan chain is traceable. This makes the ATPG tool to fail in the *drc* phase as the LTG-based technique assumes known values '0' and '1' in FF1 and FF2 respectively to generate the scan enable signal internally. In order to get around the problem of the *drc* phase of the ATPG tool failing due to the internal nature of the scan enable signals of the flip-flops, a different model of the LTG cell is used for ATPG process.

In a commercial ATPG tool, the loss in coverage due to undetectability of stuck-at-1 faults on the scan enable signals can be recovered by declaring

```
01: "load_unload" {
02:     W "_slow_WFT_";
03:     V { "CLK1"=0; "CLK2"=0;"GSEN"=1;}
04:     Shift {
05:         W "_slow_WFT_";
06:         V { "CLK1"=P; "CLK2"=P; "GSEN"=1; "_so"=#; "_si"=#; }
07:     }
        //ADDING DEAD CLOCK CYCLE
08:     V { "CLK1"=0; "CLK2"=0; "GSEN"=0; }
09:     W "_fast_WFT_";//Nth SHIFT CYCLE
10:     V { "CLK1"=P; "CLK2"=P; "GSEN"=0; "_so"=#; "_si"=#; }
11: }
```

Fig. 3.17. TetraMAX ATPG protocol file.

the scan enable signal as a clock signal. Figure 3.17 shows the *load_unload* procedure of the test protocol file in *Standard IEEE Test Interface Language (STIL) format*. Each vector (*V*) statement is a tester clock cycle. The waveform table (*W*) statement determines the clock period of the tester clock cycle defined in the timing waveform procedure of the test protocol file. The waveform table *_slow_WFT_* is the slow shift clock cycle and *_fast_WFT_* is the at-speed clock cycle. The GSEN signal is high till the (n-1)th shift and is made low in the dead cycle before the last shift which provides enough time for GSEN to go low. The last shift is done at-speed and it can be seen that the waveform timing is changed to *_fast_WFT_* before the last *V* statement.

3.5 Experimental Results

This chapter has argued in favor of the 'launch-off-shift' transition delay ATPG methodology and presented a technique that can ease the implementation of this technique using low speed testers. First, it is explained how the LTG cells are inserted using a commercial tool and the procedure following during pattern generation. Figure 3.18 shows the list of steps required to insert LTG cells during scan chain insertion. Here, it is assumed that one LTG cell is inserted per scan chain. The scan insertion tool needs to recognize the LTG cell as a scan register containing two scan cells in order to stitch it into the scan chain. This requires it to be defined as a new library cell with scan cell attributes. A workaround is to design the LTG cell as a module and declare it as a scan segment (*line 04*), e.g. *set_scan_segment* command in Synopsys DFT Compiler [32]. The tool then identifies LTG cell as a scan segment of length 2. The GSEN signal is connected to all the LTG cell's SEN_{in} input port and the clock is connected to CLK input. The scan insertion tool is then directed to stitch each individual scan chain c_i including LTG_i cell and hookup the LTG_i/SEN_{out} port to the scan enable port of the remaining scan cells in the chain (*line 09*). Only the LTG cell needs to be specified in the scan path, as the tool will stitch the rest of the cells including the LTG cell and balances the scan chain depending on the longest scan chain length parameter.

```
DFT:
01: for i = 0 to no_chains
02: {
03:     Instantiate LTG_i cell
04:     Declare LTG_i cell as scan segment
05:     Make connections to LTG cell
06:         Connect GSEN
07:         Connect clock
08:     Stitch scan chain c_i including LTG_i cell
09:     Hookup LSEN from LTG_i to all scan cells in the chain
10: }

ATPG:
01: foreach pattern
02: {
03:     N − 1 slow shifts
04:     Dead Cycle for GSEN to settle low
05:     Nth fast shift
06:     Fast capture cycle
07: }
```

Fig. 3.18. Scan Insertion and ATPG Process.

The ATPG methodology is no different from conventional when the LTG-based solution is used. The scan enable signal for the flip-flops now comes from an internally generated signal. An ATPG tool must be able to place the scan enable signal in the active mode during scan shift operation. Notice that the LTG cell generates the local scan enable signal through a logical OR of the global scan enable and the Q output of the flip-flop FF1 (see Figure 3.6). The $1 \rightarrow 0$ transition of the local scan enable is generated by the pattern shift. During "design rule check (drc)" phase, the ATPG tool performs a single sensitization shift where it assumes a don't-care in each individual scan flip-flop and shifts a known value to observe it at the output of each individual scan flip-flop. The tool starts from the scan output and performs this process till it reaches the scan-in pin and checks whether the scan chain is traceable. This makes the ATPG tool to fail in the *drc* phase as the LTG-based technique assumes known values '0' and '1' in FF1 and FF2 respectively to generate the scan enable signal internally. In order to get around the problem of the *drc* phase of the ATPG tool failing due to the internal nature of the scan enable signals of the flip-flops, a different model of the LTG cell is used during ATPG process.

Table 3.1. Designs Characteristics [15]

Design	Scan Cells	# Chains	Clocks	LTG cells	TF
A	10477	16	6	16	384998
B	40342	16	4	48	2124502
C	104539	8	1	96	4116092

Table 3.2. Results for three industrial designs obtained from conventional LOS.

Design	FC(%)	# Patt	*Conventional LOS*		
			Scan Cells	Max. Length	Test Cycles
A	81.44	1093	10477	655	0.715M
B	84.41	4160	40342	2521	10.487M
C	84.16	5958	104539	13067	77.853M

Table 3.3. Results for three industrial designs obtained from LTG-based LOS.

Design	FC(%)	# Patt	*LTG-based LOS*		
			Scan Cells	Max. Length	Test Cycles
A	81.44	1093	10509	657	0.718M
B	84.41	4160	40438	2527	10.512M
C	84.16	5958	104731	13091	77.996M

The LTG-based method have been experimented on three industrial designs, A, B and C, and Table 3.1 shows the characteristics of these designs. In design A each scan chain is inserted with one LTG cell, while in designs B and C, three and twelve LTG cells are inserted per scan chain, respectively, to meet scan enable timing requirements. The total transition faults are shown in column *TF*. The test strategy is to get the highest possible test coverage for the transition faults. When generating test patterns for the transition faults, only the faults in the same clock domain are targeted. During pattern generation, only one clock is made active during the capture cycle. Hence, only faults in that particular clock domain are tested. All PIs remain unchanged and all POs are unobservable while generating the test patterns for the transition faults. This is because the tester is not fast enough to provide the PI values and strobe POs at speed.

The ATPG results and comparison of conventional LOS with LTG-based LOS is shown in Tables 3.2 and 3.3. TetraMAX [31] tool was used for ATPG. The FC and # Patt columns show the fault coverage and the number of patterns generated, respectively, for LOS method. The chain length for LTG-based increases by twice the number of LTG cells inserted per scan chain because each LTG cell contains two flip-flops. The table also shows the area and time overhead of LTG-based LOS method over conventional LOS method (see columns 7 and 9). Notice that the increase in the number of flip-flops and the test application time by LTG insertion is not very significant.

The LTG-based solution has the following advantages.

1. The technique can be practiced using existing commercial tools for DFT insertion and ATPG.
2. The technique is applicable to all scan-based DFT techniques, including Deterministic BIST techniques that are based on scan [33].
3. The method is least intrusive and can be incorporated in existing physical design flows.
4. In the experiments, it was demonstrated the use of the technique using DFT Compiler and TetraMAX tools from Synopsys. The area overhead and impact on the functional timing due to inclusion of LTG cells is negligible.

Inserting LTG cells *within* scan chain has the following advantages:

1. It increases the scan chain controllability.
2. LTG cells can be easily tested using flush patterns during testing scan cells.
3. LTG cells can be used for both LOS and LOC methods.
4. The LTG-based method can be practiced along with other techniques such as pipelined scan.

3.6 Summary

In this chapter, a method was presented to enable the design teams to practice launch-off-shift (LOS) transition delay testing using low-cost testers. LOS testing is known to provide better quality results, both in terms of pattern count and fault coverage, but design teams may not use launch-off-shift due to the challenge of routing the scan enable signal and design effort. The solution shown in this chapter generates local scan-enable signals that can switch at functional speeds; for this purpose, the method relies on embedding some control information in the patterns. A special cell called the LTG cell is designed and used for the generation of the local scan enable signal. This cell is simple to design and layout, and its area overhead is comparable to that of a scan flip-flop. A complete analysis is also provided for finding the optimum number of LTG cells required. The LTG-based solution improves testability of the added hardware, provides greater flexibility and fewer constraints to the back-end flow during place and route step. The DFT insertion and ATPG can be easily performed using the commercial ATPG tools; therefore the LTG-based solution is easy to practice.

References

1. International Technology Roadmap for Semiconductors 2001 (http://public.itrs.net).
2. S. Natarajan, M.A. Breuer, S.K. Gupta, "Process Variations and Their Impact on Circuit Operation," in Proc. *IEEE Int. Symp. on Defect and Fault Tolerance in VLSI Systems*, pp. 73-81, 1998.
3. R. Wilson, "Delay-Fault Testing Mandatory, Author Claims," *EE Design*, Dec. 2002.
4. G. Aldrich and B. Cory, "Improving Test Quality and Reducing Escapes," in Proc. *Fabless Forum, Fabless Semiconductor Assoc.*, pp. 34-35, 2003.
5. X. Lin, R. Press, J. Rajski, P. Reuter, T. Rinderknecht, B. Swanson and N. Tamarapalli, "High-Frequency, At-Speed Scan Testing," *IEEE Design & Test of Computers*, pp. 17-25, Sep-Oct 2003.
6. V. Jayaram, J. Saxena and K. Butler, Scan-Based Transition-Fault Test Can Do Job, *EE Times*, Oct. 2003.
7. K. Cheng, "Transition Fault Testing for Sequential Circuits," *IEEE Transactions on Computer-Aided Design of Integrated Circuits and Systems*, vol. 12, no. 12, pp. 1971-1983, Dec. 1993.
8. T. M. Mak, A. Krstic, K. Cheng, L. Wang, "New challenges in delay testing of nanometer, multigigahertz designs," *IEEE Design & Test of Computers*, pp. 241-248, May-Jun 2004.
9. M. Bushnell, V. Agrawal, *Essentials of Electronics Testing*, Kluwer Publishers, 2000.
10. J. Savir, "Skewed-Load Transition Test: Part I, Calculus," in Proc. *Int. Test Conf. (ITC'92)*, pp. 705-713, 1992.
11. J. Savir and S. Patil, "On Broad-Side Delay Test," in Proc. *VLSI Test Symp. (VTS'94)*, pp. 284-290, 1994.
12. B. Dervisoglu and G. Stong, "Design for Testability: Using Scanpath Techniques for Path-Delay Test and Measurement," in Proc. *Int. Test Conf. (ITC'91)*, pp. 365-374, 1991.
13. J. Saxena, K. M. Butler, J. Gatt, R. Raghuraman, S. P. Kumar, S. Basu, D. J. Campbell, J. Berech, "Scan-Based Transition Fault Testing-Implementation and Low Cost Test Challenges," in Proc. *International Test Conference (ITC'02)*, pp. 1120-1129, Oct. 2002.
14. N. Ahmed, C.P. Ravikumar, M. Tehranipoor and J. Plusquellic, "At-Speed Transition Fault Testing With Low Speed Scan Enable," in proc. *IEEE VLSI Test Symposium (VTS'05)*, pp. 42-47, 2005.
15. N. Ahmed, M. Tehranipoor, C.P. Ravikumar and K. Butler, "Local At-Speed Scan Enable Generation for Transition Fault Testing Using Low-Cost Testers," *IEEE Transactions on Computer-Aided Design of Integrated Circuits and Systems (CAD/ICAS)*, vol. 26, no. 5, pp. 896-906, May 2007.
16. T. M. Mak, A. Krstic, K. Cheng, L. Wang, "New Challenges in Delay Testing of Nanometer, Multigigahertz Designs," *IEEE Design & Test of Computers*, pp. 241-248, May-Jun 2004.
17. K. Roy, K. T. Cheng and T.M. Mak, "Test Consideration for Nanometer-Scale CMOS Circuits," *IEEE Design & Test of Computers*, pp. 128-136, 2006.
18. K. S. Kim, S. Mitra, P.G. Ryan, "Delay defect characteristics and testing strategies, IEEE Design & Test of Computers, vol. 20,no. 5,Sept.-Oct. 2003, pp. 816.

19. A. Krstic, J. Liou, K. T. Cheng, L. C. Wang, "On structural vs. functional testing for delay faults, in Proc. *Int. Symp. On Quality Electronic Design*, pp. 438441, 2003.
20. M. Chen Chi and S. Huang, "A reliable clock tree design methodology for ASIC designs," in Proc. *International Symposium on Quality Electronic Design (ISQED'00)*, pp. 269-274, 2000.
21. S. Wang, X. Liu, S.T. Chakradhar, "Hybrid Delay Scan: A Low Hardware Overhead Scan-Based Delay Test Technique for High Fault Coverage and Compact Test Sets," in Proc. *Design, Automation and Test in Europe (DATE'03)*, pp. 1296-1301, 2004.
22. Synopsys Application Note, "Tutorial on Pipelining Scan Enables," 2004.
23. S. Bhunia, H. Mahmoodi, A. Raychowdhury and K. Roy, "A Novel Low-Overhead Delay Testing Technique for Arbitrary Two-Pattern Test Application," in Proc. *Design, Auto. and Test in Europe (DATE'05)*, pp. 1136 - 1141, 2005.
24. M. Abadir and J. Zhu, "Transition Test Generation Using Replicate-and-Reduce Transform for Scan-Based Designs," in Proc. *VLSI Test Symp. (VTS'03)*, pp. 22-27, 2003.
25. Y. Shao, I. Pomeranz and S. M. Reddy, "On Generating High Quality TeSts for Transition Faults," in Proc. *Asian Test Symp. (ATS'02)*, pp. 1-8, 2002.
26. X. Liu and M. Hsiao, "Constrained ATPG for Broadside Transition Testing," in Proc. *IEEE Int. Symp. on Defect and Fault Tolerance in VLSI Systems (DFT'03)*, pp. 175-182, 2003.
27. K. Yang, K. T. Cheng and L. Wang, "TranGen: A SAT-Based ATPG for Path-Oriented Transition Faults," in Proc. *Asia and South Pacific Design Auto. Conf. (ASP-DAC'04)*, pp. 92-97, 2004.
28. J. Rearick, "Too Much Delay Fault Coverage Is a Bad Thing," in Proc. *IEEE Int. Test Conf. (ITC'01)*, pp. 624-633, 2001.
29. X. Liu, M. S. Hsiao, S. Chakravarty and P. Thadikaran, "Techniques to Reduce Data Volume and Application Time for Transition Test," in Proc. *Intl Test Conf. (ITC'02)*, pp. 983-992, 2002.
30. B.W. Kernighan and S. Lin, "An efficient heuristic procedure for partitioning graphs," *Bell System Tech. Journal*, vol. 49, pp. 291-307, 1970.
31. Tetramax ATPG Users Guide, "SYNOPSYS Toolset Version 2004.06," Synopsys, Inc., 2004.
32. Synopsys DFT Compiler, "User Manual for SYNOPSYS Toolset Version 2004.06," Synopsys, Inc., 2004.
33. Synopsys SoCBIST DBIST Users Guide, "User Manual for SYNOPSYS Toolset Version 2004.06," Synopsys, Inc., 2004.

4

Enhanced Launch-Off-Capture

In launch-off-capture method, the at-speed constraint on the scan enable signal is relaxed (low-cost tester compliant) and the transition is launched from the functional path. The controllability of launching a transition at a target gate is less when compared to launch-off-shift method, as it depends on the functional response of the circuit under test to the initialization vector. In addition, the low-cost tester interface requirements such as no primary input changes and primary outputs being masked during launch to capture cycle impact the LOC coverage. It has become a common practice to include primary inputs and primary outputs in scan chain during scan insertion. This allows test engineers to test transition delay faults for the paths between primary inputs and internal flip-flops. Similarly, the paths between internal flip-flops and primary outputs are tested as well. However, the paths between chip pins and the cell wrapper inserted on primary inputs and primary outputs cannot be tested using low-cost testers.

This chapter focuses on improving the quality of test patterns generated using launch-off-capture method. A new enhanced launch-off-capture method is presented in which a transition can be launched either from the scan path or the functional path. The technique improves the controllability of transition fault testing, increases transition delay fault coverage, reduces the pattern count, and it still does not require the scan enable to change at-speed, i.e. it is implementable using low-cost testers. The scan enable control information to launch a transition through scan path or functional path is encapsulated in the test data and transferred during the scan operation to generate the local scan enable signals during the launch and capture cycles. A new scan cell, referred to as local scan enable generator (LSEG), is inserted in the scan chains to generate the local scan enable signals. This enhanced launch-off-capture technique can be easily practiced with no additional design effort and it is suitable for designs targeted for very low-cost testers.

4.1 Introduction

Transition fault testing is widely practiced in industry mainly due to its manageable fault count (two faults for each gate terminal) and availability of commercial tools. To perform a scan-based transition fault test, a pattern pair $(V1, V2)$ is applied to the circuit under test (CUT). Pattern $V1$ is termed as the initialization pattern and $V2$ as the launch pattern, to launch the signal transition ($0 \rightarrow 1$ or $1 \rightarrow 0$) at a desired node. Pattern $V2$ also helps propagate the transition at a target gate to any output of CUT (scan flip-flops or primary outputs). The response of the CUT to the pattern $V2$ must be captured at functional speed (rated clock period). The whole operation can be divided into 3 cycles: 1) Initialization Cycle (IC), where the CUT is initialized to a particular state ($V1$ is applied), 2) Launch Cycle (LC), where a transition is launched at the target gate terminal ($V2$ is applied) and 3) Capture Cycle (CC), where the transition is propagated and captured at an observable point. Various scan-based transition fault testing methods were proposed in literature. The two widely used methods in industry are called *launch-off-shift* (*skewed-load*) [12] and *launch-off-capture* (*broad-side*) [13]. There is a third method, called *enhanced-scan* [14], which offers highest fault coverage and lowest pattern count, however its area overhead is signifantly high.

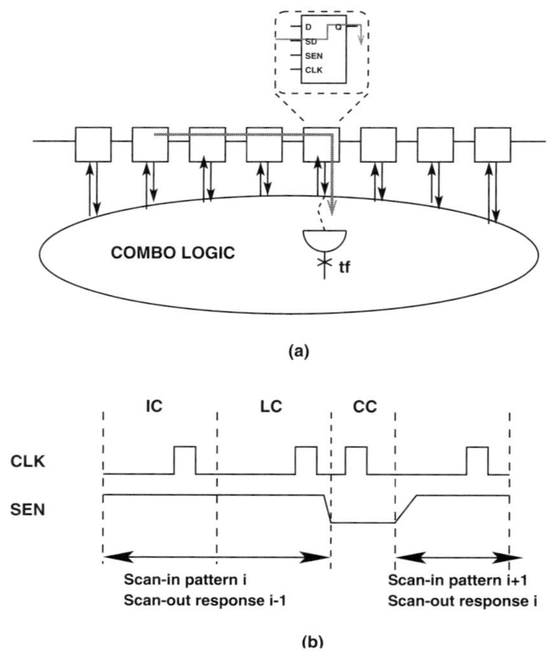

Fig. 4.1. Launch-off-shift transition delay fault pattern application [17].

In the first method, referred to as *launch-off-shift* (LOS) [12], the transition at a gate output is launched in the last shift cycle during the shift operation. Figure 4.1(a) shows the path of transition launch in LOS method. The transition is launched from the scan-in pin (SD) of any flip-flop in a scan chain. Figure 4.1(b) shows the waveforms during the different cycles of LOS operation. The LC is a part of the shift operation and is immediately followed by a fast capture pulse. The time period for SEN to make this $1 \rightarrow 0$ transition corresponds to the functional frequency. Hence, LOS requires the SEN signal to be timing critical.

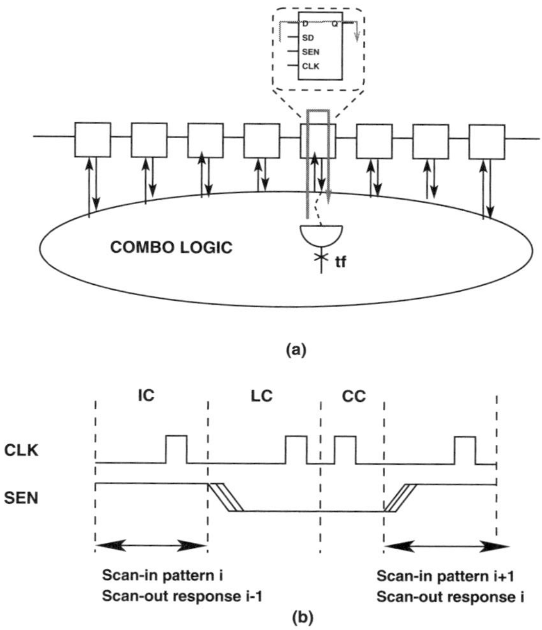

Fig. 4.2. Launch-off-capture transition delay fault pattern application [17].

Figure 4.2(a) shows the transition launch path of the second approach, referred to as *launch-off-capture* (LOC) [13]. In this method, the transition is launched and captured through the functional pin (D) of any flip-flop in the scan chain. Since, the launch pattern $V2$ depends on the functional response of the initialization vector $V1$, the launch path is less controllable due to which the test coverage is normally lower than LOS method.

Figure 4.2(b) shows the waveforms of the LOC method in which the launch cycle is separated from the shift operation. At the end of scan-in (shift mode), pattern $V1$ is applied and CUT is set to an initialized state. A pair of at-speed clock pulses is applied to launch and capture the transition at the target gate terminal. This relaxes the at-speed constraint on the SEN signal and dead cycles are added after the last shift to provide enough time for the SEN signal to settle low.

The LOS method is preferable from ATPG complexity and pattern count view points when compared to LOC method. In case of LOC, a high fault coverage cannot be guaranteed due to the correlation between the two patterns, $V1$ and $V2$, note that $V2$ is the functional response to $V1$ pattern.

As the design size increases, the SEN fanout exceeds any other net in the design. The LOS method constraints SEN to be timing critical which makes it difficult to implement using low-cost testers and on designs where the turn-around-time is critical [16]. This is the main reason that LOC method has been widely practiced, especially on low-cost testers [6]. Note that no at-speed SEN signal is required for LOC method [17] [18]. However, new techniques are required to improve the LOC method's quality.

Several techniques have been proposed to improve the transition delay fault coverage [19] [20] [21] but there has not been much work on the LOC method. The implementation of LOS method using low-cost testers is presented in [16]. An on-chip at-speed scan enable signal is generated using a slow scan enable signal generated by a low-cost tester. An on-chip scan enable generator cell is designed; it can be inserted anywhere in a scan chain and the launch and capture information are encapsulated in the test data and transferred into the scan chain. The proposed technique in [16] focuses only on LOS and its implementation on low-cost testers. The technique has no impact on fault coverage and pattern count.

In [19], a hybrid scan architecture is proposed which controls a small subset of selected scan cells by LOS and the rest are controlled by LOC approach. A fast scan enable signal generator is designed which drives only the LOS-controlled scan flip-flops. The ATPG method used is complex and current commercial tools do not support such a technique. The selection criteria of the LOS-controlled scan flip-flops determines the effectiveness of the method. In some cases, the number of patterns generated by the hybrid method exceeds the LOC pattern count [19]. Moreover, the LOS controlled flip-flops cannot be used in LOC mode.

An automatic test pattern generator (ATPG)-based scan path test point insertion technique is proposed in [20] to achieve high fault coverage for scan designs. This technique breaks the shift dependency between adjacent flip-flops by inserting dummy gates or flip-flops. The proposed technique uses a special ATPG to identify pairs of adjacent flip-flops between which test points are inserted. The authors in [22] proposed topology-based latch correlation measures and applies companion latch arrangement algorithm to guide the placement of latches in a scan-based design to minimize the effect of correlation and maximize the fault coverage of delay faults.

A restricted scan chain reordering is proposed in [21] to improve the fault coverage of skewed-load (or LOS) approach. This technique restricts the distance by which a scan flip-flop can be moved to create the new scan chain order. The scan flip-flops are reordered to minimize the number of undetectable faults due to test pattern dependency. Achieving high coverage path delay fault testing requires the application of scan justified test vector pairs, coupled with careful reordering of scan flip-flops and/or insertion of dummy flip-flops in the scan chain [23]. The authors in [23] proposed a technique considering both the number of dummy flip-flops and wirelength costs to improve path delay fault coverage.

In [24] different scan cell architectures have been proposed to improve the LOC coverage. The proposed technique uses multiple scan enable signals from the external ATE (automatic test equipment) to control the different subset of scan cells. This increases the pin count and also each scan cell is associated with an area overhead.

4.1.1 Overview of Enhanced LOC Method

In this chapter, a transition fault pattern generation technique, called *Enhanced LOC* (ELOC), is presented to improve fault coverage and reduce the pattern count [17]. In this technique, the transition launch path is deterministically determined either through a *functional path* or a *scan path*. This improves the controllability of scan chains, increases the fault coverage, and reduces the pattern count without requiring an at-speed SEN signal. A new scan cell, called local scan enable generator (LSEG), generates the local scan enable signals (not at-speed), used to control each scan chain mode of operation. A subset of scan chains are configured to work in functional-launch mode (conventional LOC) and the rest are configured in shift-launch mode. This is controlled by separate scan enable signal for each scan chain generated by LSEG. Note that the LSEG cell is ATPG controllable and the value shifted in it determines the local scan enable control signal during the launch and capture cycles and the scan chain will be controlled either in functional-launch mode or shift-launch mode only. The ELOC technique is suitable for very low-cost testers and it can be easily practiced with no additional design effort.

In the functional-launch mode, similar to LOC method, pattern $V2$ is generated using the functional response of pattern $V1$. In shift-launch mode, the pattern $V2$ is generated using the shift operation (local scan enable is held high) but the responses are not captured in the respective scan chain. The scan enable control information for the launch and capture cycle is embedded in the test data itself. The LSEG cell has the flexibility to be inserted anywhere in the scan chain and the hardware area overhead of LSEG cells is negligible with no impact on the layout of the design. The ELOC technique provides as high as 14% (approx.) coverage for ISCAS'89 and ITC'99 benchmark circuits and up to 2% (approx.) for industrial designs compared to LOC technique. For the highest achievable LOC coverage, ELOC provides up to 74% (approx.) pattern reduction for benchmark circuits and 56% (approx.) for industrial designs compared to LOC technique.

4.2 Enhanced Launch-off-Capture

The LOC method utilizes the functional response of the circuit to launch the transition at a target gate terminal and propagates the fault effect to an observable point. Launching a transition through functional response is less controllable as it depends topologically on the response of the circuit. In addition, the low-cost tester interface requirements such as no primary input changes (No-PI changes) and primary outputs being masked (No-PO measures) might impact the LOC coverage. Let's start with explaining the controllability of LOC and describe how the LOC's test coverage is impacted due to low-cost tester constraints. Then, a technique is shown to improve the test coverage by increasing its controllability and observability. Note that launching a transition through functional response is difficult due to controllability issues. The controllability of launching a transition through the functional response is less and it is a design dependent factor. Now, let's explain the controllability of LOC and describe how the LOC's test coverage can be improved by increasing its controllability.

In this chapter, controllability (observability) is referred to as the ability to launch (propagate) the transition at the target gate. Figure 4.3(a) shows a small example with two scan chains, $SC1$ and $SC2$, each consisting of two scan cells. The scan-in ports are $SI1$ and $SI2$ and the scan-out ports are $SO1$ and $SO2$, respectively. In this particular example, the two scan chains have independent scan enable signals $SEN1$ and $SEN2$, respectively. Also, there are two primary inputs (PIs) A and B, and two primary outputs (POs) Y and Z, respectively.

Consider a slow-to-rise transition fault at the target node a (see Figure 4.3(a)). The fault is non-robustly testable (off-path input of gate G5 requires $0 \rightarrow 1$ transition) using conventional LOC technique without any constraints on primary input changes during ATPG. It can be noticed that the transition is launched through the functional path and the values at each gate output and PIs represent the logic states during pattern pair V1 and V2, respectively. The scan enable signals are high during the shift operation and low during the launch and capture cycles (see Figure 4.3(b)). However, the same slow-to-rise fault at a is untestable when low-cost tester constraints (No-PI changes) are applied (see Figure 4.4(a)).

To launch a $0 \rightarrow 1$ transition at the target fault site (a) using LOC, the scan cell $SC2_2$ (suffix indicates the position from the scan-in port) must contain a logic 0 at the end of the scan shift operation (V1 applied). The functional response of the circuit must be logic 1 at the output of gate $G1$ which is required to launch a $0 \rightarrow 1$ transition at a during the launch cycle through the functional path as shown. To propagate the transition, the inputs of gate $G5$ other than the input a must have a non-controlling value (1). However, logic 1 at the output of the gate $G1$, which is required to launch the transition ($G4 = 0$), blocks the propagation of the transition to an observable point. Therefore, the slow-to-rise transition fault at the target node a is untestable

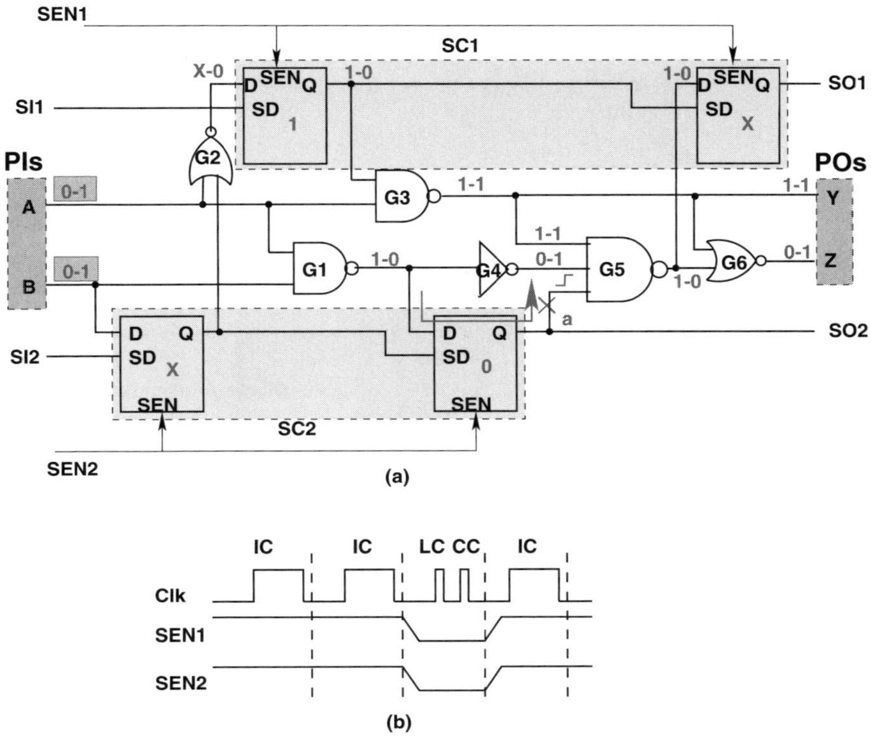

Fig. 4.3. 'a' slow-to-rise fault non-robustly testable with PI-changes using conventional LOC method [17].

by the conventional LOC method with the low-cost tester requirement (No-PI changes).

In some cases, the controllability can be improved by using the scan path to launch a transition instead of the functional path. Figure 4.5(a) shows the same example in which the slow-to-rise transition fault at node a, untestable using conventional LOC method with No-PI changes ATPG constraint, becomes robustly testable by controlling the launch path of the target transition fault using the scan enable signals. The $0 \rightarrow 1$ transition at a is launched through the scan path instead of the functional path. The remaining inputs of the gate $G5$ other than the target fault site a are controllable to non-controlling value 1 to propagate the transition. This method is referred to as *enhanced launch-off-capture (ELOC)*. Figure 4.5(b) shows the scan enable signals $SEN1$ and $SEN2$ during the shift (IC), launch (LC) and capture (CC) cycles. In this method, the scan enable signal of the second scan chain $SEN2$ is kept constant at 1 during both launch and capture cycles. In other words, the scan chain $SC2$ is used only to shift bits, i.e, to launch transitions

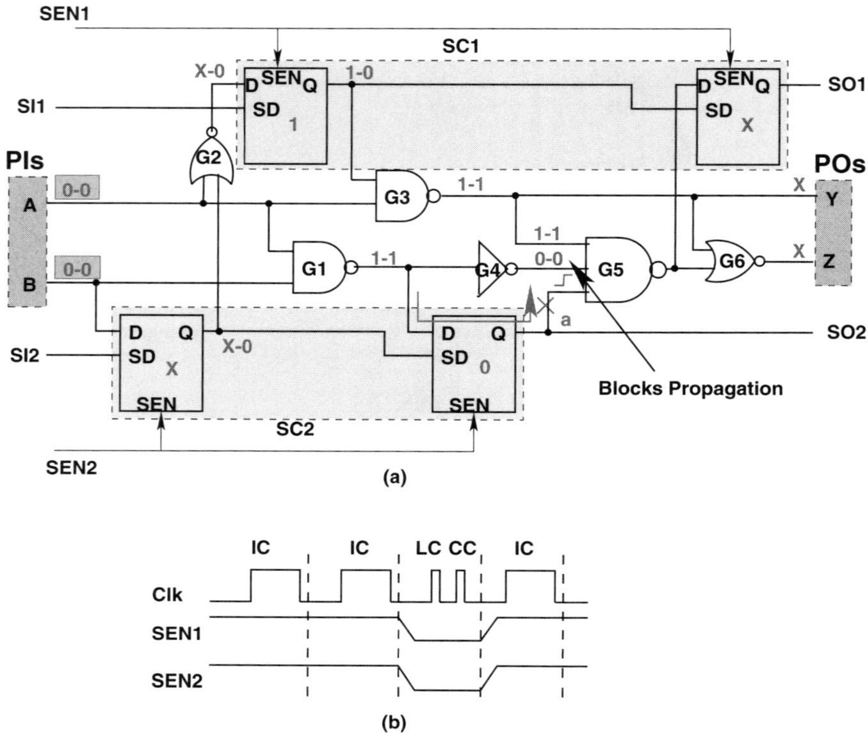

Fig. 4.4. 'a' slow-to-rise fault un-testable with No-PI-changes using conventional LOC method [17].

in the circuit. It acts like a shift register and does not capture any functional response of the circuit.

The conventional LOC method may be viewed as a special condition of the ELOC method, where the scan enable signals of all the chains are 0 during the launch and capture cycles. ELOC provides more controllability of launching the transition either through the scan path or the functional path. Note that, the scan enable (SEN) signals do not change between the launch and capture cycles and any scan enable transition is at shift frequency. Figure 4.6 shows a circuit with two scan chains, one acting as a shift register (shift-launch mode) and the other in the functional-launch mode. The transitions in the first scan chain (Scan Chain 1) are launched through the functional path while the transitions from the second scan chain (Scan Chain 2) are launched from the scan path.

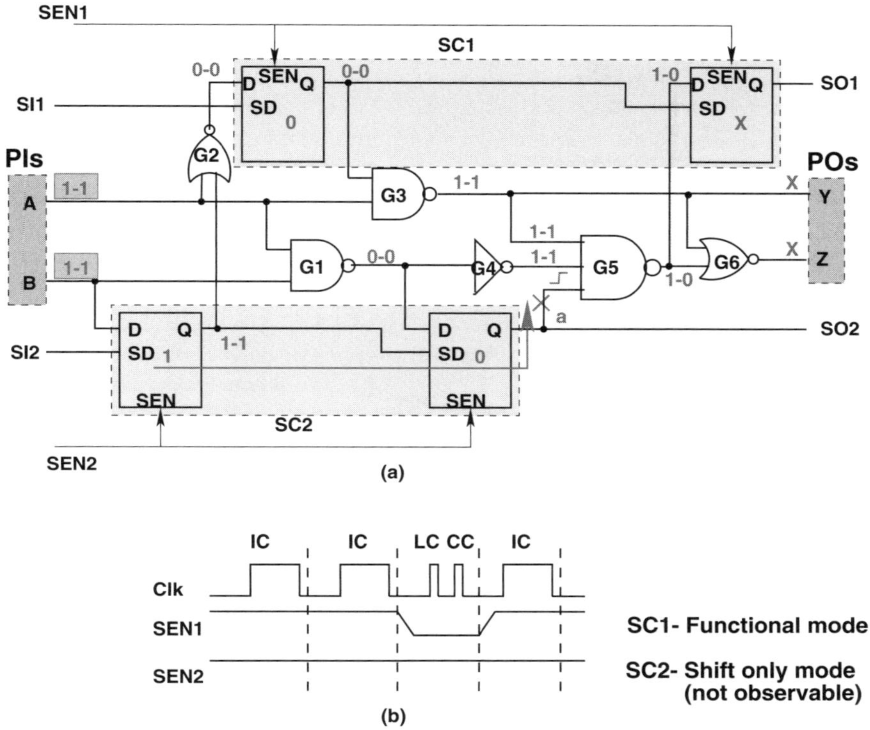

Fig. 4.5. 'a' slow-to-rise fault robustly testable with No-PI-changes using **Enhanced LOC** (ELOC) method [17].

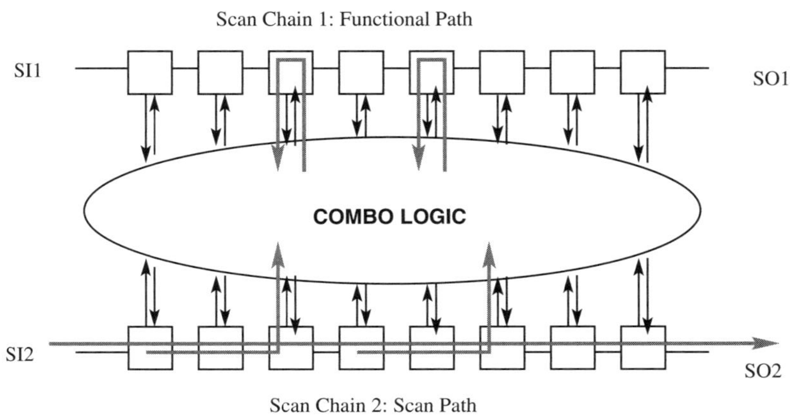

Fig. 4.6. Controllability of Enhanced LOC [17].

4.3 Local Scan Enable Signal (LSEN) Generation

The ELOC method provides more controllability to launch a transition but requires independent scan enable signal for each scan chain. Multiple SEN ports can be used, but this increases the number of pins. The scan enable control information for all the scan chains differ only during the launch and capture cycles of the pattern. Hence, the scan enable signal from the external tester can be utilized for the scan shift operation and the scan enable control information for only the launch and capture cycles can be generated internally. The local scan enable generator cells are inserted within the scan chains. Therefore, the control information is to be passed as part of the test data. The scan enable control information will be part of each test pattern and is stored in the tester's memory.

The normal scan architecture with a single scan enable signal from the external tester is shown in Figure 4.7(a). In general, the design can contain multiple scan chains but here only one scan chain is shown for the purpose of explaining local scan enable generation. There are eight scan flip-flops in the scan chain and the test pattern shifted is 10100110. The external scan enable signal from the tester is referred to as the global scan enable (GSEN). Figure 4.7(b) shows the same circuit in which a local scan enable signal is generated from a combination of GSEN and some information in the test pattern data for the enhanced LOC method. The internally generated scan enable signal is termed as local scan enable (LSEN). The main objective is to de-assert GSEN after the entire shift operation and then generate the LSEN signal during the launch and capture cycles from the test data. In this case, the pattern shifted is modified to [C]10100110, where C is the scan enable control bit which is stored in scan flip-flop A at the end of the scan operation.

One extra scan flip-flop *(A)* and an OR-gate are added for the generation of LSEN signal. The output of A is ORed with GSEN to generate the LSEN signal (see Figure 4.7(b)). Note that GSEN is not an at-speed signal. The GSEN signal asynchronously controls the shift operation. GSEN is de-asserted after the nth shift (IC) cycle, where n=9. n is the length of scan chain after inserting new cell A. After the GSEN signal is de-asserted at the end of the shift operation, the scan enable control during the launch and capture cycles is the control bit C stored in A. At the end of the capture cycle, the LSEN signal is asynchronously set to 1 by GSEN for scanning out the response. Figure 4.7(c) shows the timing waveforms during the test pattern application. In general:

$$LSEN = (GSEN + A) = \begin{cases} 1 & \text{if GSEN=1} \\ A & \text{if GSEN=0} \end{cases}$$

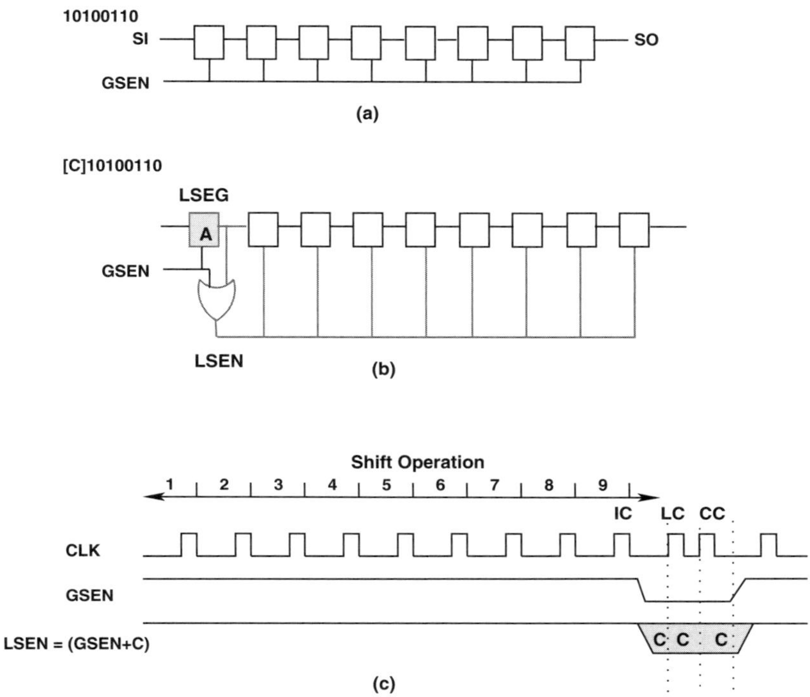

Fig. 4.7. (a) Scan chain architecture, (b) Local scan enable (LSEN) generation and (c) LSEN generation process.

4.3.1 Local Scan Enable Generator (LSEG)

As explained earlier, during the launch and capture cycles of the pattern, the control bit shifted into scan flip-flop A is used as the scan enable control. Figure 4.8 shows the LSEG cell architecture. It consists of a single flip-flop which is used to load the control information required for the launch and capture cycles. The port definition and architecture of LSEG cell is similar to a scan cell and the output of the scan flip-flop is fed back to the functional input port of the scan flip-flop. It consists of a scan-in (SEN_{in}) pin which takes GSEN signal as input. An additional scan-out (SEN_{out}) pin represents the LSEN signal ($GSEN + Q$). Therefore, after going to a control state *(C)* at the end of the shift operation (GSEN is de-asserted), LSEN remains in this state as long as it is asynchronously set to 1 by GSEN.

Table 4.1 shows the different modes of operation of LSEG cell. $GSEN = 1$ represents the normal shift operation of the pattern. When $GSEN = 0$ and $C = 1$, $LSEN = 1$ and the scan chain acts in the shift mode to launch the transitions (*Shift-Launch* mode). The scan chain acts in the conventional LOC mode when $GSEN = 0$ and $C = 0$ (*Functional-Launch* mode). Note that the

LSEG cell can be inserted anywhere in the scan chain and it is not connected to the circuit-under-test (CUT). Hence, it has no impact on the functional timing and the CUT fault coverage.

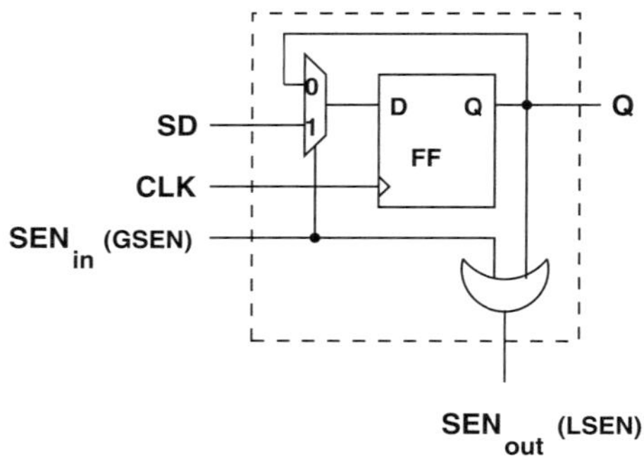

Fig. 4.8. Local scan enable generator (LSEG) cell [17].

Table 4.1. LSEG Operation [17].

GSEN	Q	LSEN	Operation
1	X	1	Shift
0	1	1	Shift-Launch
0	0	0	Functional-Launch

The LSEG cell provides a simple mechanism to generate the local internal scan enable signals. But, it has a shift dependency for the following scan flip-flop in the shift register mode. If $C = 1$, the LSEG flip-flop is constant at 1 for the launch and capture cycles and the scan flip-flop following the LSEG cell in the scan chain can generate only a $0 \rightarrow 1$ transition at its output. This may result in slight loss of coverage for faults which are in the logic cone and require a $1 \rightarrow 0$ transition on the scan flip-flop following the LSEG cell. In order to avoid loss of coverage without significant change in the architecture, the LSEG cell is modified such that the LSEG cell when loaded with the control bit, the cell will remain in this state and it is not in the shift path during the launch and capture cycles for $C = 1$. Figure 4.9 shows the modified LSEG cell architecture. It consists of an additional multiplexer and it does not impact the functional path timing. When $GSEN = 0$, the LSEG cell is bypassed and the SD pin is directly connected to Q. In the *Shift-Launch mode*, the value from the previous scan flip-flop of LSEG cell is shifted directly into its following scan flip-flop.

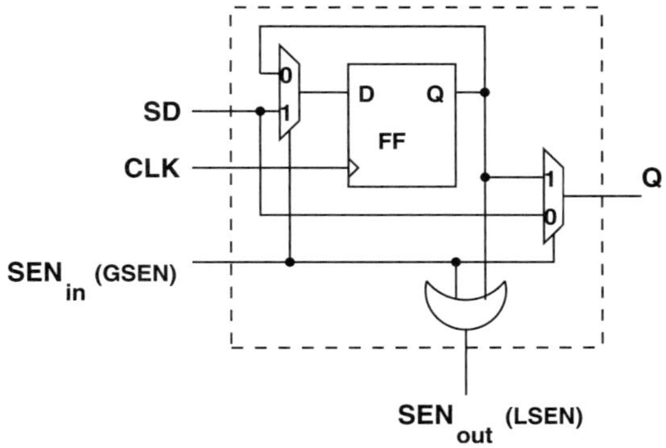

Fig. 4.9. Modified local scan enable generator (LSEG) cell [17].

4.3.2 Operation of LSEG Cell

Figure 4.10(a) shows the previous example with the LSEG cell inserted in the scan chain. The value loaded in the LSEG flip-flop is represented as C ([C]10100110) in the test pattern. Figure 4.10(b) shows the pattern and the timing waveform for the conventional LOC method. The SEN_{out} (LSEN) signal is generated by the boolean equation $SEN_{out} = Q(FF) + SEN_{in}$. The GSEN signal is high during the entire shift operation. At the end of the shift operation, the GSEN signal is asynchronously de-asserted and the LSEN signal must be logic 0. Hence, the LSEG cell flip-flop must be constrained to 0 during ATPG. After the capture cycle, the LSEN signal is asynchronously asserted high by GSEN.

For ELOC, similar to conventional LOC, the GSEN signal is high during the entire shift operation. At the end of the shift operation, the LSEN signal is determined by the control bit *(C)* shifted into the LSEG cell flip-flop during the pattern shift. Figure 4.10(c) shows the pattern and the timing waveform for ELOC. The LSEN signal is constant at logic value C during the launch and capture cycle. It can be noticed that the LSEN transitions are not at-speed. After the capture cycles, the LSEN is asserted to 1 asynchronously by GSEN.

4.4 Scan Insertion and ATPG Flow

4.4.1 Test Architecture

The LSEG-enabled solution explained in Section 4.3 provides a method of generating internal local scan enable signals from the pattern data and global scan enable signal from the external ATE. The overhead of generating the local scan enable signal is the addition of an LSEG cell in the scan chain. The area overhead of a LSEG cell is a few extra gates, which is negligible in modern designs. The methodology is applicable to designs with multiple clock

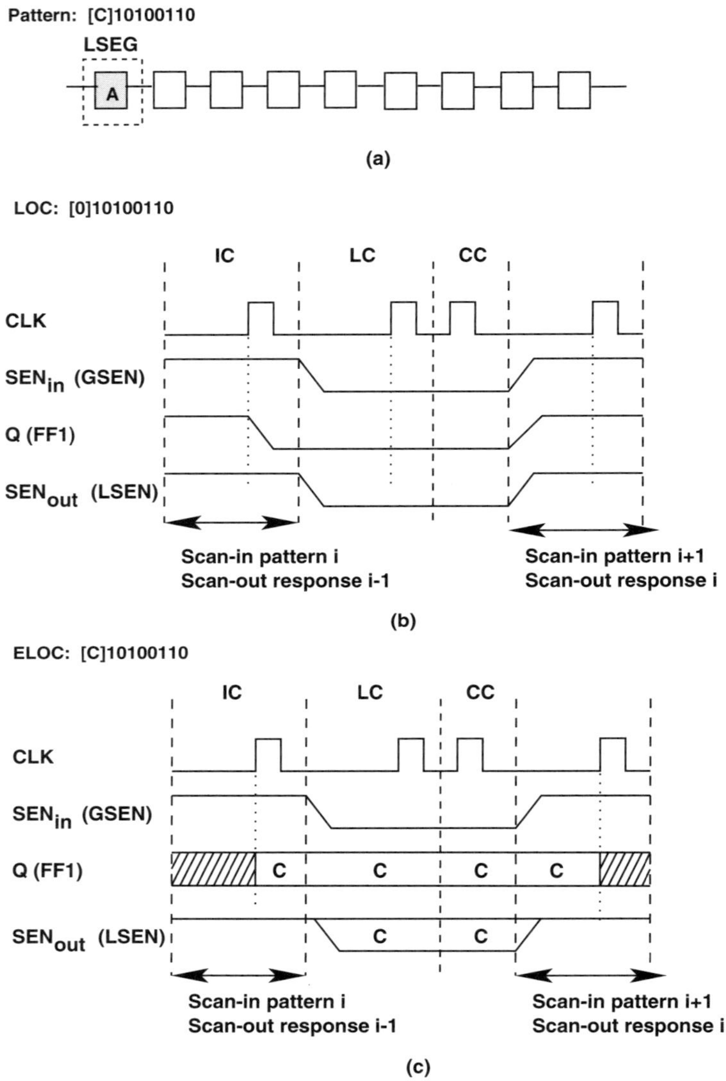

Fig. 4.10. Operation of LSEG cell, (a) Scan chain, (b) Conventional LOC and (c) Enhanced LOC [17].

domains. In general, there can be multiple scan chains in a design to reduce the test application time. The test application time is directly proportional to the longest scan chain length in the design. Figure 4.11 shows a multiple scan chain architecture with n scan chains. Each scan chain i, where $1 \leq i \leq n$, consists of an LSEG cell which generates the local scan enable signal $LSEN_i$ for the respective scan chain. The GSEN signal connects only to the SEN_{in} port of LSEG cells.

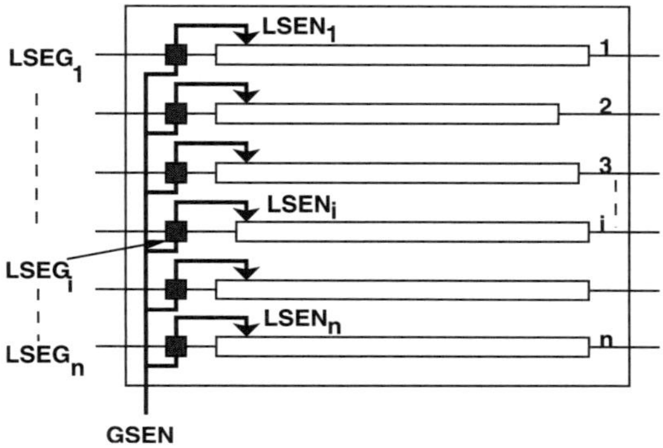

Fig. 4.11. Test Architecture [17].

4.4.2 Test Synthesis and ATPG

Synopsys DFTCompiler [26] was used for scan chain insertion in the design. One LSEG cell per scan chain is inserted. To insert the LSEG cells, additional commands are required during the scan chain configuration. The test synthesis tool must recognize the LSEG cell as a scan cell in order to stitch it into the scan chain. This requires it to be defined as a new library cell with the scan cell attributes. A workaround is to design the LSEG cell as a module and declare it as a scan segment of length 1.

The GSEN signal is connected to all the LSEG cells SEN_{in} input pin. To insert the LSEG cell in each respective scan chain, the scan insertion tool requires explicitly to declare the scan path including the LSEG cell for each scan chain. Only the LSEG cell is specified in the scan path, as the tool will stitch the rest of the cells including the LSEG cell and balance the scan chain depending on the longest scan chain length parameter. Finally, during scan enable routing, the LSEG cell's SEN_{out} port in each chain is hooked up to all the scan enable input port of the scan flip-flops in the respective chain.

The ELOC method provides better controllability to launch a transition fault either through the scan path or the functional path. The ATPG tool must be able to understand the local scan enable generation methodology and deterministically decide the transition fault activation path. By default, when generating test patterns for transition fault LOC method, the commercial ATPG tools uses an optimized two-clock ATPG algorithm that has features of both the combinational ATPG (only one capture clock between pattern scan-in and scan-out of response) and sequential ATPG engine (sequential time-frame ATPG algorithm). The tool understands the local scan enable generation technique using LSEG cells and is capable to decide the launch path for the target transition fault deterministically. Hence, there is no fundamental difference in the ATPG methodology when the LSEG-based solution is used.

The scan enable signal for the scan flip-flops during the launch and capture cycles is an internally generated signal. Notice that the OR gate in the LSEG cell generates the local scan enable signal through a logical OR of the global scan enable and the Q output of the flop FF (see Figure 4.9). The global scan enable signal is in active mode during scan shift operation. The tool determines the local scan enable for each chain and shifts the control value into the LSEG cell during pattern shift, used for launch and capture. It also deterministically decides the combination of scan chains to work in *Shift-Launch* mode, to activate a transition fault.

As explained in Section 4.2, in the conventional LOC method the scan enable signal is zero during the launch and capture cycles. The LSEG cell must be loaded with logic 0 at the end of the shift operation for conventional LOC method. Cell constraints can be used to control the load values allowed on LSEG cells. The ATPG tool creates only patterns that satisfy the cell constraints. The enhanced LOC method does not require any cell constraints. The ATPG tool generates patterns based on the controllability of the target fault site. During ELOC pattern generation, the ATPG tool deterministically decides the value of LSEG cell which determines the LSEN signal depending on the transition fault activation path.

The available commercial ATPG tools are mostly timing unaware. During ELOC, if a scan chain is used in a *Shift-Launch* mode, the ATPG tool will shift data in the scan chain during the LC at functional frequency and captures (CC) response (shifted response) at functional frequency. This will be a problem as the scan chain is usually not timing closed for functional frequency and the captured response through the scan-in pin will be erroneous. Since the scan chain is used only to launch the transitions in *Shift-Launch* mode, it does not contribute to the fault coverage. Hence, all the scan chains that are operating in *Shift-Launch* mode can be masked for the respective pattern. The operation mode of the scan chain can be determined by the value of the LSEG cell in the pattern. The generated patterns are post-processed to mask the particular scan chain with the LSEG cell value equal to 1. In future, if the ATPG tool is enhanced to mask the scan chains that are not observable, the post processing step can be skipped.

4.5 Case Study

In this case study, a subchip of an industrial-strength design is used that had the following characteristics (Table 4.2). The design has 16 scan chains and approximately 10K scan cells. There are 13 non-scan cells and six internal clock domains. One LSEG cell is inserted per scan chain. The test strategy is to get the highest possible test coverage for the transition faults. When generating test patterns for the transition faults, only the faults in the same clock domain are targeted. During pattern generation, only one clock is active during the launch and capture cycle. Hence, only faults in that particular clock domain are tested. All PIs remain unchanged and all POs are unobservable while generating the test patterns for the transition faults. This is because the very low cost ATEs are not fast enough to provide the PI values and strobe POs at speed.

Table 4.2. Design Characteristics [17]

Clock Domains	6
Scan Chains	16
Scan Flops	10477
Non-scan Flops	13
Transition Delay Faults	320884

The results for conventional LOC, enhanced LOC and LOS transition-delay ATPG on this design are shown in the Table 4.3. It is seen that LOS methodology gave approximately 3% higher fault coverage than the conventional LOC methodology. The ELOC method gave approximately 1.9% higher fault coverage compared to LOC method. The number of patterns generated is also less due to better controllability in ELOC method. However, the CPU time of ELOC method is greater than LOC as the ATPG tool does more processing to determine the possible combinations of the scan chains to work in shift register (scan path) mode or in functional mode.

Table 4.3. ATPG Results

	Conventional LOC	*Enhanced LOC*	*LOS*
Detected faults	282658	288681	292342
Test Coverage	88.27	90.15	91.30
Fault Coverage	88.09	89.96	91.11
Pattern Count	2145	2014	1112
CPU Time *[sec]*	896.96	924.74	329.30

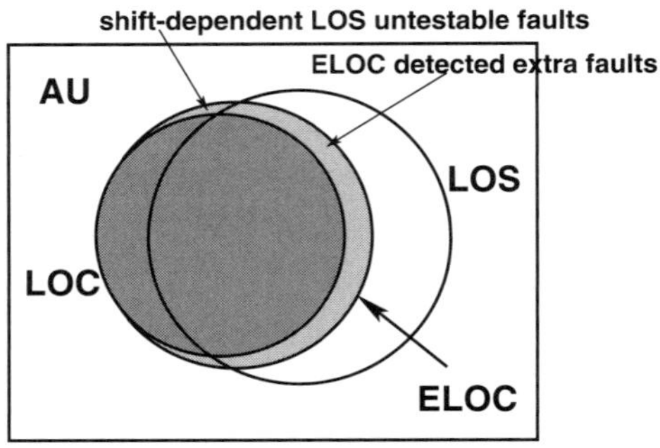

Fig. 4.12. Fault Coverage Analysis [17].

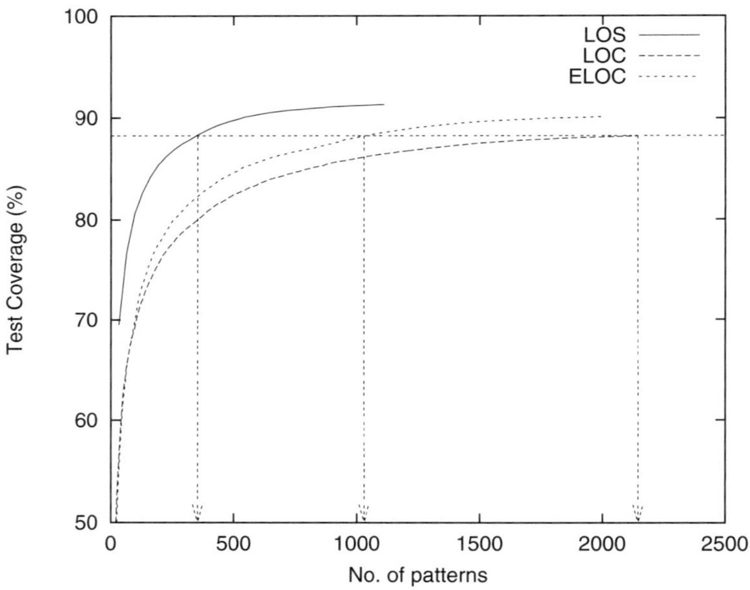

Fig. 4.13. Transition Coverage for LOC, ELOC and LOS [17].

Figure 4.12 shows the fault coverage analysis for the three different transition fault methods. There is a common set of transition faults which are covered by both LOS and LOC and there are some faults in the LOC transition fault set which are not covered by LOS. However, ELOC covers the entire transition fault set of LOC and also detects some extra faults which fall in the LOS set. This is because, LOC is a special case where all the LSEG cells are constrained to 0 during ATPG. In addition, ELOC also detects some faults which are not detected by LOS and LOC due to its hybrid nature of launching a transition. One class of such faults are referred as *shift-dependent untestable faults* [20] which are not detected by LOS. The new ELOC method provides an intermediate fault coverage point between LOS and the conventional LOC method.

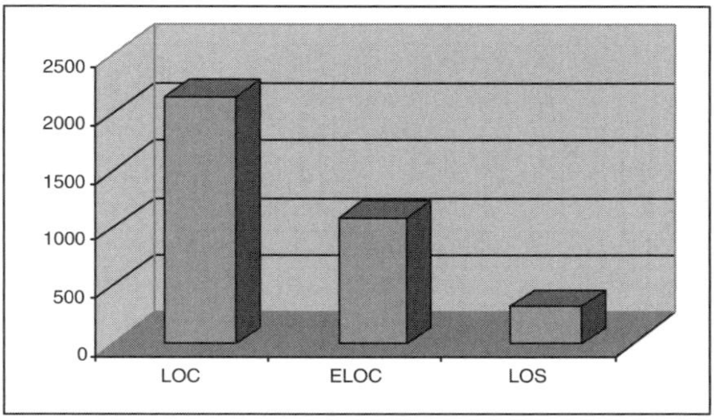

Fig. 4.14. Comparison of no. of patterns in LOC, ELOC and LOS [17].

The test coverage curves for LOS, LOC and ELOC are shown in Figure 4.13. The improvement in ELOC coverage can be seen due to greater controllability. Figure 4.14 shows the number of patterns required by LOC, ELOC and LOS methods for the highest coverage achieved for LOC method in all three cases. For the design under consideration, it is observed that only 1092 patterns of ELOC method offered the same coverage as conventional LOC method. This represents a pattern reduction of about 52% compared to conventional LOC.

4.6 Analysis of ELOC Detected Additional Faults

In [27], potential yield loss problem due to detection of faults not exercised during functional operation was discussed. An analysis is performed on the extra faults detected by ELOC compared to conventional LOC method. To determine the nature of these extra faults, an ATPG is performed using Synopsys TetraMAX [28] on the largest ISCAS'89 and ITC'99 benchmark circuits (the industrial design used for case study was not available to us at the time of performing these experiments) in three different cases: *Case1)* conventional LOC with both PI-changes and PO-measures, *Case2)* conventional LOC with no PI-changes and no PO-measures and *Case3)* enhanced LOC with no PI-changes and no PO-measures. *Case1* represents the state of a high-speed ATE capable of delivering at-speed PI-changes and PO-measures. However, not all companies can afford such a high-cost test equipment. Similarly, *Case2* and *Case3* represents the state of a low-cost tester usage.

Tables 4.4 and 4.5 show the fault classification by the ATPG tool into the following four categories: 1) *DT-detected:* This fault class includes faults which

Table 4.4. Fault class analysis for ISCAS'89 and ITC'99 benchmark designs (Cases1 and Case2).

Design	*LOC (Case1)*				*LOC* (Case2)*			
	PI changes, PO measures				*No-PI changes, No-PO measures*			
	DT	*AU*	*NC*	*NO*	*DT*	*AU*	*NC*	*NO*
s13207	7168	33	0	1051	5957	729	0	1566
s15850	9158	27	0	1597	6907	657	27	3191
s38417	27881	37	0	620	27410	281	0	847
s38584	35851	29	0	1736	28469	1061	0	8086
b17	94535	8	8	20811	93106	856	8	21392
b18	328896	8	41	76017	314307	10484	41	80130
b19	560196	8	167	161513	540169	13584	167	167964

Table 4.5. Fault class analysis for ISCAS'89 and ITC'99 benchmark designs (Case3).

Design	*ELOC (Case3)*			
	No-PI changes, No-PO measures			
	DT	*AU*	*NC*	*NO*
s13207	6978	707	0	567
s15850	8479	641	4	1658
s38417	27432	255	0	851
s38584	31666	1043	0	4907
b17	100510	856	0	13996
b18	349879	10484	4	44595
b19	544486	13584	58	163756

can be initialized and propagated to an observable point fault effect value. 2) *AU-ATPG untestable:* The AU fault class includes faults which neither fall into DT nor PT classes under the current ATPG conditions. One example is a fault being untestable due to an ATPG constraint (such as No-PI changes or No-PO measures) which is in effect. 3) *NC-not controlled:* This class contains faults that the ATPG algorithm could not control to achieve both a logic 0 and a logic 1 state. Nodes that are always at an X state are classified as NC because ATPG cannot achieve either a logic 0 or a logic 1. 4) *NO-not observed:* Faults that could be controlled, but could not be propagated to an observable point are placed in this class. It can be clearly noticed that the application of low-cost tester ATPG constraints to LOC reduces the number of detected (*DT*) faults (*Case2*) and some of the detectable faults are transferred into *AU*, *NC* and *NO* categories. For example, in case of benchmark s38584, the number of detected faults decreased from $DT_{LOC} = 35851$ to $DT_{LOC^*} =$

28469 and correspondingly the AU and NO faults increased from $AU_{LOC} = 29$ to $AU_{LOC^*} = 1061$ and $NO_{LOC} = 1736$ to $NO_{LOC^*} = 8086$, respectively.

Table 4.6. Analysis of additional faults detected by ELOC for ISCAS'89 and ITC'99 benchmark designs.

Design	*ELOC (Case3)*				
	$\Delta DT = DT_{ELOC} - DT_{LOC^*}$	DT_{LOC}	AU_{LOC}	NC_{LOC}	NO_{LOC}
s13207	1021	303	0	0	718
s15850	1572	305	0	0	1267
s38417	22	22	0	0	0
s38584	3197	1845	0	0	1352
b17	7404	172	0	0	7232
b18	35572	695	0	15	34862
b19	4317	472	0	2	3843

The ELOC detects some of these detectable faults turned untestable (transferred to AU, NC and NO) due to low-cost ATPG constraints and improves the fault coverage due to increased controllability to launch and propagate a transition. However, the important question is how many of these additionally detected faults by ELOC fall under DT category of conventional LOC without any ATPG constraints on PIs and POs (*Case1*). Table 4.6 shows further analysis of the extra faults ($\Delta DT = DT_{ELOC} - DT_{LOC^*}$) detected by ELOC in presence of low-cost tester ATPG constraints. The Columns 3 to 6 show the number of these additionally detected ELOC faults in each fault class of LOC without low-cost ATPG constraints (*Case1*) DT, AU, NC and NO, respectively. For benchmark s38584, 1845 faults out of $\Delta DT = 3197$ extra faults detected by ELOC fall under DT fault class of conventional LOC (*Case1*) without any low-cost ATPG constraints and the rest of the extra detected faults were from the NO category. It can be noticed that in majority of the cases, a large portion of the detected faults were from the NO category than the DT category of *Case1*.

It can be argued that these NO detected faults might result in yield loss as some of them can be functionally untestable. However, a NO fault need not necessarily be functionally untestable because of the over simplifications assumed during transition fault test such as single-cycle defect size, zero-delay model, etc. For example, in [25], it was illustrated that a functionally untestable NO fault might affect the functionality if the defect size is greater than single-cycle. With technology scaling and increasing operating frequencies, detection of such faults might become important and more than two vectors are required to detect the multi-cycle delay faults and requires a sequential ATPG engine for pattern generation. The ELOC method can be advantageous as it eases the ATPG and detects such multi-cycle faults using a two vector pair. In addition, ELOC also detects some faults which are classified as *NO* due to low-cost ATPG constraints.

4.7 Experimental Results

The ELOC method has been experimented on the largest ISCAS'89, ITC'99 benchmark circuits and eight industrial designs. In all these designs each scan chain is inserted with one LSEG cell. During ATPG, the faults related to clocks, scan-enable and set/reset pins, referred to as untestable faults, are not added to the fault list. The faults related to clocks can only be detected by implication and the remaining faults (scan-enable/set/reset) are untestable as the signals remain unchanged during the launch and capture cycles. These faults contribute approximately 10-15% of the total transition faults.

Tables 4.7, 4.8, 4.9 and 4.10 show the ATPG results comparing LOC and ELOC methods for several ISACS'89, ITC'99 and industrial designs. For both methods, DT, FC, # Patt columns shows the detected faults, the fault coverage percentage and the number of patterns generated, respectively. Note that the CPU time for ELOC method is greater than LOC method since the tool has to do additional processing to find the transition launch activation path. The ΔDT are the extra faults detected by the ELOC method. The LSEG-based solution provides better fault coverage and pattern count with a simple addition of a LSEG cell controlling the scan path. As shown in Tables 4.7, 4.8, 4.9 and 4.10, the ELOC gives up to 14% (ΔFC) higher fault coverage for benchmark circuits and up to 2% higher fault coverage for industrial designs compared to conventional LOC method. It had been noticed that the ELOC fault coverage improvement depends on the fault coverage difference between the LOC and LOS method as ELOC provides an intermediate coverage between LOC and LOS. For example, in the experiments on the industrial designs, it is noticed that the fault coverage difference between LOC and LOS was not very significant. Also, for a given fault coverage, the number of patterns generated by ELOC is less in all designs. The ($\Delta Patt$) column is the percentage pattern reduction for the maximum fault coverage achieved by LOC method. Figure 4.15 shows the number of patterns in both methods for the highest LOC fault coverage achieved. In all cases the ELOC method generates a smaller pattern set as the controllability of transition launch has increased. $\Delta Patt$ is calculated by:

$$\Delta Patt = \{\frac{LOC - ELOC}{LOC}\}_{FC_{LOC_{max}}}$$

Capture-Dependency Untestable

In ELOC, in case a scan chain is used as a shift-register to launch a target transition fault and if the fault is observable at a scan flip-flop in the same

Table 4.7. ATPG Results for ISCAS'89 and ITC'99 benchmark circuits using LOC method.

Design	*LOC*			
	DT	FC (%)	# Patt	CPU Time *[sec]*
s13207	5957	72.14	167	1.15
s15850	6907	64.02	106	1.86
s38417	27410	96.05	248	4.76
s38584	28469	75.55	360	32.89
b17	93106	80.71	1132	1000.11
b18	314307	77.00	1339	1000.74
b19	540169	69.27	1152	1001.60

Table 4.8. ATPG Results for ISCAS'89 and ITC'99 benchmark circuits using ELOC method and comparison with the results obtained from LOC method.

Design	*ELOC*						
	DT	FC (%)	# Patt	CPU Time *[sec]*	Δ DT	Δ FC (%)	Δ Patt (%)
s13207	6978	84.50	250	1.48	1021	12.36	58.08
s15850	8479	78.60	172	4.98	1572	14.58	56.60
s38417	27432	96.12	277	7.77	22	0.07	1.11
s38584	31666	84.04	353	18.66	3197	8.49	74.72
b17	100510	87.13	1240	810.25	7404	6.42	56.44
b18	349879	85.71	1327	1001.12	35572	8.71	69.23
b19	544486	69.83	1088	1000.89	4317	0.55	14.23

Table 4.9. ATPG Results for the first four industrial designs A through D using LOC method.

Design	*LOC*			
	DT	FC (%)	# Patt	CPU Time *[sec]*
A	282658	88.09	2145	896.96
B	770043	87.14	8360	7800.47
C	2052330	95.16	13463	76867.39
D	1640185	87.43	12401	12949.53
E	1727905	87.53	12527	22930.56
F	3662034	91.88	28762	55436.58
G	2810269	95.09	37124	81618.71
H	3192991	91.56	10219	11635.25

Table 4.10. ATPG Results for the next four industrial designs E through H using ELOC method and comparison with the results obtained from LOC method.

Design	*ELOC*						
	DT	FC (%)	# Patt	CPU Time *[sec]*	Δ DT	Δ FC (%)	Δ Patt (%)
A	288681	89.96	2014	924.74	6023	1.87	52.7
B	773821	87.57	8539	8702.29	3778	0.43	17.5
C	2072388	96.09	9515	26082.12	20058	0.93	55.6
D	1645459	87.71	11304	13447.33	5274	0.28	22.1
E	1746713	88.48	11583	25642.80	18808	0.95	43.4
F	3677805	92.28	24666	61098.98	15771	0.40	31.9
G	2818722	95.38	34350	90853.44	8453	0.29	24.9
H	3217952	92.28	12505	47788.05	24961	0.72	56.4

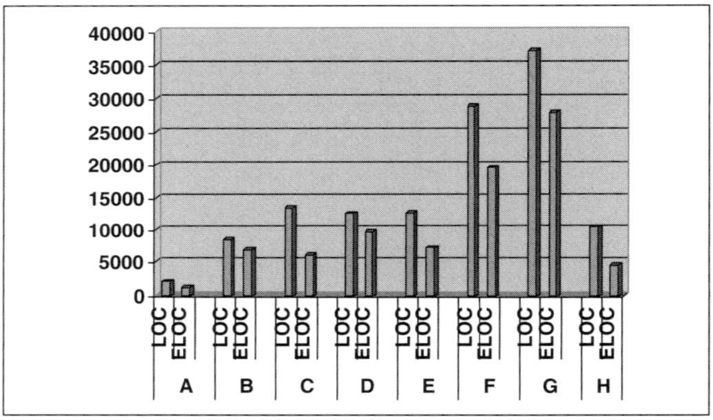

Fig. 4.15. Comparison of no. of patterns in LOC and ELOC for industrial designs.

chain, then the fault cannot be captured and is referred to as *capture dependency untestable*. Figure 4.16 shows an example of such a type of untestable fault. Since, the scan chain is only controllable and not observable while it is acting as a shift register, the fault cannot be captured.

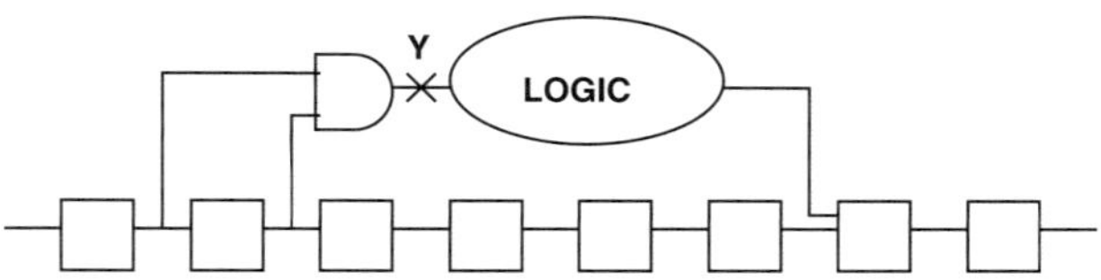

Fig. 4.16. Capture dependent untestable [17].

Fault Coverage Improvement

The fault coverage of ELOC method can be increased further by careful scan insertion to reduce capture-dependent untestable faults. Presently, the scan insertion tool stitches the scan cells based on the lexical naming convention of the scan flops. The best results in terms of fault coverage can be obtained, if the scan insertion tools can stitch the chains such that the inter-scan chain faults are increased. However, irrespective of scan chain order, the pattern volume reduction is achieved due to improved controllability.

Table 4.11. Comparison of Methodologies

Methodology	*SEN Effort*	*FC*	*Pattern Count*
LOC	Low	Low	High
LOS + LOC	High	High	Low
LOS + LOC(LTG)[16]	Medium	High	Low
ELOC	Low	Medium	Medium

Qualitative Analysis of Various Methods

Table 4.11 shows a qualitative analysis of the different transition fault methodologies. The parameters compared are the scan enable effort *(SEN Effort)* during physical design, fault coverage (FC) and the pattern count. The conventional LOC and the ELOC method require the least *SEN* effort as it is not at-speed. The LOS method requires the *SEN* signal to work at functional speed and takes high *SEN* effort. In [16], the entire scan enable tree is divided into sub-trees and the local scan enable for each sub-tree is generated internally. This reduces the *SEN* effort during the physical design step. The ELOC method provides an intermediate fault coverage and pattern count between the conventional LOC method and LOS method while the SEN effort is still kept low.

4.8 Summary

In this chapter, a new method of transition fault testing, referred as *enhanced launch-off-capture* (ELOC), was presented which provides better controllability than the conventional launch-off-capture (LOC) method even in the presence of low-cost tester ATPG constraints (No-PI changes and No-PO measures). LOC testing is known to provide less quality results, both in terms of pattern count and fault coverage, but design teams may not use launch-off-shift (LOS) due to the challenge of routing the scan enable signal. The solution shown in this chapter generates local scan-enable signals that control the transition launch path; for this purpose, the method relies on embedding some scan enable control information in the patterns. A special cell called the LSEG cell is used for the generation of the local scan enable signal. This cell is simple to design and layout, and its area overhead is comparable to that of a scan flip-flop. The number of LSEG cells inserted in the design will be small, thereby making the area overhead due to ELOC technique negligible. The LSEG-based solution provides greater flexibility and controllability for transition fault pattern generation. The ELOC technique provides as high as 14% (approx.) fault coverage and upto 74% (approx.) pattern reduction compared to LOC technique. The DFT insertion and ATPG can be easily performed using the commercial ATPG tools; therefore ELOC solution is easy to practice.

References

1. International Technology Roadmap for Semiconductors 2003 (http://public.itrs.net).
2. S. Natarajan, M.A. Breuer, S.K. Gupta, "Process Variations and Their Impact on Circuit Operation," in Proc. *IEEE Int. Symp. on Defect and Fault Tolerance in VLSI Systems*, pp. 73-81, 1998.
3. R. Wilson, "Delay-Fault Testing Mandatory, Author Claims," *EE Design*, Dec. 2002.
4. G. Aldrich and B. Cory, "Improving Test Quality and Reducing Escapes," in Proc. *Fabless Forum, Fabless Semiconductor Assoc.*, pp. 34-35, 2003.
5. X. Lin, R. Press, J. Rajski, P. Reuter, T. Rinderknecht, B. Swanson and N. Tamarapalli, "High-Frequency, At-Speed Scan Testing," *IEEE Design & Test of Computers*, pp. 17-25, Sep-Oct 2003.
6. V. Jayaram, J. Saxena and K. Butler, Scan-Based Transition-Fault Test Can Do Job, *EE Times*, Oct. 2003.
7. K. Cheng, "Transition Fault Testing for Sequential Circuits," *IEEE Transactions on Computer-Aided Design of Integrated Circuits and Systems*, vol. 12, no. 12, pp. 1971-1983, Dec. 1993.
8. T. M. Mak, A. Krstic, K. Cheng, L. Wang, "New challenges in delay testing of nanometer, multigigahertz designs," *IEEE Design & Test of Computers*, pp. 241-248, May-Jun 2004.
9. M. Bushnell, V. Agrawal, *Essentials of Electronics Testing*, Kluwer Publishers, 2000.
10. N. Devtaprasanna, A. Gunda, P. Krishnamurthy, S.M. Reddy, I. Pomeranz, "A novel method of improving transition delay fault coverage using multiple scan enable signals," in Proc. *Int. Conf. on Computer Design*, 2005.
11. I. Pomeranz and S.M. Reddy, "At-speed delay testing of synchronous sequential circuits," in Proc. *Design Automation Conference*, pp.177181, 1992.
12. J. Savir, "Skewed-Load Transition Test: Part I, Calculus," in Proc. *Int. Test Conf. (ITC'92)*, pp. 705-713, 1992.
13. J. Savir and S. Patil, "On Broad-Side Delay Test," in Proc. *VLSI Test Symp. (VTS'94)*, pp. 284-290, 1994.

14. B. Dervisoglu and G. Stong, "Design for Testability: Using Scanpath Techniques for Path-Delay Test and Measurement," in Proc. *Int. Test Conf. (ITC'91)*, pp. 365-374, 1991.
15. X. Liu, M. S. Hsiao, S. Chakravarty and P. Thadikaran, "Techniques to Reduce Data Volume and Application Time for Transition Test," in Proc. *Intl Test Conf. (ITC'02)*, pp. 983-992, 2002.
16. N. Ahmed, C.P. Ravikumar, M. Tehranipoor and J. Plusquellic, "At-Speed Transition Fault Testing With Low Speed Scan Enable," in Proc. *IEEE VLSI Test Symp. (VTS'05)*, pp. 42-47, 2005.
17. N. Ahmed, M. Tehranipoor and C.P. Ravikumar, "Enhanced Launch-off-Capture Transition Fault Testing," in Proc. *Int. Test Conf. (ITC'05)*, pp -, 2005.
18. J. Saxena, K. M. Butler, J. Gatt, R. Raghuraman, S. P. Kumar, S. Basu, D. J. Campbell, J. Berech, "Scan-Based Transition Fault Testing - Implementation and Low Cost Test Challenges," in Proc. *International Test Conference (ITC'02)*, pp. 1120 - 1129, Oct. 2002.
19. S. Wang, X. Liu, S.T. Chakradhar, "Hybrid Delay Scan: A Low Hardware Overhead Scan-Based Delay Test Technique for High Fault Coverage and Compact Test Sets," in Proc. *Design, Automation and Test in Europe (DATE'03)*, pp. 1296-1301, 2004.
20. S. Wang and S. T. Chakradhar, "Scalable Scan-Path Test Point Insertion Technique to Enhance Delay Fault Coverage for standard scan designs," in Proc. *Int. Test Conf. (ITC'03)*, pp. 574-583, 2003.
21. W. Li, S. Wang, S. T. Chakradhar and S. M. Reddy, "Distance Restricted Scan Chain Reordering to Enhance Delay Fault Coverage," in Proc. *Int. Conf. on VLSI Design*, pp. 471-478, 2005.
22. W. Mao and M. D. Ciletti, "Reducing Correlation to Improve Coverage of Delay Faults in Scan-Path Design," *IEEE Trans. on Computer-Aided Design of Integrated Circuits and Systems*, vol. 13, no. 5, pp. 638-646, May 1994.
23. P. Gupta, A. B. Kahng, I. Mandoiu, P. Sharma, "Layout-Aware Scan Chain Synthesis for Improved Path Delay Fault Coverage," in Proc. *Int. Conf. on Computer Aided Design (ICCAD'03)*, pp. 754-759, 2003.
24. N. Devtaprasanna, A. Gunda, P. Krishnamurthy, S. M. Reddy and I. Pomeranz, "Methods for Improving Transition Delay Fault Coverage Using Broadside Tests," in Proc. *Int. Test Conf. (ITC'05)*, pp. -, 2005.
25. W. C. Lai, A. Krstic and K. T. Cheng, "On Testing the Path Delay Faults of a Microprocessor Using its Instruction Set," in Proc. *VLSI Test Symp. (VTS'00)*, pp. 15-20, 2000.
26. Synopsys DFT Compiler, "User Manual for SYNOPSYS Toolset Version 2004.06," Synopsys, Inc., 2004.
27. K. J. Rearick, "Too Much delay Fault Coverage is a Bad Thing," in Proc. *Int. Test Conf. (ITC'01)*, pp. 624-633, 2001.
28. Tetramax ATPG Users Guide, "SYNOPSYS Toolset Version 2004.06," Synopsys, Inc., 2004.

5

Hybrid Scan-Based Transition Delay Test

Launch-off-shift (LOS) method provides higher fault coverage and lower pattern count when compared to launch-off-capture (LOC) method. Investigations have proven that some faults can be detected using LOC but not LOS and vice-versa. In LOS, the second pattern (i.e. pattern $V2$) is generated using the last shift while in LOC method, the second pattern is the functional response of the first pattern (i.e. pattern $V1$). To take advantage of both methods' ability in detecting different faults, the patterns generated using one method are applied on top of the other method's pattern set. This method, in industry, is often called LOS+LOC. It provides a fault coverage higher than that of LOS but the design effort still remains high since the scan enable to all scan chains must be timing closed because of using LOS method.

This chapter presents a hybrid scan-based transition delay fault test. The hybrid technique controls a small subset of scan cells by launch-off-shift (LOS) method and the rest by enhanced launch-off-capture (LOC) method. An efficient ATPG-based controllability and observability measurement approach is presented to select the scan cells to be controlled by launch-off-shift or enhanced launch-off-capture which was presented in the previous chapter. In this technique, local scan enable signals are generated on-chip using two local scan enable generator cells (fast scan enable signal for launch-off-shift and slow scan enable signal for enhanced launch-off-capture). The cells can be inserted anywhere in a scan chain and the area overhead of cells is negligible. The launch and capture control information of scan enable signals are embedded into test patterns and transferred into the scan chain during scan-in process. This hybrid scan-based technique improves the fault coverage, reduces the pattern count and scan enable design effort. The technique is practice-oriented, specially suitable for low-cost testers since no external at-speed scan enable signal is required, and implemented using current commercial ATPG tools.

5.1 Introduction

As device frequencies become higher, the ability to test the device at-speed becomes limited by the capabilities of the production test equipment. Several on-chip design-for-test alternatives are deployed in industry to using a more expensive tester such as on-chip clock generator for at-speed clock, pipeline scan enable generation and on-chip at-speed scan enable generation [16] for launch-off-shift transition fault test. The LOS method is preferable from ATPG complexity and pattern count view points, compared to LOC method. However, due to increasing design sizes, the SEN fanout exceeds any other net in the design. The launch-off-shift constraints SEN to be timing critical requiring high design-effort which makes it difficult to implement on designs where the turn-around-time is critical. That is the main reason that LOC method has been widely practiced, especially on very low cost testers [14]. In this chapter, a technique presented to use both LOS and LOC methods in scan-based designs with low design effort for SEN signal providing higher fault coverage and lower pattern count.

In a related work [15], a hybrid scan architecture is proposed which controls a small subset of selected scan cells by LOS and the rest are controlled by LOC technique. A fast scan enable signal generator is designed which drives the LOS-controlled scan flip-flops. The selection criteria of the LOS-controlled scan flip-flops determines the effectiveness of the method. In some cases, the number of patterns generated by the hybrid method exceeds the LOC pattern count. Moreover, the LOS-controlled flip-flops cannot be used in LOC mode. Figure 5.1(a), generally, shows the SEN signal waveforms of the hybrid technique that uses both LOS and LOC methods.

A new scan-based at-speed test referred to as *enhanced launch-off-capture* (ELOC) was presented in the previous chapter, in which the ATPG tool deterministically determines the transition launch path either through a functional path or the scan path. The technique improves the controllability of transition fault testing, improves fault coverage and it does not require the scan enable to change at-speed. Figure 5.1(b) shows SEN signal waveforms in ELOC technique. SEN of a subset of scan chains stays at 1 (SEN1) during launch and capture cycles to launch the transition only. The remaining scan chains were controlled using the second SEN signal (SEN2) to launch a transition through the functional path during launch cycle and capture the response during the capture cycle. Figure 5.1(c) shows a circuit with two scan chains, $Chain1$ acting as a shift register and $Chain2$ in the functional mode. The conventional LOC method may be viewed as a special condition of the enhanced LOC (ELOC) method, where the scan enable signals of all the chains are '0' (SEN2) during the launch and capture cycles.

In [22][23], techniques are proposed to improve the LOS fault coverage by reducing the shift dependency. The scan flip-flops are reordered in [22] to minimize the number of undetectable faults and the technique restricts the distance by which a scan flip-flop can be moved to create the new scan chain order. The authors in [23] proposed a technique to insert dummy flip-flops and re-order scan flip-flops taking wire length costs into consideration, to improve path delay fault coverage. In [24], the proposed technique uses a special ATPG to identify pairs of adjacent flip-flops between which test points (dummy gates or flip-flops) are inserted.

5.1.1 Overview of the Hybrid Method

In this chapter, a new hybrid scan-based technique is presented to improve the fault coverage of transition fault test [20] [21]. The hybrid scan architecture controls a small subset of selected scan cells by LOS and the rest by ELOC method. An efficient ATPG-based controllability/observability measurement approach is presented to select scan cells to be controlled by LOS or ELOC method. The selection criteria improves the fault coverage of transition test and reduces the overall pattern count. As mentioned above, a small number of scan cells are LOS-controlled, hence only a small subset of scan chains' scan enable signals need to be timing closed resulting in reduced SEN design effort. The method is robust, practice-oriented and implemented using existing commercial ATPG tools [18], i.e. no special ATPG is required. To control the scan chain mode of operation (LOS or ELOC), two new cells, called local scan enable generators (LSEG), are presented to generate on-chip scan enable signals. The scan enable control information for the launch and capture cycles are embedded in the test data itself. The LSEG cells have the flexibility to be inserted anywhere in the scan chain and the hardware area overhead is negligible. This hybrid technique is suitable for use by low-cost testers since no external at-speed scan enable is required.

5.2 Motivation

Enhanced launch-off-capture (ELOC) improves controllability of launching a transition either through the scan path or the functional path [19]. However, it has less observability compared to LOS as a scan chain working in shift-mode to launch a transition is not observable at the same time (scan enable held high during LC and CC cycles, SEN1 shown in Figure 5.1(c)). Therefore, the fault coverage of ELOC is lesser than LOS but greater than LOC. Figure 5.2(a) shows the fault coverage analysis for the three different transition fault methods, LOS, LOC and ELOC. There is a common set of transition faults which are detected by both LOS and LOC and there are some faults in the LOC transition fault set which are not detected by LOS, e.g. shift-dependency untestable faults [15] [24]. However, ELOC covers the entire transition fault set of LOC and also detects some extra faults which fall in the LOS detected fault set. This is because, LOC is a special case where all the local SEN signals are held 0 during the launch and capture cycles. The ELOC method provides

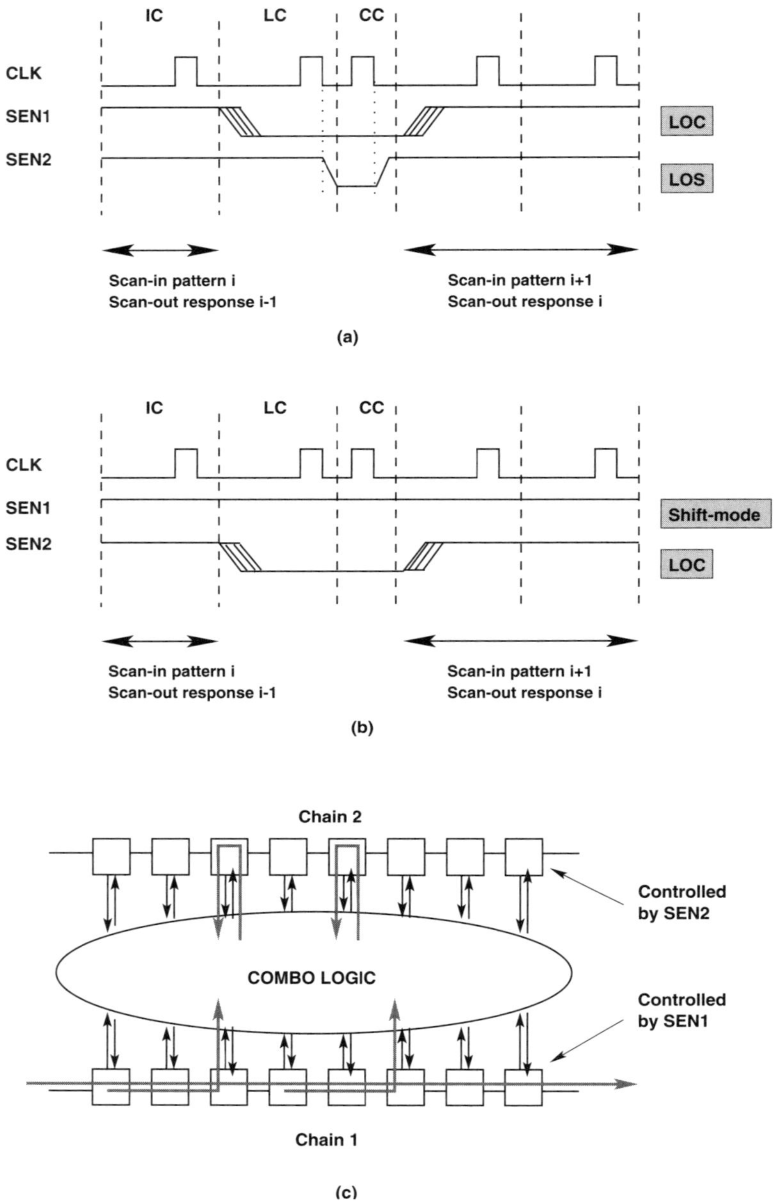

Fig. 5.1. (a) SEN waveforms in the hybrid-scan technique (b) SEN waveforms in ELOC technique and (c) Controllability of ELOC (*Chain1* used in shift mode and *Chain2* in functional mode) [21].

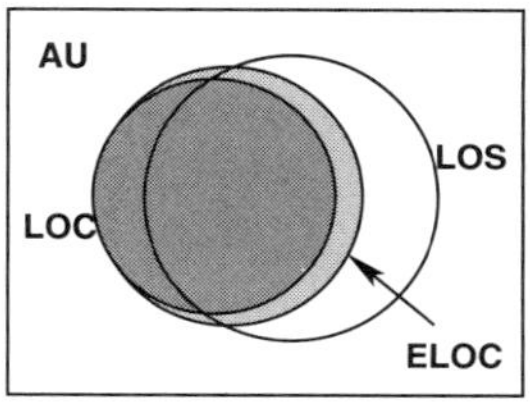

(a)

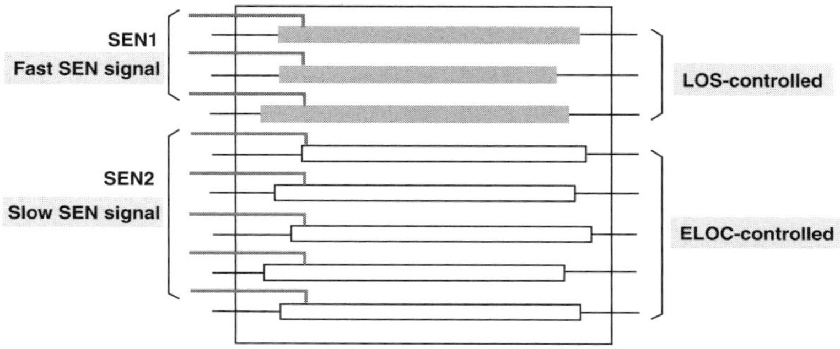

(b)

Fig. 5.2. (a) Fault analysis of LOS, LOC and ELOC techniques and (b) Hybrid-Scan Architecture: LOS-controlled scan chains using fast SEN signal and ELOC controlled scan chains using slow SEN signal [21].

an intermediate fault coverage point between LOS and the conventional LOC method [19].

To improve the fault coverage and detect the union of fault sets detected in both LOS and ELOC mode, the scan cells must be controllable in both LOS and ELOC mode. Also, to reduce the design effort for at-speed scan enable required for LOS, it is intended to determine the minimum number of scan cells which requires very high controllability and observability during pattern generation and control the resulting smaller subset of scan cells in LOS mode and the remaining scan cells in ELOC mode. This reduces the design effort to timing close SEN signal at-speed required for LOS-controlled scan flip-flops. Figure 5.2(b) shows an example of a hybrid-scan architecture with eight scan chains. The LOS-controlled scan flip-flops are stitched in separate scan chains. The first three scan chains containing LOS-controlled flip-flops are controlled using a fast scan enable signal and the remaining scan chains are controlled in ELOC mode using a slow scan enable signal. Moreover, it is also required to configure the LOS-controlled scan chains in functional mode as there are some faults detected only in LOC mode and not by LOS.

5.3 Local Scan Enable Signal (LSEN) Generation

The new hybrid-scan transition fault testing method described in Section 5.2 provides more controllability to launch a transition but requires independent scan enable signal (SEN) for each scan chain. Multiple SEN ports can be used, but this increases the number of pins. Note that two types of SEN signals need to be generated on-chip. The scan enable control information for the scan flip-flops differ only during the launch and capture cycles of the pattern. Hence, the low-speed scan enable signal from the external tester can be utilized for the scan shift operation and the scan enable control information for only the launch and capture cycles can be generated internally.

5.3.1 Local Scan Enable Generator (LSEG) Cells

Here, two local scan enable generator (LSEG) cells are presented to generate on-chip local scan enables using low-speed external scan enable generated by a low-cost tester. Since the hybrid technique uses both LOS and enhanced LOC techniques, both fast and slow local SEN signals must be generated. In the following, the two LSEG cells and their operations are presented.

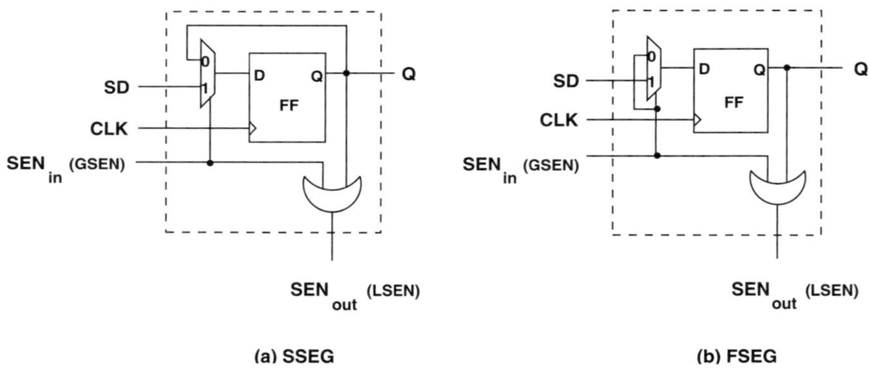

Fig. 5.3. LSEG cells: (a) Slow scan enable generator (SSEG) cell and (b) Fast scan enable generator (FSEG) cell [21].

5.3.2 Slow Scan Enable Generator (SSEG)

A local scan enable generator is designed in [19], to control the transition launch path of a scan flip-flop. In the rest of this chapter, this cell will be referred to as the slow scan enable generator (SSEG), as the local scan enable signal does not make any at-speed transition. Figure 5.3(a) shows the SSEG cell architecture. It consists of a single flip-flop which is used to load the control information required for the launch and capture cycles. The input scan enable (SEN_{in}) pin which is connected to the external scan enable signal from the tester is referred to as the global scan enable (GSEN). An additional output scan enable ($SEN_{out} = GSEN + Q$) pin represents the local scan enable (LSEN) signal. Therefore, after going to a control state *(Q)* at the end of the shift operation (GSEN is de-asserted), LSEN remains in this state as long as it is asynchronously set to 1 by GSEN. The SSEG cell essentially holds the value 0(1) for the launch and capture cycles, which is loaded at the end of the shift operation (GSEN=1).

$$LSEN = (GSEN + Q) = \begin{cases} 1 & \text{if GSEN=1} \\ Q & \text{if GSEN=0} \end{cases}$$

Table 5.1. SSEG Operation [20]

GSEN	Q	LSEN	Operation
1	X	1	Shift
0	1	red1 → 1	Shift-Launch (No Capture)
0	0	red0 → 0	Functional-Launch-Capture

Table 5.1 shows the different modes of operation of SSEG cell. $GSEN = 1$ represents the normal shift operation of the pattern. When $GSEN = 0$ and $Q = 1$, $LSEN = 1$ and the controlled scan flip-flops act in the shift mode to launch the transitions only (*Shift-Launch (No Capture)* mode) and there is no capture as the LSEN signal is '1' ($LSEN = 1 \rightarrow 1$ at launch edge). The capture is performed at the other observable scan flip-flops. The LSEN controlled scan flip-flops act in the conventional LOC method when $GSEN = 0$ and $Q = 0$ (*Functional-Launch-Capture* mode).

5.3.3 Fast Scan Enable Generator (FSEG)

A new at-speed local scan enable generator architecture, referred to as the fast scan enable generator (FSEG), is shown in Figure 5.3(b). Table 5.2 shows the different modes of operation of FSEG cell. Similar to the SSEG cell operation, $GSEN = 1$ represents the normal shift operation of the pattern. When $GSEN = 0$ and $Q = 1$, $LSEN = 1$ and the scan flip-flops act in the (*Shift-Launch-Capture* mode) to launch the transition from the scan path and capture the response at the next capture cycle (conventional LOS method). The LSEN from the FSEG cell, makes a $1 \rightarrow 0$ at-speed transition at the launch cycle. The LSEN controlled scan flip-flops act in the conventional LOC method when $GSEN = 0$ and $Q = 0$ (*Functional-Launch-Capture* mode).

Table 5.2. FSEG Operation [20]

GSEN	Q	LSEN	Operation
1	X	1	Shift
0	1	red1 $\rightarrow$ 0	Shift-Launch-Capture
0	0	red0 $\rightarrow$ 0	Functional-Launch-Capture

5.3.4 Operation of LSEG cells

The local scan enable generator cells are inserted within the scan chains and the control information is passed as part of the test data. The scan enable control information will be part of each test pattern and is stored in the tester's memory. The normal scan architecture with a single scan enable signal from the external tester is shown in Figure 5.4(a). There are eight scan flip-flops in the scan chain and the test pattern shifted is 10100110. Figure 5.4(b) shows the same circuit in which a local scan enable signal is generated from the test pattern data for the hybrid transition fault test method. The internally generated scan enable signal is termed as local scan enable (LSEN). The main objective is to de-assert the external scan enable signal (GSEN) after the entire shift operation and then generate the LSEN signal during the launch and capture cycle from the test data. In this case, the pattern shifted is modified to [C]10100110, where C is the scan enable control bit which is stored in LSEG cell at the end of the scan operation.

The GSEN signal asynchronously controls the shift operation. GSEN is de-asserted after the nth shift (IC) cycle, where n=9. n is the length of scan chain after inserting LSEG cell. After the GSEN signal is de-asserted at the end of the shift operation, the scan enable control during the launch and capture cycles is the control bit C stored in LSEG. At the end of the capture cycle, the LSEN signal is asynchronously set to 1 by GSEN for scanning out the response. Figure 5.4(c) shows the process of test pattern application and also the timing waveforms for the two different local scan enable generator cells, i.e. SSEG and FSEG.

5.4 Flip-Flop Selection: ATPG-Based Controllability/Observability Measurement

In LOS technique, the fault activation path (scan path) is fully controllable from the scan chain input compared to the functional path used in LOC method. Hence, in most cases, for the same detected fault, a LOS pattern will require fewer number of care bits than a LOC pattern. The controllability measure of a scan flip-flop is defined as the percentage of patterns, in the entire pattern set *(P)*, for which a care-bit is required in the scan flip-flop, to enable activation or propagation of a fault effect. Figure 5.5 shows a scan

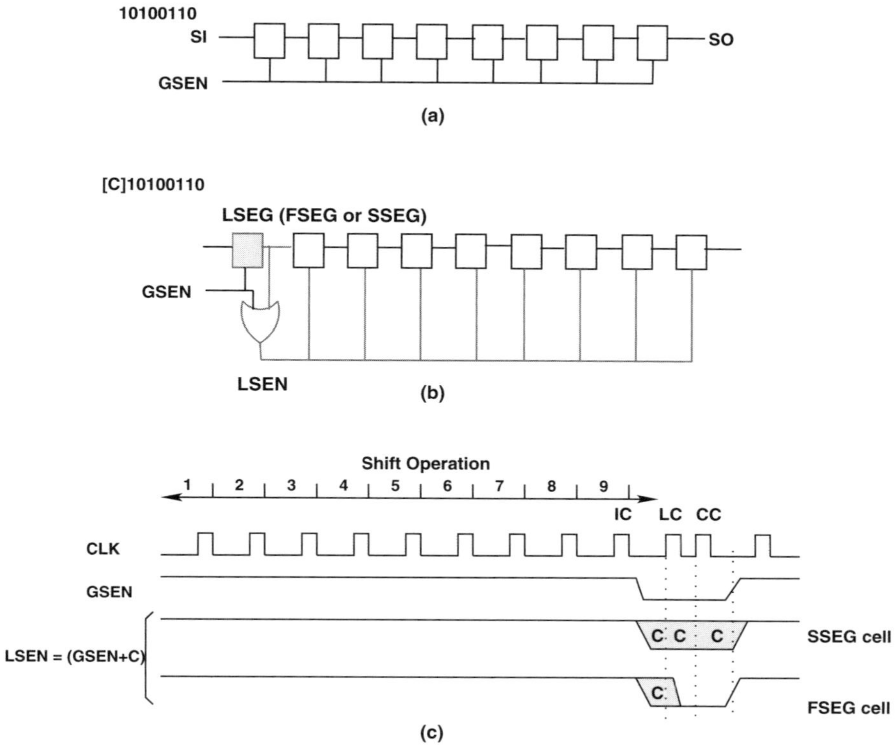

Fig. 5.4. (a) Scan chain architecture, (b) Local scan enable (LSEN) generation using LSEG and (c) LSEN generation process and waveforms [20].

flip-flop with an input (observability) and output (controllability) logic cone. A large output logic cone implies that the scan flip-flop will be used to control more number of faults, i.e. a care-bit will be required in their activation or propagation. Similarly, the observability of scan flip-flop is determined by the input logic cone and it is defined as the percentage of patterns, in the entire pattern set *(P)*, for which a valid care-bit is observed in the scan flip-flop.

In a transition fault test pattern pair (V_1, V_2), the initialization pattern (V_1) is essentially an IDDQ pattern to initialize the target gate to a known state. In the next time frame, pattern V_2 is a stuck-at-fault test pattern to activate and propagate the required transition at the target node to an observable point. Therefore, to find the controllability/observability measure of a scan flip-flop, an ATPG tool is used to generate stuck-at patterns and force the tool to fill don't-care (X) value for scan flip-flops which do not affect the activation/propagation of any fault. The controllability of the ith scan flip-flop is measured as $C_i = \frac{p_c}{P}$, where p_c is the number of patterns with a care-bit in the scan flip-flop during scan-in and P is the total number of stuck-at patterns.

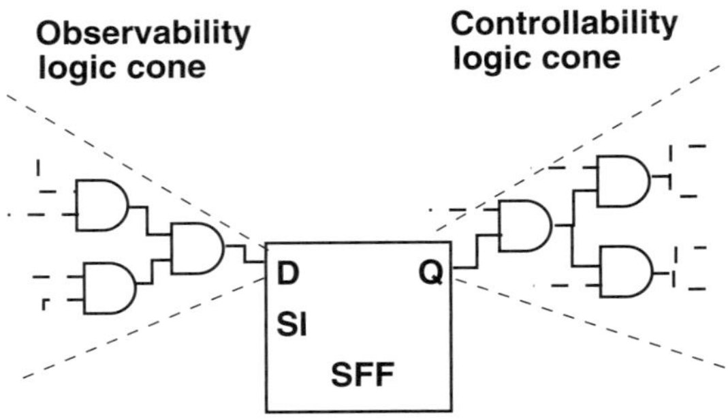

Fig. 5.5. Scan flip-flop controllability and observability measure [20].

Similarly, the observability is measured as $O_i = \frac{p_o}{P}$, where p_o is the number of patterns with a care-bit in the scan flip-flop during scan-out. We, then, use the measured controllability and observability factors of each scan flip-flop and determine the cost function ($CF_i = C_i \times Oi$). The scan flip-flops are arranged in decreasing order of cost function and a subset with very high cost function are selected as LOS-controlled flip-flops. The ATPG-based controllability/observability measure technique overcomes the limitation of SCOAP [26] based method used in [15], in which there is a possibility to select a scan flip-flop whose 0(1) controllability is high but not controlled to 0(1) during pattern generation by the ATPG tool.

5.5 CASE Study: DFT Insertion, ATPG Flow and Fault Analysis

5.5.1 Test Architecture

The local scan enable generator based solution presented in this chapter provides a method of generating the internal local scan enable signals from the pattern data and global scan enable signal from the tester. The overhead of generating the local scan enable signal is the addition of a LSEG (SSEG or FSEG) cell in the scan chain. The area overhead of an LSEG cell is a few extra gates, which is negligible in modern designs. The area of the buffer tree required to drive all the LOS-controlled scan flip-flops through LSEG cells is assumed to be equal to the case of applying an at-speed GSEN signal from external ATE. Figure 5.6 shows a multiple scan chain architecture with n scan chains. The LOS-controlled scan flip-flops determined using the controllability and observability measurement are stitched in separate scan chains. Each scan chain i, where $1 \leq i \leq n$, consists of an LSEG (FSEG or SSEG) cell which generates the local scan enable signal $LSEN_i$ for the respective scan chain. The GSEN signal connects only to the SEN_{in} port of LSEG cells.

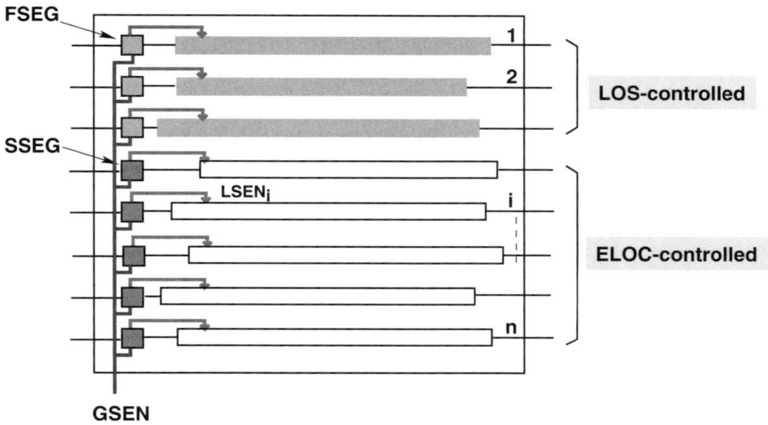

Fig. 5.6. Hybrid-Scan Test Architecture. FSEG cells driving LOS-controlled scan chains and SSEG cells driving ELOC-controlled scan chains [20].

5.5.2 Case Study

In this case study, the hybrid method was experimented with a subchip of an industrial-strength design that had the following characteristics (Table 5.3). The design has 16 scan chains and approximately 10K scan cells. There are 13 non-scan cells and six internal clock domains. One LSEG cell is inserted per scan chain. The test strategy is to get the highest possible test coverage for the transition faults. When generating test patterns for the transition faults, only the faults in the same clock domain are targeted. During pattern generation, only one clock is active during the launch and capture cycle. Hence, only faults in that particular clock domain are tested. All PIs remain unchanged and all POs are unobservable while generating the test patterns for the transition faults. This is because the very low cost testers are not fast enough to provide the PI values and strobe POs at-speed.

Table 5.3. Design Characteristics [21]

Clock Domains	6
Scan Chains	16
Scan flip-flops	10477
Non-scan flip-flops	13
Transition Delay Faults	320884

5.5.3 DFT Insertion Based on Controllability/Observability Measure

The cost function (*controllability* × *observability*) of a scan flip-flop is measured using the ATPG-based technique explained in Section 5.4. Figure 5.7 shows the cost function of each scan flip-flop in the design. It can be noted that approximately only 20-30% of the entire flip-flops require very high controllability and observability. Hence, scan enable need not be at-speed for all scan flip-flops.

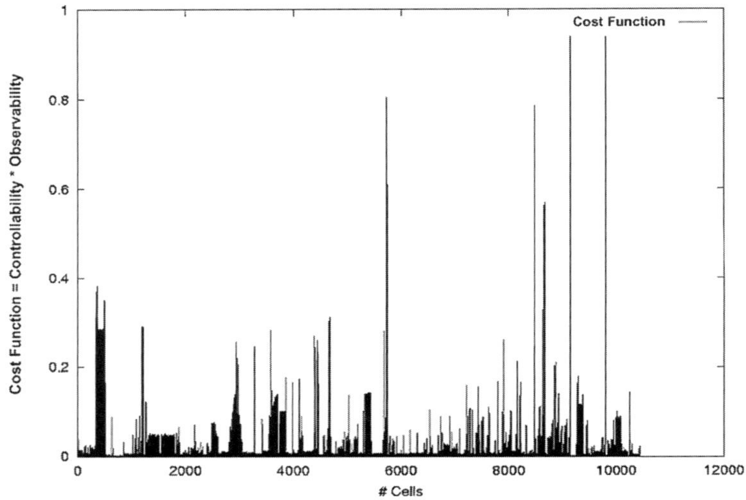

Fig. 5.7. Scan flip-flop controllability/observability measure [20].

The scan flip-flops are then arranged in decreasing order of cost function and this order is used during scan insertion. In the new order of scan chains, the initial few chains consists of very high controllability/observability flip-flops and they are selected for LOS based on their average cost function. The average cost function of a scan chain was measured as $\Sigma\ CF_i/N$, where $CF_i = C_i \times O_i$ is the cost function of ith scan flip-flop in the chain as explained in Section 5.4 and N is the number of flip-flops in the scan chain. Figure 5.8 shows the average cost function of each chain for normal scan insertion and after controllability/observability-based scan insertion. It can be noticed that after controllability/observability based scan insertion technique, the average cost function of the first 5 scan chains, as shown in Figure 5.8(b) is very high (due to very high cost function scan flip-flops) and it is very low for the rest of the chains. Therefore, the first 5 chain's scan enable signal can be designed to

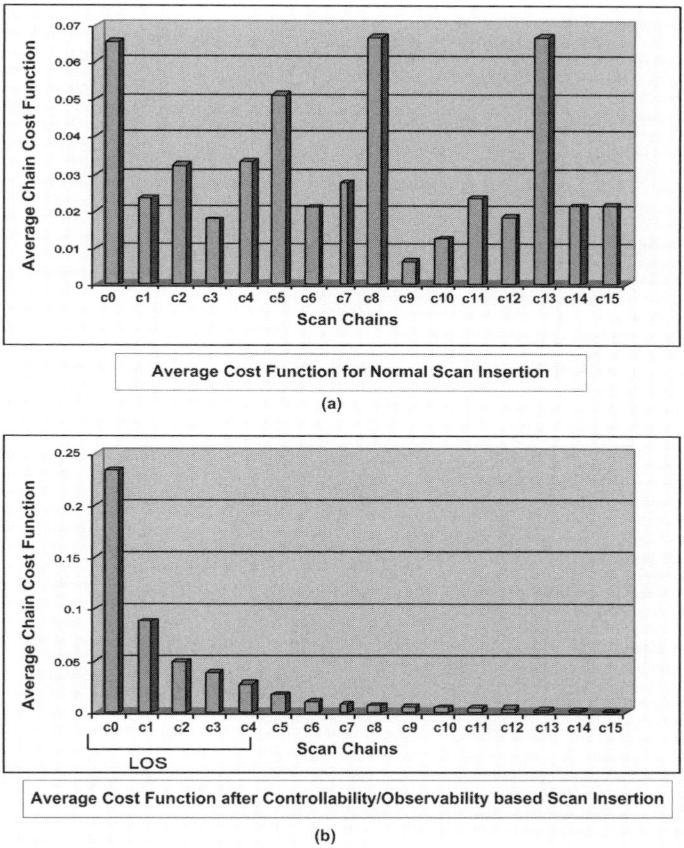

Fig. 5.8. Average cost function before and after controllability/observability-based scan insertion [20].

be at-speed (controlled using FSEG cell) and the rest of the scan chains can be controlled using the slow speed scan enable (controlled using SSEG cell).

Synopsys DFT Compiler [18] was used for scan chain insertion in the design. To insert the LSEG cells, the synthesis tool must recognize the LSEG cell as a scan cell in order to stitch it into the scan chain. This requires it to be defined as a new library cell with the scan cell attributes. A workaround is to design the LSEG cell as a module, instantiate it and declare it as a scan segment of length 1. The GSEN signal is connected to all the LSEG cells SEN_{in} input pin. During scan insertion, only the LSEG cell is specified in the scan path, as the tool will stitch the rest of the cells including the LSEG cell and balance the scan chain depending on the longest scan chain length parameter. Additionally, the tool provides the flexibility to hookup each LSEG cell's SEN_{out} port in a particular chain to all the scan enable input port of the scan flip-flops in the respective chain.

Figure 5.9 shows the algorithm used for the controllability measure based scan insertion.

```
01: Normal Scan Insertion
02: ATPG-based controllability measure of flip-flops
03: Arrange flip-flops in non-increasing order of controllability measure
04: Assign the new order of scan cells to each scan chain
05: Re-stitch scan chains with LSEG-based control for each chain
```

Fig. 5.9. Algorithm for controllability based scan insertion.

5.5.4 ATPG

The ATPG tool must be able to understand the local scan enable generation methodology and deterministically decide the transition fault activation path. A commercial ATPG tool was used for test pattern generation (Synopsys TetraMax [18]). The test pattern generation tool supports two ATPG modes: 1) Basic-Scan and 2) Sequential. Basic-Scan ATPG is a combinational-only mode with only one capture clock between pattern scan-in and scan-out of response while the Sequential mode uses sequential time-frame ATPG algorithm. By default, when generating test patterns for transition fault model in functional launch mode, the commercial ATPG tools use a two-clock ATPG algorithm that has some features of both the Basic-Scan and Sequential engines. The tool understands the local scan enable generation technique using LSEG cells and is able to decide the launch path for the target transition fault deterministically. Hence, there is no fundamental difference in the ATPG methodology when the LSEG-based solution is used.

The scan enable signal for the flip-flops for the launch and capture cycles now comes from an internally generated signal. Notice that the OR gate in

Table 5.4. ATPG results [21]

	LOS	*LOC*	*ELOC[19]*	*Hybrid*
Detected faults	292342	282658	288681	295288
Test Coverage	91.30	88.27	90.15	91.92
Fault Coverage	91.11	88.09	89.96	91.74
Pattern Count	1112	2145	2014	1799
CPU Time *[sec]*	329.30	896.96	924.74	1014.60

the LSEG cell generates the local scan enable signal through a logical OR of the global scan enable and the Q output of the flip-flop FF (see Figures 5.3(a) and (b)). The global scan enable signal is active high during scan shift operation. The tool determines the local scan enable for each chain and shifts the control value into the LSEG cell during pattern shift, used for launch and capture. It also deterministically decides the combination of scan chains to work in *Shift/Functional-Launch* mode, to activate a transition fault.

The results for conventional LOS and LOC (*Normal Scan Insertion*), Enhanced LOC (ELOC) [19] and hybrid transition-delay ATPG on this design are shown in the Table 5.4. It is seen that LOS methodology gave approximately 3% higher fault coverage than the LOC methodology. The Enhanced LOC method gives approximately 1.9% higher fault coverage compared to LOC method. The hybrid transition fault technique gives a better fault coverage when compared to LOS, LOC and ELOC methods and it also provides a better pattern count compared to LOC and ELOC methods. The pattern count is more than LOS but at the advantage of less scan enable design effort, only 5 scan chains being timing closed for at-speed scan enable. The hybrid scan technique proposed in [15] gives, in some cases, more pattern count compared to LOC technique. The CPU time of hybrid method is greater than all other techniques because for the hard-to-detect faults, the ATPG tool has to do more processing to determine the possible combinations of the SSEG controlled scan chains to work in shift register mode or in functional mode.

5.5.5 Analysis of Extra Detected Faults

In [27], potential yield loss problem due to detection of functionally untestable faults was discussed. An analysis is performed on the extra faults detected by hybrid-scan architecture over the conventional LOC technique. To determine the nature of these extra faults, a conventional LOC ATPG is performed on them. For example, for design *b17*, there were 17926 extra faults detected by hybrid-scan while LOC ATPG on these faults showed all of them as non-observable (NO) faults. *NO* faults are those that could be controlled, but could not be propagated to an observable point.

It can be argued that some of these *NO* detected faults might result in yield loss as some of them can be functionally untestable. However, some of these

(NO) faults are actually functionally testable but, for instance, due to low-cost tester ATPG constraints, such as no primary input changes and no primary output measures, become non-observable. For example, of the 17926 extra faults detected by hybrid scan in the *NO* class, 1155 faults were detectable without the low-cost tester constraints. Also, in [28], it was illustrated that the functionally untestable *NO* faults may not need to be tested if the defect does not cause the delay to exceed twice the clock period. With technology scaling and increasing operating frequencies, detection of such faults might become important and more than two vectors are required to detect the multi-cycle delay faults [28]. The hybrid scan can be advantageous as it eases the ATPG and detects such multi-cycle faults using a two vector pair.

5.6 Experimental Results

The hybrid-scan technique was experimented on four more industrial designs with design sizes ranging from 10K-100K flip-flops and three largest ITC'99 benchmark circuits. The # FFs column in Tables 5.5 and 5.6 show the number of scan flip-flops in each design. In all designs, sixteen scan chains were inserted. For LOS and LOC techniques, normal scan insertion was performed using Synopsys DFT Compiler [18]. While, for ELOC and hybrid techniques controllability/observability based scan insertion was performed and each scan chain is inserted with one LSEG (SSEG or FSEG) cell. In case of ELOC, only SSEG cells were inserted in each scan chain. In hybrid technique, only the first four scan chains are selected to be LOS-controlled (FSEG) after controllability/observability measurement and the remaining were ELOC-controlled (SSEG) cell. This reduces the at-speed scan enable design effort significantly as the scan enable to only one fourth of the scan flip-flops needs to be timing closed.

Table 5.5. ATPG results for seven designs (four ITC benchmarks and three industrial designs) using LOS and LOC methods [21].

Design	# FFs	*LOS*			*LOC*		
		FC (%)	# Patt	CPU Time *[sec]*	FC (%)	# Patt	CPU Time *[sec]*
b17	1.4K	95.09	1088	95.4	81.02	1190	1000.8
b18	3.3K	92.67	1451	279.7	77.50	1309	1020.9
b19	6.6K	85.98	2280	645.3	69.21	1153	1050.4
A	10K	91.11	1112	329	88.09	2145	896
B	30K	87.94	4305	3569	85.14	8664	7800
C	50K	81.10	6869	8415	79.42	12073	22930
D	104 K	92.15	5933	6559	91.56	10219	12088

Table 5.6. ATPG results for seven designs (four ITC benchmarks and three industrial designs) using ELOC and Hybrid methods [21].

Design	# FFs	*ELOC [19]*			*Hybrid*		
		FC (%)	# Patt	CPU Time *[sec]*	FC (%)	# Patt	CPU Time *[sec]*
b17	1.4K	94.29	1328	325	96.50	1179	187.9
b18	3.3K	93.01	1876	726	95.18	1334	336.6
b19	6.6K	84.81	1422	1000	88.33	1590	1000.9
A	10K	89.96	2014	924	91.74	1799	1014
B	30K	86.57	8539	8702	88.03	8062	6611
C	50K	80.48	11583	25642	83.29	8134	14451
D	104K	92.28	12505	47788	94.83	9674	18410

During ATPG, the faults related to clocks, scan-enable and set/reset pins, referred to as untestable faults are not added to the fault list. Tables 5.5 and 5.6 show the ATPG results comparing LOS, LOC, ELOC and hybrid methods. The FC and # Patt columns shows the fault coverage percentage and the number of patterns generated, respectively for each method. The ELOC method provides higher fault coverage compared to LOC method (up to 15.6% for design *b19*) and in most cases an intermediate fault coverage and pattern count between LOS and LOC. The hybrid method provides a better coverage compared to all other methods, as it has the flexibility to use combinations of functional and scan path for launching a transition. It provides higher fault coverage up to 2.68% (*design D*) and 19.12% (*design b19*) compared to LOS and LOC, respectively.

Based on a worst-case analysis, the lower bound for ELOC is LOC with no extra faults being detected over LOC and the upper bound is LOS. Similarly, for the hybrid technique, the lower bound is ELOC and upper bound can be greater or equal to LOS due to its hybrid nature. However, in the worst case scenario, for a given fault coverage, the hybrid method will still benefit in test pattern count reduction compared to LOC, thereby reducing test time, with minimum scan enable design effort. Note that, in some cases the CPU time for hybrid and ELOC method is greater than LOC method due to a larger search space for the ATPG tool to find the transition launch activation path for hard-to-detect faults.

Typically, in an ASIC design flow, the scan insertion is performed in a bottoms-up approach and it is independent of physical synthesis step. The DFT insertion tool stitches the scan chains based on the alpha-numeric order of scan flip-flop name in each module. The resulting scan chains are then re-ordered during physical synthesis step to reduce the scan chain routing area. At the top-level, the module level scan chains are stitched together.

Similarly, in the bottoms-up scan insertion flow, the scan chains in each module are stitched based on the scan flip-flops decreasing order of cost function and the resulting scan chains are re-ordered during physical synthesis to reduce the scan chain routing area. Hence, the new scan insertion method will not impact significantly as the scan insertion and physical synthesis is performed in a bottoms-up approach and not for the entire chip. Although, it can be argued that the controllability- and observability-based scan chain stitch might slightly increase the scan chain routing area in some cases, the decrease in scan enable design effort and area overhead compared to LOS are quite significant. Moreover, the technique still has the flexibility to shuffle and re-order the different groups of scan chains (LOS-controlled and ELOC-controlled) if there is any scan chain routing problem.

5.7 Summary

A new hybrid technique is presented to improve the transition delay fault coverage and reduce the pattern count. The technique uses a subset of scan flip-flops in LOS mode and the remaining in ELOC mode. Two local scan enable generator cells are used, fast scan enable generator (FSEG) cell for LOS-controlled scan flip-flops and slow scan enable generator (SSEG) cell for ELOC-controlled scan flip-flops. Both cells are easily tested using chain test patterns during test (using flush patterns). The technique reduces the design effort for at-speed scan enable signal considerably as only a few scan flip-flops are LOS-controlled. The technique was implemented on a number of industrial designs and ITC'99 benchmark circuits. The experimental results show up to 19.12% increase in fault coverage over LOC (ITC'99 design b19) achieved with lower number of patterns.

References

1. S. Natarajan, M.A. Breuer, S.K. Gupta, "Process Variations and Their Impact on Circuit Operation," in Proc. *IEEE Int. Symp. on Defect and Fault Tolerance in VLSI Systems*, pp. 73-81, 1998.
2. R. Wilson, "Delay-Fault Testing Mandatory, Author Claims," *EE Design*, Dec. 2002.
3. S. Wang, K. Peng, K. Li, "Layout-Aware Scan Chain Reorder for Skewed-Load Transition Test Coverage," in Proc. *Asian Test Symposium*, 2006.
4. R. D. Eldred, "Test Routing Based on Symbolic Logical Statement," *Journal ACM*, vol. 6, pp. 33-36, Jan 1959.
5. G. Aldrich and B. Cory, "Improving Test Quality and Reducing Escapes," in Proc. *Fabless Forum, Fabless Semiconductor Assoc.*, pp. 34-35, 2003.
6. X. Lin, R. Press, J. Rajski, P. Reuter, T. Rinderknecht, B. Swanson and N. Tamarapalli, "High-Frequency, At-Speed Scan Testing," *IEEE Design & Test of Computers*, pp. 17-25, Sep-Oct 2003.
7. K. Cheng, "Transition Fault Testing for Sequential Circuits," *IEEE Transactions on Computer-Aided Design of Integrated Circuits and Systems*, vol. 12, no. 12, pp. 1971-1983, Dec. 1993.

8. T. M. Mak, A. Krstic, K. Cheng, L. Wang, "New challenges in delay testing of nanometer, multigigahertz designs," *IEEE Design & Test of Computers*, pp. 241-248, May-Jun 2004.
9. M. Bushnell, V. Agrawal, *Essentials of Electronics Testing*, Kluwer Publishers, 2000.
10. V. Jayaram, J. Saxena and K. Butler, Scan-Based Transition-Fault Test Can Do Job, *EE Times*, Oct. 2003.
11. J. Savir, "Skewed-Load Transition Test: Part I, Calculus," in Proc. *Int. Test Conf. (ITC'92)*, pp. 705-713, 1992.
12. J. Savir and S. Patil, "On Broad-Side Delay Test," in Proc. *VLSI Test Symp. (VTS'94)*, pp. 284-290, 1994.
13. B. Dervisoglu and G. Stong, "Design for Testability: Using Scanpath Techniques for Path-Delay Test and Measurement," in Proc. *Int. Test Conf. (ITC'91)*, pp. 365-374, 1991.
14. J. Saxena, K. M. Butler, J. Gatt, R. Raghuraman, S. P. Kumar, S. Basu, D. J. Campbell, J. Berech, "Scan-Based Transition Fault Testing - Implementation and Low Cost Test Challenges," in Proc. *International Test Conference (ITC'02)*, pp. 1120 - 1129, Oct. 2002.
15. S. Wang, X. Liu, S.T. Chakradhar, "Hybrid Delay Scan: A Low Hardware Overhead Scan-Based Delay Test Technique for High Fault Coverage and Compact Test Sets," in Proc. *Design, Automation and Test in Europe (DATE'03)*, pp. 1296-1301, 2004.
16. N. Ahmed, C.P. Ravikumar, M. Tehranipoor and J. Plusquellic, "At-Speed Transition Fault Testing With Low Speed Scan Enable," in Proc. *IEEE VLSI Test Symp. (VTS'05)*, pp. 42-47, 2005.
17. J. Savir and S. Patil, "Scan-Based Transition Test," *IEEE Transactions on Computer-Aided Design of Integrated Circuits and Systems*, vol. 12, no. 8, pp. 1232-1241, Aug. 1993.
18. Synopsys DFT Compiler, "User Manual for SYNOPSYS Toolset Version 2005.06," Synopsys, Inc., 2005.
19. N. Ahmed, M. Tehranipoor and C.P. Ravikumar, "Enhanced Launch-off-Capture Transition Fault Testing," to appear in *Int. Test Conf. (ITC'05)*, 2005.
20. N. Ahmed and M. Tehranipoor, "Improving Transition Delay Fault Coverage Using Hybrid Scan-Based Technique," in Proc. *International Symposium on Defect and Fault Tolerance in VLSI Systems (DFT'05)*, 2005.
21. N. Ahmed and M. Tehranipoor, "Improving Quality of Transition Delay Test Using Hybrid Scan-Based Technique," in Proc. *IEEE Design and Test of Computers*, 2006.
22. W. Li, S. Wang, S. T. Chakradhar and S. M. Reddy, "Distance restricted scan chain reordering to enhance delay fault coverage," in Proc. *Int. Conf. on VLSI Design*, pp. 471-478, 2005.
23. P. Gupta, A. B. Kahng, I. Mandoiu, P. Sharma, "Layout-aware scan chain synthesis for improved path delay fault coverage," in Proc. *Int. Conf. on Computer Aided Design (ICCAD'03)*, pp. 754-759, 2003.
24. S. Wang and S. T. Chakradhar, "Scalable scan-path test point insertion technique to enhance delay fault coverage for standard scan designs," in Proc. *Int. Test Conf. (ITC'03)*, pp. 574-583, 2003.
25. W. Mao and M. D. Ciletti, "Reducing correlation to improve coverage of delay faults in scan-path design," *IEEE Trans. on Computer-Aided Design of Integrated Circuits and Systems*, vol. 13, no. 5, pp. 638-646, May 1994.

26. L. H. Goldstein and E. L. Thigpen, “SCOAP: Sandia Controllability/Observability Analysis Program,” in Proc. *Design Automation Conf. (DAC’80)*, pp. 190-196, 1980.
27. K. J. Rearick, “Too Much delay Fault Coverage is a Bad Thing,” in Proc. *Int. Test Conf. (ITC’01)*, pp. 624-633, 2001.
28. W. C. Lai, A. Krstic and K. T. Cheng, “On Testing the Path Delay Faults of a Microprocessor Using its Instruction Set,” in Proc. *VLSI Test Symp. (VTS’00)*, pp. 15-20, 2000.

6

Avoiding Functionally Untestable Faults

Modern Very Large Scale Integration (VLSI) designs require extensive testing to ensure the shipped product will function correctly when it reaches the consumer. Transition delay fault testing is one of many commonly used testing techniques. However, it is a structural-based test and the chip may suffer from overtesting. A simple method to avoid overtesting is to identify the faults that are functionally untestable and omit the faults from the fault list but automatic test pattern generation (ATPG) tool may incidentally detect these faults when filling in don't-care states. The percentage of don't-care bits in a test pattern can be very high in large designs, about 99%. These don't-care bits are either filled randomly to provide higher defect coverage by increasing the chance of detecting non-modeled faults or filled by compression tools to obtain the highest compression to reduce test data volume and test time. However, filling these don't-care bits without considering the functionally untestable faults can cause yield loss.

In this chapter, a technique is presented that is be able to enhance current commercial ATPG tools such that it avoids detection of functionally untestable faults during both the initial pattern generation phase and while the tool fills in don't-care states. Previous attempts at functionally untestable fault avoidance have required modification to a custom ATPG tool or development of a new ATPG entirely. Rather than altering the tool, the framework presented in this chapter modifies the netlist to restrict the invalid states the ATPG tool can generate, allowing the use of a commercial tool that is known to work well. The results show that the test coverage of functionally testable faults is minimally affected while not significantly increasing the effort needed by the ATPG tool.

6.1 Introduction

Chip scaling and market demands continue to push designers to fit more complex designs into smaller areas. Not only does a higher design density correlate to a higher probability of defects, but as the feature size shrinks, the effective length of the interconnects become longer due to a shrinking width. This creates the potential for severe signal integrity problems in the chip, which often presents itself as signal delay. Delay fault testing detects many of these defects that cause gate and interconnect delay. However, design-for-testability (DFT) techniques, like scan design, allow for better delay fault coverage while also allowing scan patterns that detect functionally untestable faults. If these faults are included during test, patterns are generated that would be applied to the chips which never occur in the field, potentially failing chips unnecessarily. As a result, overtesting can occur, potentially producing a significant yield loss [1].

Current transition fault testing techniques already avoid a portion of the functionally untestable faults based on the test application methodology. Compared to launch-off-shift (LOS) [10] and enhanced scan [11], launch-off capture (LOC) [12] testing assists in avoiding the largest percentage of such faults due to the functional dependence between the initializing and transition launching patterns. However, functionally untestable faults can still be stimulated and detected by initializing the circuit-under-test (CUT) with a pattern that is not the functional response of the design. Avoiding these patterns during pattern generation will be the key to avoiding the respective untestable fault it stimulates.

Since it is possible to determine functionally untestable faults separately [9] [3] [2] before ATPG, a naive solution to avoid these faults is to exclude them from the active fault list during pattern generation. However, a small proportion of the scan cells in patterns are care-bits, forcing ATPG tools to fill don't-care bits with some value that may produce a pattern that will incidentally detect these faults. In order to ensure only functionally testable faults are detected, additional steps must be taken to constrain the ATPG tool from creating patterns with functionally invalid states.

Previous techniques that avoid functionally untestable faults during pattern generation have required custom modifications to ATPG tools or designing a new ATPG [4] [5]. This makes the immediate application of such techniques to commercial ATPG tools rather difficult. The work in [4] constraints LOC by selecting a random sample of patterns, determines the patterns which detect functionally untestable faults, and uses such patterns in conjunctive normal form (CNF) to constrain their custom ATPG tool. In [5], the custom ATPG tool contains a list of illegal state cubes that would detect functionally untestable faults. During LOC pattern generation, the ATPG checks the list of illegal states to avoid stimulation of such combinations in the generated pattern. There has also been work that uses a sequential SAT solver to validate the results of an ATPG tool and reject any patterns that cannot be validated as a functional response of the circuit [13]. Although this work allows the use of a commercial tool, since it is a post ATPG validation technique, it still relies on random fill of any don't-care bits in the pattern, potentially leading to incidental detection.

In this chapter, a framework is presented to be used as a wrapper around any commercial ATPG tool. Using a functionally untestable transition fault identification tool, the faults, which need to be avoided during ATPG, are identified. These faults are then used as constraints by altering the netlist instead of modifying the ATPG algorithm. For constraint generation, the ATPG is used to identify the initializable circuit states to test such functionally untestable faults. Two cases can happen: 1- ATPG cannot generate a pattern for a fault, i.e. the fault is undetectable by ATPG. 2- ATPG can generate a pattern for a functionally untestable fault which it basically proves the fact that the current ATPGs cannot avoid detecting such faults. A pattern generated for a functionally untestable fault is considered as an invalid state for the circuit under test and such states must be avoided during pattern generation.

Constraining an ATPG will allow test engineer to avoid such faults only and if only the final pattern contains care bits. Since this is not usually the case, therefore filing don't-cares either randomly or using a compression tool could result in incidentally detection of such faults. These are the main reasons why modifying netlist is a better alternative than modifying ATPG. After modifying the netlist, transition fault pattern generation is performed without the functionally untestable faults in the active fault list, but since the constraints are in place, patterns generated will not incidentally detect functionally untestable faults. The constraints ensure that the filled don't-care bits do not create an invalid state for the circuit under test. The result of the ATPG will be a test pattern set that will only detect functionally testable faults.

6.2 Overview of the Framework

The framework for functionally untestable transition fault avoidance consists of a four step process. Figure 6.1 outlines the general flow of the framework. First, a functionally untestable fault list (FUFL) is generated, which only needs to be performed once for each design. This list is generated using the technique described by Liu et al. [2], who extended FIRES [3], a sequential stuck-at redundant fault detection technique, towards application to transition faults.

Using the functionally untestable transition fault list, one can use existing ATPG tools to generate LOC patterns for each fault in the list without filling don't-care bits. The advantage of using LOC as the basis for pattern generation is that any fault that cannot be detected during LOC can be assumed to be functionally untestable due to the functional constraint LOC inherently applies to all test patterns. This pattern generation for functionally untestable faults will result in two lists (see Figure 6.1: *LOC detectable functionally untestable test patterns* (DFUTP) and *LOC undetectable functionally untestable fault list* (UDFUFL). Any faults that LOC ATPG can successfully

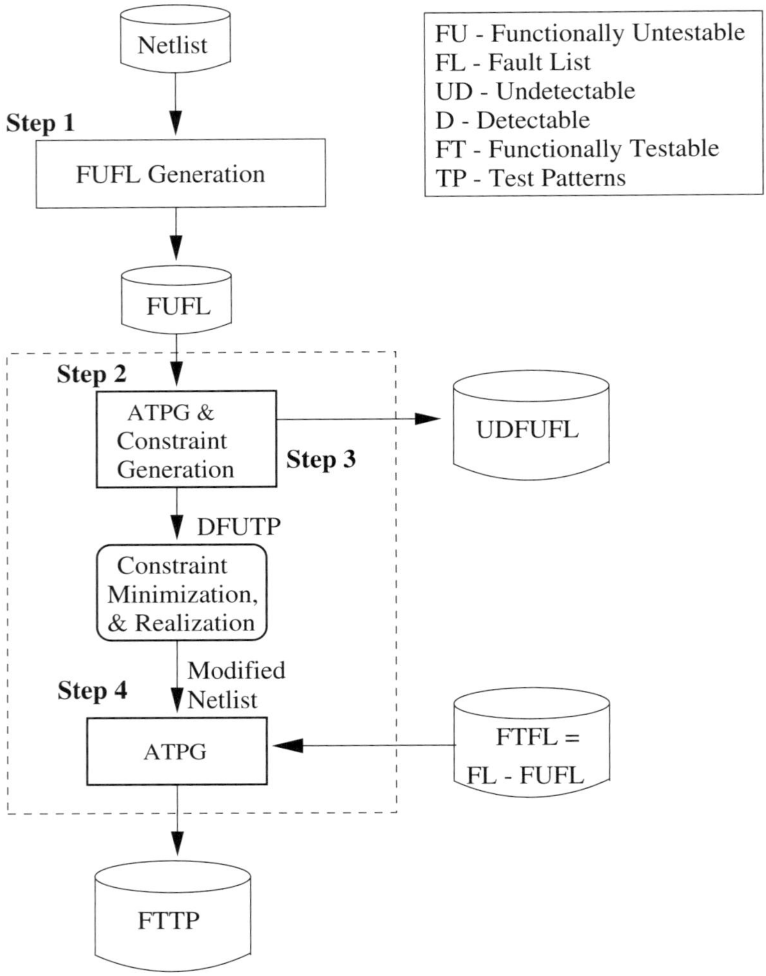

Fig. 6.1. Flow of the framework for functionally untestable fault avoidance.

generate a pattern for are LOC detectable and any fault that does not have a corresponding pattern are LOC undetectable. The UDFUFL are the existing 15–20% (according to the experimental results on ISCAS'89 benchmarks) of faults undetectable by LOC due to the inability to functionally stimulate and propagate the appropriate transition to an observable output. The remaining LOC detectable faults are a result of initializing the CUT with a pattern that is not a functional response during normal operation and cause chip overtesting.

The patterns generated for the LOC detectable faults (DFUTP) are then used in the next step, *Constraint Generation, Minimization, and Realization*, which will be used for ATPG. This step is done to make sure that running

ATPG on functionally testable faults will not incidentally detect functionally untestable ones when filling in don't-care bits in the final patterns. Unlike previous techniques, the constraints are applied to the design netlist instead of modifying the ATPG tool to handle an entirely new set of constraints. For each care bit of a LOC detectable fault pattern, there is a corresponding reachable input. A simple combinational logic tree can be added to the design on each of these reachable inputs for each pattern, which can then be ORed together with the trees of the other LOC detectable patterns. The single OR gate will be constrained by the ATPG to prevent generation of a pattern that will detect any faults in the FUFL. Most ATPG tools (e.g. Synopsys TetraMax [6]) can easily constrain a single net to a zero or a one. This will be further discussed in Section 6.4.

For designs that have a large number of DFUTPs, a form of constraint minimization must be used to reduce some of the burden that may be placed on the ATPG tool. To alleviate this problem, in this framework, the size of constraints are reduced such that if many of the patterns have similar bit combinations, one logic tree is used to apply many of the constraints.

The modified netlist will then be used for ATPG. The fault list input into the ATPG tool consists of the functionally testable faults only, i.e. the original fault list minus those faults determined to be functionally untestable generated in Step 1 (FTFL = FL-FUFL). Constraining the output of the OR gate, that was mentioned above, to a logic zero prevents the ATPG tool from generating a pattern that will detect any faults in the FUFL since the don't-care bits of the final patterns are only filled with values that will not incidentally detect a functionally untestable transition fault.

6.3 Functionally Untestable Fault Identification

The first step of the framework is based on the functionally untestable transition fault identification technique in [2]. This technique uses static logic implication (static learning) to expand on the work in [9], which developed a technique called FIRE.

The efficiency of the identification technique is determined by the number of implications performed for each circuit. As was done in [2], the program developed for functionally untestable fault identification was limited to direct, indirect, and extended backward implications. These implications are stored as a graph for efficient searching and easy traversal through the circuit. Direct implication is straight forward and can be learned based on the function of the gate. Assume that a simple circuit is given as in Figure 6.2. The implications can be immediately connected for $C = 1$ as shown in Figure 6.3(a). Indirect implications are derived based on the direct implications. For example, if one implies a 1 on net C, the direct implications imply nets A and B are both 1, which each directly imply $D = 1$. So an indirect implication can be made with $C = 1$ and $D = 1$, implying $E = 0$, which itself has a set of direct implications

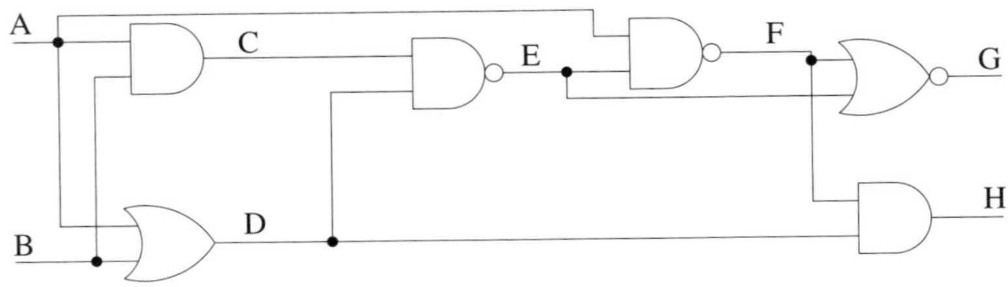

Fig. 6.2. A simple combinational circuit used to demonstrate static logic implication.

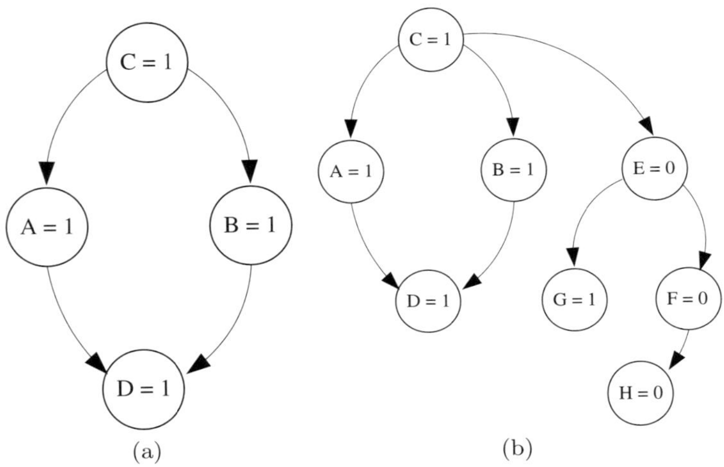

Fig. 6.3. (a) Direct implications graph for C = 1. (b) Graph after direct, indirect, and backward implications have been performed.

and becomes part of the implication graph of $C = 1$. Extended backward implication is used in cases where the output of a gate is known, but the inputs cannot be directly implied based on the current output. One example of this is when the output of an AND gate is 0, since it is not directly implied as to whether one input is 0 or both are 0. Extended backward implication will determine whether there are any common implications when implying the dominant value on each of the inputs of the current gate. After direct, indirect and backward implications are completed, the final implication graph for net D implied to 0 for the circuit shown in Figure 6.2 will be as shown in Figure 6.3(b).

There are many additional implications that can be performed to more fully describe the circuit [8], which result in a more complete description of the circuit and potentially yield a larger set of identified functionally untestable faults. However, this increases the complexity of the program and the time required to perform the implications. To reduce the time requirement of the program, one can limit the number of implications.

The implication graphs learned from static learning help to easily identify functionally untestable faults, which will become the FUFL. In order to identify these faults, a single-line conflict technique similar to FIRE is used to identify any functionally untestable faults. However, the original FIRE must first be extended to include static learning [7]. Extending FIRE basically results in the intersection of the implication graphs of a net implied to 0 and 1. If the intersection yields a non-empty set, any implications in this non-empty set can be identified as functionally untestable faults. This technique determines the faults that are sequentially uninitializable and sequentially uncapturable. Sequentially uninitializable faults cannot be detected due to redundancies in the design that prevent nets from being set to a logic one or zero. Sequentially uncapturable faults are due to an inability to propagate a transition to an observable output.

In order to extend FIRE further for application towards functionally untestable transition faults, sequentially constrained faults must also be considered. Sequentially constrained faults can be individually initialized and captured, but are functionally constrained by LOC such that the two faults cannot occur sequentially. Similar to the sequentially uninitializable and sequentially uncapturable faults, this relation is dependent on an intersection operation. This set considers the union of the initializable set with the capturable set of a net implied to 0 intersected by the union of the initializable set with the capturable set of the same net implied to 1. Any implications in the non-empty set are considered sequentially constrained and functionally untestable.

6.4 Constraint Generation, Minimization, and Realization

Constraint generation is based on the faults that were identified in the first step of the functionally untestable fault avoidance framework described in Section 6.3. To determine these constraints, a commercial ATPG tool is used to generate LOC transition fault patterns for the FUFL, which is then minimized and then transformed into constraints that are temporarily added to the final design solely for the purpose of transition fault pattern generation in the final step of the framework.

Under the assumption that the faults identified by Step 1 of the framework are functionally untestable transition faults, it is also assumed that those faults will never be sensitized during normal operation and are only sensitizable due to states that are initialized during test mode through the scan chain. So, it is only during the initialization phase of LOC that functionally untestable fault patterns must be avoided since the launch pattern is a functional response of the circuit and is based on the initialization pattern. In other words, one needs to make sure the initialization pattern is a valid, reachable states, which makes the functional response state valid.

6.4.1 Constraint Generation

In order to prevent the detection of these faults during LOC ATPG, the scan cells must be constrained from any state that would place an LOC initialization pattern for any faults in the FUFL into the scan chain. To realize these constraints, LOC ATPG is performed on the FUFL (Step 2 of framework). During this ATPG step, don't-care bits must remain unfilled in order to isolate only those cells that are necessary for detection of functionally untestable faults. This is easily done in Synopsys' TetraMax by removing the random fill option during ATPG.

LOC ATPG will determine test patterns for a fraction of all faults in the FUFL, leaving the remaining faults as undetectable functionally untestable faults (UDFUF). Since LOC in general leaves approximately 20% of the total faults of a design undetected, the set of faults declared as UDFUF should be the same faults as the 20% of total faults that are LOC undetectable.

6.4.2 Constraint Minimization

Although the UDFUFs make up a majority of the functionally untestable faults, for larger designs, the number of patterns generated during the second step of the framework (DFUTP) can become rather large. If constraints were based solely from this large set of patterns, it could unnecessarily constrain the ATPG tool too much and require significantly more effort and time to avoid the faults in the FUFL. Therefore, a constraint minimization strategy is used to minimize the time the ATPG tool spends on constraints.

The method used to minimize the number of constraints is based on *pattern dominance*. For example, assume there is a design with 8 scan cells and with four patterns in the format $b_7 b_8 b_6 b_5 b_4 b_3 b_2 b_1 b_0$ that detect some functionally untestable faults: $XXX1X0XX$, $XX11X0XX$, $XX0XXX0X$, $XXX100XX$. There are clearly similarities in the first, second, and fourth pattern. Assuming each of these patterns detect different, but intersecting fault sets, the pattern with the fewest number of care bits can be used to determine whether the other patterns can be eliminated from constraint consideration. In this example, the first and third patterns have the fewest number of care bits. Starting with the first pattern, searching for all other patterns that also contain the same care bits yield the second and fourth patterns, so these two patterns can then be eliminated from constraint consideration. Since the third pattern does not have the same care bits in the same positions as the first pattern, it cannot be removed from consideration. As a result only $XXX1X0XX$ and $XX0XXX0X$ are used as constraints during ATPG.

The reason the second and fourth patterns could be eliminated from the pattern set is due to pattern dominance. The first pattern will detect at least one functionally untestable fault with the given care bits. The second pattern will also detect that same faults as the first pattern plus those detected due to the additional care bits in the pattern. However, a necessary condition to detect those additional faults are the care bits of the first pattern. So, if the first pattern is constrained from occurring, then those faults detected by the second and fourth patterns will not be detected either. Because of this constraint, the ATPG tool will not generate a pattern that contains a 1 at b_4 and a 0 at b_2.

6.4.3 Constraint Realization

After constraint minimization, the constraints are finally realized into a netlist that will be used for the final design. Since the position of the care bit in the pattern stream and the scan cell order is known, a one to one mapping between care bit and scan cell can be made. For every pattern that remains after generation and minimization, a single behavioral equation can be formulated based on the sequence of care bits.

Using the first pattern from the example above, $XXX1X0XX$, the constraint created by the pattern can be realized as $pattern_1 = cell_4 \cdot \overline{cell_2}$, where the dot represents the logical AND operation. The equation represents the case when b_4 and b_2 are 1 and 0, respectively, excluding the remaining bits since they are don't-care states. Equation 6.1 is a generalized form of the constraints created from the DFUTP, where n is the total number of scan cells in the design.

$$pattern_j = \prod_{i<n} c_i,\ c_i = \begin{cases} cell_i & \text{if } b_i = 1 \\ \overline{cell_i} & \text{if } b_i = 0 \end{cases} \tag{6.1}$$

$$constraint = \sum_{j<k} pattern_j \tag{6.2}$$

To ease the application of the ATPG constraints, rather than constraining each of the *pattern* signals to 0, if all the signals are ORed together, a single net will have to be constrained to a 0 to ensure the ATPG will not generate patterns that contain a state that will incidentally detect a functionally untestable fault. Equation 6.2 shows the general form for performing the logical OR operation between all of the *pattern* signals, where k is the total number of patterns after minimization.

Once constraint generation, minimization, and realization is performed, there is now a functional description of states to avoid during LOC ATPG. It must now be integrated into the current design in order to be effective. A more detailed implementation of constraint integration with the targeted design is described in the following section.

6.5 Framework Implementation

The framework is implemented to easily wrap around existing commercial synthesis and test tools. The framework was implemented using a C program to fully integrate each of the four steps together. Synopsys DFT Compiler and TetraMax [6] were the commercial tools used to synthesize and generate patterns. The framework implementation shown below is an easy to follow flow.

Functionally Untestable Fault Avoidance Flow

---

1. Synthesize design and insert scan chains using DFT Compiler.
2. Run functionally untestable fault identification program on synthesized netlist. (Framework Step 1)
   - Functionally untestable fault list (FUFL) is generated by the program.
3. Use FUFL as fault list in TetraMax and perform LOC ATPG on synthesized design. (Framework Step 2)
   - TetraMax generates patterns to detect faults that are LOC detectable but functionally untestable (DFUTP).
4. DFUTP are extracted and minimized
   - Behavioral model of constraints is generated. (Framework Step 3)
5. Constraint model is synthesized and optimized using DFT Compiler and connected to already synthesized design.
6. The final netlist is used for LOC ATPG in TetraMax with a functionally testable fault list. (Framework Step 4)

The functionally untestable fault identification program was implemented in C and followed the techniques explained in Section 6.3. A Perl script was also used to extract the DFUTP from the STIL file, minimize the number of patterns, and generate the behavioral model.

DFT Compiler was used to incorporate the generated constraints into the synthesized design. By passing the constraints netlist through the synthesis tool as a separate module, many redundancies in the constraint netlist were removed and the constraints were reduced to a structural netlist that maintained the original function of the behavioral model. Adding the constraint module to the design was straightforward and maps the the output net of all the scan cells to the input ports of the constraint module. Since the STIL file that extracted the patterns contained the order of the scan cells, the constraint module inputs were placed in the same order as the scan cells. The output signal of the constraint module is tied to the output of the final logical

Table 6.1. Identified Functionally Untestable Faults.

Bench Name	Total # of Faults	FUFL based on [2]
s1423	3028	5
s5378	6822	118
s13207	15534	25
s15850	18240	324
s38417	56490	3010

OR operation of all the *pattern* signals to simply searching for the signal when applying the single constraint during ATPG.

When using this final netlist with the included constraints with TetraMax, the ATPG is constrained to always hold the single output of the constraint logic to 0. Although this is functionally equivalent to constraining the scan chain to individually prevent the generation of functionally untestable fault states, it is significantly easier to apply the constraint on a single net as opposed to constraining the scan cell values individually or on a per pattern basis.

6.6 Analysis

The framework was run on the ISCAS'89 benchmarks using a 3.2 GHz Pentium 4 with 1 GB of memory running the Linux Operating System. Table 6.1 lists the number of functionally untestable faults that were identified by the first step of the framework. The first column lists the benchmark name and the second column lists total number of faults identified by TetraMax. Finally, the third column lists the number of faults identified as functionally untestable by the technique referenced in Section 6.3.

The number of functionally untestable faults found by the implementation of the technique have found significantly fewer faults than those discovered in [2]. This is probably due to a combination of circuit optimizations performed by DFT Compiler and the possibility of the developed tool performing fewer implications than what was done in [2].

The results of the presented framework are shown in Table 6.2. Columns 2 and 3 show the results of the ATPG using TetraMax for conventional LOC and the coverage when applying the framework. For both LOC ATPG and ATPG with the framework shown in this chapter, the entire fault list of the design is included to show the constraints filtering out those faults in the FUFL. Column 4 shows the number of patterns used after minimization to generate the constraints for the final ATPG. Finally, columns 5 and 6 are the overall time of conventional LOC ATPG and the time of the framework, respectively. The time of the framework does not include the time taken to identify the functionally untestable faults since it is considered a preprocessing step and

Table 6.2. Functionally Untestable Fault Avoidance Framework Results

Benchmark Name	LOC Fault Coverage(%)	Framew. Fault Coverage(%)	# of Constraints	Conv. Time (s)	Framew. Time (s)
s1423	95.72	95.55	2	4.7	5.9
s5378	89.06	70.62	41	6.4	8.8
s13207	89.99	88.04	10	28.2	33.9
s15850	89.46	71.50	94	27.7	28.3
s38417	96.80	49.14	1105	155	301

ideally one would already have a full list of functionally untestable faults to be used as an input to the framework.

The number of constraints listed in Column 4 of Table 6.2 correlates closely with the number of functionally untestable faults identified in Column 3 of Table 6.1. Since the FUFL identification program did not identify as many functionally untestable faults for each of the benchmarks as desired, the number of constraints were quite limited. The number of constraints grows linearly with the number faults in the FUFL and, for the cases shown here, remains roughly half the number of faults identified as functionally untestable.

As can be compared between Columns 2 and 3, the fault coverage between LOC and the presented framework is directly related to the number of constraints used. For s1423 and s13207, since the program was not able to identify many functionally untestable faults, the number of constraints were few, and very few functionally untestable faults were filtered out during pattern generation.

For the remaining three cases, there was a sufficient number of constraints to clearly show a substantial drop in fault coverage. For s5378 and s15840, there was almost a 20% drop in coverage. In each case, the fault identification tool only indicated approximately 2% of the total faults as functionally untestable. However, if the percentages are compared with the percentage of functionally untestable faults for those two benchmarks in [2], one can assume the additional faults excluded during the framework ATPG are also functionally untestable.

Since the framework essentially determines invalid/unreachable states, it is assumed that any fault that requires initialization with an invalid state is functionally untestable. Due to this, by first finding a subset of functionally untestable faults and the invalid states that would detect them, additional faults will also potentially be filtered out by this process since these additional faults also require the same functionally unreachable states to be detected. So, with a greater number of constraints, the more states that can be concluded as invalid.

However, as can be seen with s38417, since there are so many constraints that still remain after minimization, the ATPG tool cannot effectively reach a high fault coverage and is obviously impaired. This problem most likely can be alleviated with a better constraint minimization technique, which will be pursued further in future work. An appropriate balance obviously must be reached between constraining the ATPG to effectively prevent detection of functionally untestable faults and the ATPG effort load.

Overall, the framework did not increase the amount of time to complete pattern generation by an exorbitant amount, even on s38417. A majority of the time for the framework was actually spent synthesizing the constraints circuit that was added to the netlist. Reducing the number of constraints will potentially reduce the generation time in addition to restoring the fault coverage to an acceptable level. Even with almost 100 constraint for s15850, the entire framework flow took less than one minute with the majority of the time spent on constraint generation, minimization, and realization instead of pattern generation with the constrained ATPG.

6.7 Summary

A framework was presented for avoiding functionally untestable faults during pattern generation that can be used in conjunction with a commercial ATPG tool. Rather than altering a custom tool as previous implementations have done, the netlist is modified to include additional logic that is constrained during ATPG. The additional constraint ensures the ATPG does not generate an LOC test pattern that will detect a functionally untestable fault. Application of the framework is straightforward and does not significantly increase pattern generation time nor hinder the ATPG from reaching reasonable coverage levels if there are a manageable number of constraints that are applied. Initial results show that a current commercial ATPG tool without significant modification can be used to avoid incidental detection of faults that have already been identified as functionally untestable and potentially identify additional faults.

References

1. Jeff Rearick, "Too Much Delay Fault Coverage Is a Bad Thing," in Proc. *International Test Conference (ITC)*, pp. 624–633, 2001.
2. Xiao Liu and Michael S. Hsiao, "On Identifying Functionally Untestable Transition Faults," IEEE Intl. High-Level Design Validation and Test Workshop, pp. 121–126, 2004.
3. Mahesh A. Iyer and David E. Long and Miron Abramovici, "Identifying Sequential Redundancies Without Search," in Proc. *Design Automation Conf. (DAC)*, pp. 457–462, 1996.
4. Xiao Liu and Michael S. Hsiao, "A Novel Transition Fault ATPG That Reduces Yield Loss," IEEE Design & Test of Computers, pp. 576–584, 2005.
5. Zhou Zhang and Sudhakar M. Reddy and Irith Pomeranz, "On Generating Pseudo-Functional Delay Fault Tests for Scan Designs," in Proc. *IEEE Intl. Symposium on Defect and Fault Tolerance in VLSI Systems*, pp. 398–405, 2005.

6. Synopsys Inc., “User Manuals for Synopsys Toolset Version 2005.09,” Synopsys Inc., 2005.
7. Jian-Kun Zhao and Jeffrey A. Newquist and Janak H. Patel, “A Graph Traversal Based Framework for Sequential Logic Implication with an Application to C-Cycle Redundancy Identification,” in Proc. *Intl. Conf. on VLSI Design*, pp. 163–169, 2001.
8. Mana Syal and Rajat Arora and Michael S. Hsiao, “Extended Forward Implications and Dual Recurrence Relations to Identify Sequentially Untestable Faults,” in Proc. *Intl. Conf. on Computer Design*, 2005.
9. Mahesh A. Iyer and Miron Abramovici, “FIRE: A Fault-Independent Combinational Redundancy Identification Algorithm,” IEEE Transactions on VLSI Systems, vol. 4, no. 2, 1996, pp. 295–301.
10. J. Savir, “Skewed-Load Transition Test: Part I, Calculus,” in Proc. *Intl. Test Conference*, pp. 705–713, 1992.
11. J. Savir and S. Patil, “On Broad-Side Delay Test,” in Proc. *IEEE VLSI Test Symposium*, pp. 284–290, 1994.
12. B. Dervisoglu and G. Stong, “Design for Testability: Using Scanpath Techniques for Path-Delay Test and Measurement,” in Proc. *Intl. Test Conference*, pp. 365–374, 1991.
13. Yung-Chieh Lin and Feng Lu and Kai Yang and Kwang-Ting Cheng, “Constraint Extraction for Pseudo-Functional Scan-based Delay Testing,” in Proc. *Asia and Sourth Pacific Design Automation Conference*, pp. 166–171, 2005.

7

Screening Small Delay Defects

As technology scales, new subtle defects are seen during fabrication which can have significant impacts on yield and reliability of the shipped products. Resistive open and short are two such defects that cause timing or logic failures in the design. Such defects can cause gross or small delay defects depending on the size of their resistance. It is proven that the population of such defects increases as technology scales, thus increasing small delay defects.

Timing unaware commercial ATPG tools mostly generate delay fault pattern set through very short paths, thereby increasing the escape chance of smaller delay defects. The small delay defects cause timing failure if activated on longer paths during functional operation and must be detected during production test. In this chapter an efficient pattern generation procedure is presented for transition fault model, which provides a higher coverage of small delay defects that lie along the long paths. The proposed procedure uses commercial no-timing ATPG tool as the basic engine and static timing analysis to identify the target fault sites. In the pre-processing step, only a subset of scan flip-flops are selected and used as observe points during pattern generation based on the least slack path terminating at each scan flip-flop. This reduces the ATPG search space and complexity as the tool targets a limited set of faults at a subset of observable endpoints. Then, pattern generation is performed and a novel pattern selection technique is applied to screen test patterns affecting longer paths. Also, a new delay defect size metric based on the affected path length and required increase in test frequency is developed. Using this technique will provide the opportunity of using existing timing unaware ATPG tools as slack-based ATPG. The resulting pattern set improves the defect screening capability of small delay defects. This timing-based ATPG will be presented in details in this chapter.

7.1 Introduction

Scan-based structural delay test has found its place in today's industry test flow, as it is considered a cost-effective alternative to the at-speed functional pattern approach [1] [2]. Transition and path delay fault models are the two prevalent fault models [3] [4]. The transition fault model targets each gate output in the design for a *slow-to-rise* and *slow-to-fall* delay fault while the path delay model targets the cumulative delay through the entire list of gates in a pre-defined path [5]. In comparison, transition fault model is widely practiced in industry and existing commercial tools have matured in test generation and debug of such tests. On the other hand, the number of faults targeted using transition fault model is manageable and the pattern generation is faster when compared to path delay fault model. Due to the ATPG complexity, path delay model is mostly applied to test critical paths. Note that critical paths are timing sensitive and small delay defects on such paths can be easily detected as the slack is very small.

The traditional transition fault tests are generated assuming a fixed cycle time for each clock domain. In general, delay tests are generated/applied one clock domain at a time. This shows that a delay defect will be detected only when it causes a transition to reach an observe point (primary output or scan flip-flop) by more than the positive slack of the affected path. Slack of a path is a measure of how close a transition on the respective path meets the timing of an observe point, relative to the test cycle time. The slack reflects the relation between the size of delay defect and length of path under test. In other words, for a path with a very small slack, a small delay defect can potentially be detected and for a path with very large slack, a small delay defect can escape the test.

There are a large number of available paths for a delay defect to be activated and propagated. For a particular defect, a pattern which affects a longer path is more efficient than a pattern which detects it through a shorter path. A small delay defect might escape, if activated through a short path during test. While the same defect might be activated on a long path during functional operation and it may cause a timing failure. Therefore, the detection of small delay defects on long paths is a *quality* issue. The detection through the longer path ensures the detection of varying sizes of the delay defect.

The detection of small delay defects on short paths is more of a *reliability* issue. A small delay defect escape on such paths during test might magnify during subsequent aging in the field and cause a failure of the device. If a manufacturer's defective part per million (DPPM) level is low, then detecting small delay defects on short paths may not be necessary. However, for zero DPPM-required products such as in automobile and space applications, it is recommended to detect such defects. On the other hand, knowing the population of such defects in the production flow can help make a decision on detecting or not detecting such faults.

In this chapter is focused only on small delay defects on long paths to improve the quality of delay test pattern set. Moreover, the definition of long paths is very important and it depends on the frequency. If the frequency is increased, the slack of not-long-paths decreases hence more paths which were not considered long enough earlier will become timing critical. To make small delay defect detectable on short and intermediate paths, the frequency can be increased to reduce the size of slack. Such method is known as faster-than-at-speed test which will be discussed in details in the next chapter.

Due to increasing population of resistive opens and shorts in nanometer technology designs, there is a growing industry concern and demand for timing-aware ATPG tools. Encounter True-Time Delay Test Tool ™ [6] is one such available commercial ATPG tool that uses actual design timing information for ATPG but increases the test frequency which might in turn increases the power and IR-drop. True time starts with targeting short paths and masking the endpoints connected to the longer paths. It then chooses next longer paths while masking the paths longer than those selected. This may result in a very large number of test patterns.

Most of the widely used commercial ATPG tools are still timing unaware and generate test patterns for gross delay defects, i.e. they generate patterns based on ease of finding an affected path, instead of a least slack path. In general, a short path can be considered an easy path, i.e. there are fewer gates on the path to be controlled for both activation and propagation of the fault effect.

To demonstrate this point, transition fault test patterns for an ISCAS'89 benchmark (s38584) have been generated using a commercial timing unaware ATPG tool. Note that results obtained by this tool did not consider any slack based options. Figure 7.1 shows the delay distribution of the pattern set. The patterns were generated using launch-off-capture technique (*broadside*) for the total transition fault list (52874 faults) and the gross delay fault coverage and pattern count were 76.92% and 372, respectively. It can be noticed from the figure that majority of the paths exercised for delay fault detection are short paths. The minimum defect size detectable depends on the path delay region affected by the pattern set, relative to the clock cycle. For instance, in this particular example, most of the paths affected are less than 30% of cycle time. A delay defect size of at least 70% cycle time is required for the faults to be detected. This figure shows that a small delay defect is likely to escape. Therefore, more robust at-speed techniques are required to improve the effectiveness of transition fault testing to affect more longer paths and screen the small delay defects better.

In the past few years, various techniques have been proposed for improving the small delay defect screening quality of a pattern set. A number of these methods such as very-low-voltage (VLV) [8] and burn-in [9], modify the operating conditions of test environment and magnifies the defect size, which may escape at nominal conditions. However, these methods don't necessarily target small delay defects. The effectiveness of VLV testing for very

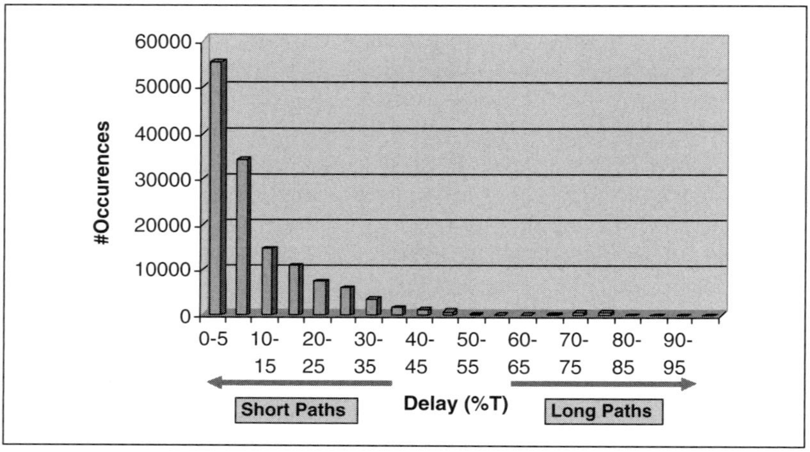

Fig. 7.1. Path delay distribution for no-timing ATPG transition fault pattern set generated using launch-off-capture method for ISCAS'89 benchmark s38584 [10].

deep submicron designs is reducing since the scaling of threshold voltage is not proportionate to supply voltage and issues like IR-drop and crosstalk are becoming more prominent. The burn-in test however is associated with considerable high costs and time which may not be suitable for today's designs due to shortened time to market.

Researchers are investigating alternative methods to detect such delay defects. In [14], a transition fault model, called As Late As Possible Transition Fault (ALAPTF) was proposed. The method tries to activate and propagate a transition fault at the target gate terminal through the least slack path possible. Although effective in identifying longer paths, the ATPG method used is complex and will be more CPU intensive compared to a no-timing ATPG. A delay fault coverage metric is proposed in [15] which tries to detect the longest path affecting a line. The technique attempts to find the longest sensitizable path passing through the target line producing a rising (falling) transition on it. In [16], the authors proposed an efficient ATPG tool to generate K longest paths per gate for transition fault test. The technique targets all the transition faults to find the longest path. A longest path does not reflect the detectable delay defect size. For example, if the least slack path of a gate is a short path then a small delay defect on such a gate output cannot be detected for the nominal frequency.

In [12], an ATPG methodology is proposed for detecting delay defects by integrating timing information, e.g. from Standard Delay Format (SDF) files, into the ATPG tool. The timing information is used to guide the test generator to detect faults through the longest paths in order to improve the ability to detect small delay defects. During fault simulation, a new fault-dropping criterion, named Dropping based on Slack Margin (DSM), is proposed to facilitate the trade-off between the test set quality and the test pattern count. The complexity of the proposed methodology appears to be high which makes its immediate use limited. The technique proposed in [17] is based on detecting a smaller delay on a shorter path by increasing the frequency of operation, instead of detecting it on a long path which requires a timing aware ATPG tool. Due to increasing the frequency, the capture edge might occur in the hazard region for some of the observation points. Such methods may also be limited by the highest possible frequency of operation which exacerbates the already well known issues of peak power during test and IR-drop. The authors in [18] showed a case study of the effects of IR-drop and explored quiet pattern (reduced transition) generation methods to reduce it. Recently, the effects of power supply noise on clock frequency during delay test was presented in [19].

In [20], multiple-detect test pattern sets are used to improve the quality of tests by maximizing the probability of detecting bridging defects but generates high pattern count compared to a single detect pattern set. To enhance the effectiveness of screening frequency dependent defects, the authors in [21] propose a pattern selection methodology to reduce the delay variation of the selected pattern set and higher frequency is used for pattern application. The method uses a multiple-detect transition fault pattern set and it uses statistical timing analysis techniques to reduce pattern delay variations.

7.1.1 Overview of the Proposed Timing-based Pattern Generation Procedure

The pattern modification or selection techniques described above assume that a single-detect or a multiple-detect pattern set is already available, respectively. Using multiple detect will increase the test length and inefficient patterns must be excluded from the pattern set. Moreover, most previously proposed methods try to improve the defect screening effectiveness by detecting small delay defects on all paths of the design at multiple higher test frequencies, which might further worsen the issues of power during test and IR-drop.

In this chapter, a novel pattern generation technique is proposed which targets the small delay defects only on the long paths of a design for the functional operating frequency. But the user can define multiple test frequencies based on his/her analysis for tolerable increase in test power and IR-drop. The proposed technique uses static timing analysis tool to divide the path lengths and their corresponding observation points into different categories (long, intermediate and short paths respectively). A new delay defect size metric is defined based on the affected path length category and required increase in test frequency. Then, multiple-detect ATPG is performed on all the fault sites along the long paths to detect small delay defects. A novel pattern selection technique is used which selects patterns activating higher percentage of long paths and masks all short paths. Using this technique, existing timing unaware ATPG tools can be used to obtain a high coverage of small delay defects along the long paths of a design. The experimental results show the effectiveness of the proposed technique in detecting small delay defects on longer paths when compared to traditional no-timing pattern generation.

7.2 Path Length and Pattern Delay Analysis

Traditionally, the transition delay ATPG methods targeted gross delay defects. A *gross* delay defect can be detected irrespective of the affected path. This delay defect model was efficient for higher technology nodes. However, the population of not-gross (very small, small or intermediate) delay defects increases as technology scales due mainly to the existence of resistive opens and shorts, crosstalk and signal integrity. A *very small* delay defect can be detected only when affected by an extremely timing sensitive path, referred to as a critical path. For such paths, the slack is very small and any kind of delay defect can cause a timing failure. Such paths are limited in number and they are used for path delay fault testing.

A *small* delay defect which cannot affect any critical paths needs to be detected through its least possible slack path. But, each fault can literally be detected through tens and hundreds of paths with varying lengths. Also, there are millions of paths in a design with different paths of varying length converging to each observation point. To differentiate between all the various paths in a design, four categories of paths are defined based on their path length and the minimum size of the delay defect that can be detected through the path.

1. **Critical Path (CP):** A critical path is very timing sensitive and very small delay defects on such path can be detected. Path delay fault model can be used for such paths.

2. **Long Path (LP):** A long path in a design is defined as a path, if affected by a small delay defect can cause a timing failure. Such paths are timing sensitive as the path's timing length is very close to the target frequency next to the critical path (CP) category.

3. **Short Path (SP):** A short path requires a significant delay defect size that will create a very large timing variability to cause a failure. Detecting small delay defects on such paths requires a very high test frequency to make the slack near zero.

4. **Intermediate Path (IP):** A path with a delay in the range other than long paths and short paths is defined as an intermediate path.

The path length range of each category is shown in Figure 7.2. The vertical axis is organized with the least slack at the top and the highest slack at the bottom. The cutoff limits of the long and short path's region are shown as

LP-cutoff and SP-cutoff, respectively. These cut-off points can be determined based on the delay defect size coverage and the increase in frequency required to make the slack of the path near zero.

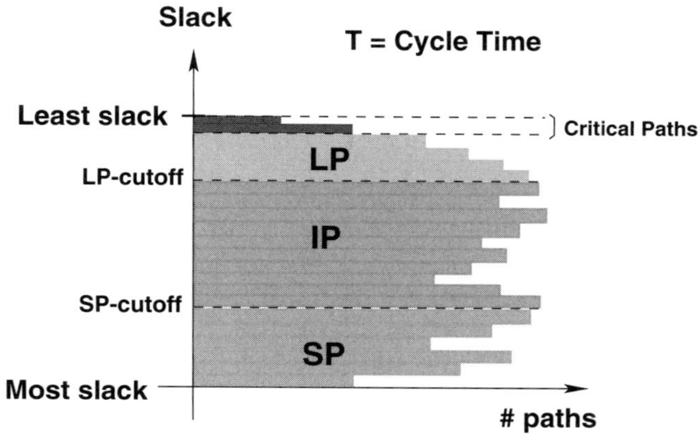

Fig. 7.2. Different categories of paths (critical, long, intermediate and short paths) [10].

The above path length analysis shows that small delay defects on some paths may only be detected by increasing the frequency. As an example, consider a case where the longest path for a detectable small delay defect might still be short therefore, it will not be detected under normal functional frequency. This will be further evaluated in the following.

Table 7.1 compares the detectable delay defect size at nominal functional frequency and the test frequency increase required compared to the functional frequency for each path length category to detect small delay defects. The critical paths (CPs) are very timing critical and even very small delay defects can be detected on such paths. To detect small delay defects on LP paths, it requires less increase (or no increase) in test frequency. As a result, the test power or IR-drop is not expected to increase. Whereas, detecting small delay defects on IP paths requires higher test frequency. The IP paths are important because as the test frequency is increased, some of the paths in IP category become long paths for the next target frequency. On short paths (SPs), only gross delay defects can be detected. To detect small delay defects on short paths, a very high test frequency (probably $\geq 2X$ increase) will be required.

The above path length analysis shows that small delay defects on intermediate and short paths can only be detected using higher test frequencies. Applying patterns at higher test frequencies require special considerations in terms of the potential excessive power and IR-drop [11]. In this chapter considers only the small delay defects on long paths for pattern generation. However, if increasing frequency during test is not an issue, our technique is flexible and it can be applied for small delay defects on all paths.

Table 7.1. Comparison of delay defect size and test frequency for different path categories [10].

Path Type	Defect Size Range	Frequency Increase
CP	very small - gross	No change
LP	small - gross	Slight - No change
IP	intermediate - gross	Medium
SP	gross	High

7.2.1 Endpoint Definition

An observation point at the end of a path (primary output or scan flip-flop) is referred to as an *endpoint*. Note that, during delay testing, the primary outputs are not measured between the launch and capture cycles. This is due to insufficient timing accuracy of increasingly used low-cost testers to strobe the primary outputs before the capture event. Therefore, in the rest of the chapter, an endpoint refers to a scan flip-flop.

Each endpoint is associated with a path delay distribution. Due to the complexity of finding all the paths to an endpoint, only the least slack path to each endpoint is considered. Figure 7.3 shows the number of endpoints with least slack divided across the entire cycle period for the benchmark s38584. A static timing analysis tool (Synopsys PrimeTime [7]) was used to find the least slack path to each endpoint. In this example, the LP-cutoff limit was set as slack less than 30% of cycle time. It can be noticed that the critical path endpoints in the LP category are included as they might contain long paths in their path distribution. Similarly, a slack greater than 70% of cycle time is considered for the SP-cutoff limit. The static timing analysis tool measures the slack of a path by $slack = T_{cycle} - T_{setup} - T_{delay}$, where T_{cycle}, T_{setup} and T_{delay} refer to cycle period, setup time and delay of path, respectively. As seen, a small percentage of endpoints (approx. 12%) have their least slack in the long path category. A majority of the endpoints with least slack fall in the intermediate category (approx. 61 %), while the remaining (approx. 27 %) fall in the short path length category.

7.3 Pattern Generation

As shown in the previous section, the delay defects on the long paths require a smaller defect size to cause a timing failure compared to intermediate and

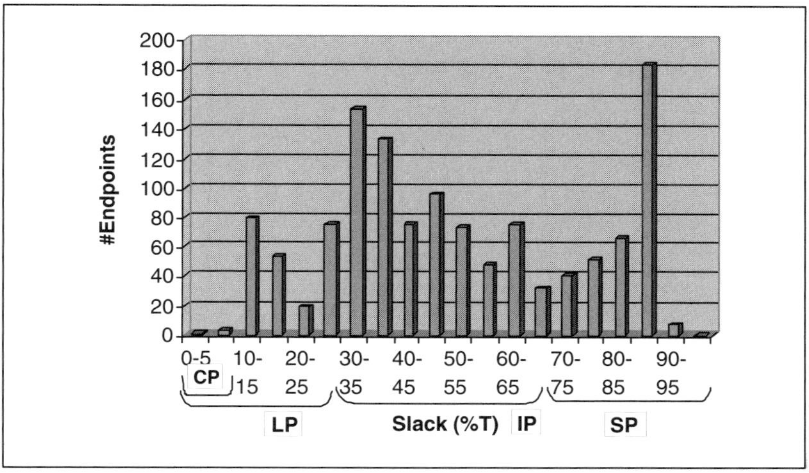

Fig. 7.3. Slack versus endpoints for benchmark s38584 [10].

short paths. Such small delay defects on long paths might escape at nominal frequency during test, since a timing unaware ATPG tool will affect short paths through them. One possible solution to detect such defects is to extract all the long paths in the design and perform path delay test pattern generation. If it could achieve 100% path delay coverage then the defect coverage of all small delay defects on long paths would be 100%. However, due to inherent robust pattern generation, path delay test gives a very small coverage of long paths. Therefore, heuristics are required to detect small delay defects on long paths using transition fault model.

Each LP-endpoint will have several short and intermediate paths converging to it other than a long path. In order to force the timing unaware ATPG tool not to exercise the short paths, the initializing points of the short/intermediate paths need to be held constant. Such logic sensitization control (holding certain logic state's constant) might be very complex. In case of launch-off-capture (broadside) method which launches a transition through the functional path will require the next time-frame information to determine the present state values to hold them constant. This requires a detailed analysis to find all the initializing endpoints and hold them constant during pattern generation. This is not possible due to the very high computational cost.

To avoid the high cost and complexity of path delay analysis, multiple-detect method is used. Multiple-detect method is an existing feature in the ATPG tools which increases the probability of a fault to be detected using a long path. Multiple-detect technique tries to activate a fault site through multiple paths. Since, most of the ATPG tools have the capability of multiple-detect pattern generation, this capability is utilized to activate the long paths. When using a single-detect method, the ATPG tool detects the faults mostly through shorter paths (see Figures 7.1 and 7.3). When using multiple-detect method, one fault will be detected through different paths and there is a high chance that one path will be long (LP).

Figure 7.4 shows the pattern delay distribution of a 15-detect transition fault pattern set for the same benchmark (s38584). The LP-endpoints are only used as observe points and the remaining endpoints (IP and SP) are masked. This is performed using cell constraints during pattern generation which forces the ATPG tool to ignore the value captured in the respective endpoints. However, an endpoint with a cell constraint can still be loaded with a valid care-bit value during scan chain shift for transition launch and propagation. The histogram excludes the endpoints that do not observe a transition since they do not contribute to fault coverage. The number of long paths affected in a 15-detect pattern set is higher compared to a 1-detect pattern set (see Figure 7.1). However, there are still a lot of short paths being affected to the LP-endpoints. Hence, the goal is to select a subset of patterns from the 15-detect pattern set with majority of long paths used to observe the delay faults. The 15-detect pattern set is larger than a single detect pattern set due to multiple combinations of activating each fault site. For this particular experiment, approximately 1000 patterns were generated for a 15-detect compared to 300 patterns for a 1-detect delay fault pattern set.

An analysis can be done to identify n in n-detect. If n is chosen very small then the ATPG may not generate efficient patterns to affect longer paths, i.e. the probability of generating patterns that affect longer paths goes down when n is small and vice versa. If n is chosen to be very large, then there may be large number of patterns in the pattern set that are not necessarily contributing to further long path delay defect detection. Consider an endpoint with only one long path and k number of short or intermediate paths. Also, assume that ATPG targets a delay fault at the input of the endpoint. To ensure that ATPG detects this fault, $n = k + 1$. Although this analysis is possible to be done for all faults to identify the maximum n to be used in n-detect ATPG, it adds another complexity to the pattern generation procedure. In this work, however, 15-detect is used for pattern generation. Note that 20-detect have also been tried for various benchmark circuits but the obtained small delay coverage was almost the same as 15-detect.

The pattern generation process is divided into four steps as listed below:

- **Step 1**: In the first step, path delay test is performed to cover some of the small delay defects on the long paths.

- **Step 2**: In the second step, only the LP-endpoints are considered to be observable for delay fault test generation. The remaining endpoints (IP and SP) are made non-observable.

- **Step 3**: In this step, the SP-endpoints are masked, since these endpoints with least slack in the SP category require a huge delay fault in order to fail.

- **Step 4**: In the final step, the above multiple-detect pattern generation procedure using only IP-endpoints as observe points is repeated. This step is optional and depends on the decision whether higher frequencies can be applied during test. The IP path range is the next range of paths close to LP. Any increase in frequency will push the IP paths into LP range for the new target frequency. Increase in frequency creates a possibility of hazards in patterns, especially to LP-endpoints as their path range might exceed the new target clock period. In such a case, the respective endpoints are required to be masked to avoid any timing failures.

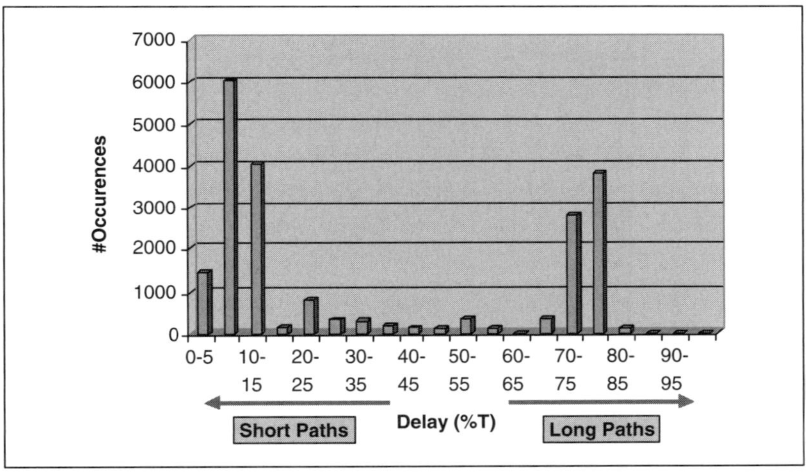

Fig. 7.4. Path Delay distribution for 15-detect pattern set (benchmark s38584) [10].

7.4 Pattern Selection

As mentioned earlier, a multiple-detect pattern set is larger than a single detect pattern set and there are still many short paths being exercised. Therefore, a subset of patterns are selected from the multiple-detect pattern set with majority of long paths being affected. This is necessary to exclude inefficient patterns from pattern set and reduce the test time. To perform the selection, The path delay distribution of each pattern in the pattern set is analyzed. Suppose, there are two patterns $P1$ and $P2$ in a delay fault pattern set where the pattern $P1$ affects multiple paths in the LP range while pattern $P2$ affects only a single long path and rest are short paths. For the same set of delay defects, pattern $P1$ would screen more small delay defects than $P2$ as it will affect more longer paths.

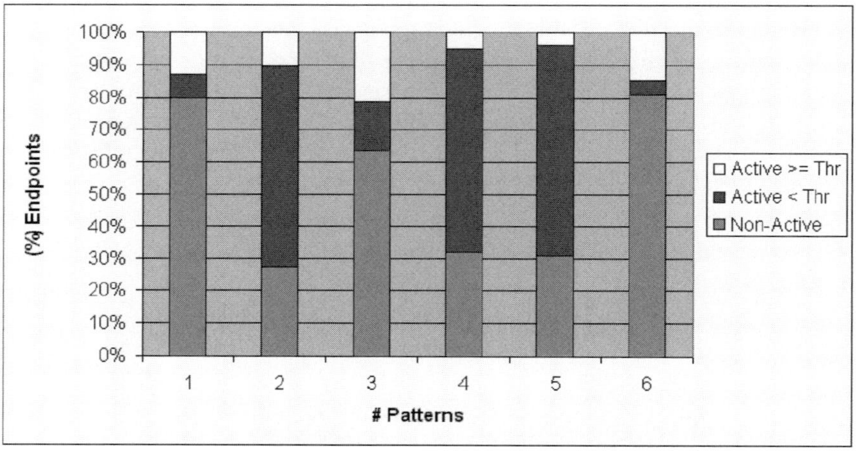

Fig. 7.5. Endpoint analysis for six patterns generated using multiple-detect for s38584 benchmark [10].

In the pattern selection process, each pattern is investigated to determine the number of active endpoints. An endpoint which observes a transition is referred to as an *active endpoint*. An endpoint which does not observe a transition, referred to as *non-active*, has no contribution to the fault coverage. Figure 7.5 shows the breakup of endpoints for six patterns, generated using multiple-detect technique, into three different categories: 1) non-active endpoints, 2) active endpoints with path length less than a threshold limit and 3) active endpoints with affected paths length greater than the threshold limit. The threshold limit is defined as the maximum slack of the path length region which also implies the minimum small delay defect size that can be detected. For this experiment, the thresholds LP-cutoff limit is fixed at 30% cycle time and SP-cutoff limit as 70% cycle time. Patterns 2, 4 and 5 have higher percentage of active endpoints. This criteria alone does not ensure that all active endpoints are observing long path delays. Consider patterns 1, 3 and 6 respectively which have less number of active endpoints. For these patterns, the percentage of active endpoints observing long paths is very high, although the overall percentage of active endpoints is low. Such patterns ensure that most of the faults are detected through long paths.

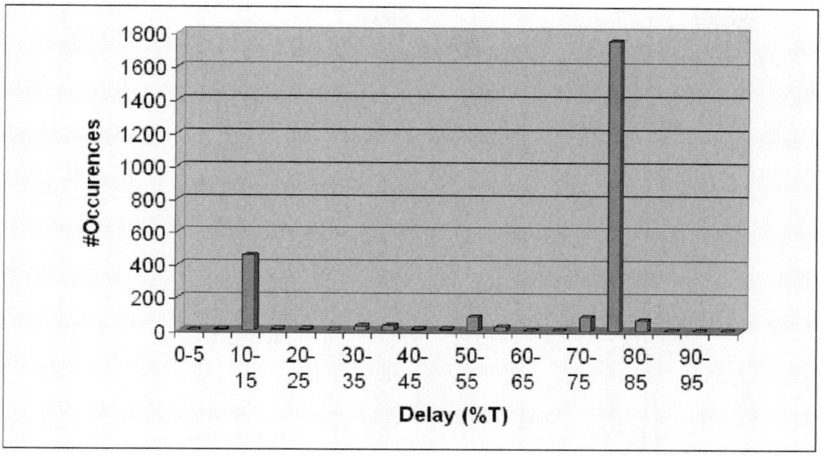

Fig. 7.6. Path delay distribution after pattern selection from a 15-detect pattern set (benchmark s38584) [10].

The path delay distribution of the resulting pattern set using the pattern selection process is shown in Figure 7.6. As seen, the variation of path delays is much smaller and concentrated in the long path range, although a small percentage of short paths are still activated. The entire n-detect pattern set is used in the pattern selection procedure to reduce the pattern count. The results are shown in the following section.

7.5 Experimental Results

The entire automation flow consisting of various steps is shown in Figure 7.7. The complete process can be divided into two phases: 1) Pre-processing phase and 2) Pattern generation and selection phase which, in the following, both will be discussed in details.

7.5.1 Pre-processing Phase

Before starting the test pattern generation process, the design is pre-processed for path delay analysis. This is performed using a static timing analysis tool (Synopsys PrimeTime [7]) and the endpoints are classified into different path length categories based on the least slack reported for each endpoint (*Step 1*). Table 7.2 shows the total number of endpoints (column 2) and the endpoints with their least slack path for five largest ISCAS'89 benchmark circuits in each of the different regions. In the experiments, the cutoff limit for LP-region (LP-cutoff) was used as slack less than 30% of cycle period. Similarly, the SP-cutoff was set to slack greater than 70% of cycle period. It can be noticed that majority of the endpoints for circuits s13207, s15850 and s35932 have their least slack path in the SP region. While for circuits s38417 and s38584, it is in the IP region. The NP-endpoints are endpoints with a path starting from the primary input. Since, the primary inputs are held constant during the delay fault test generation due to low-cost tester speed limitations, *NP* endpoints do not contribute to delay fault coverage.

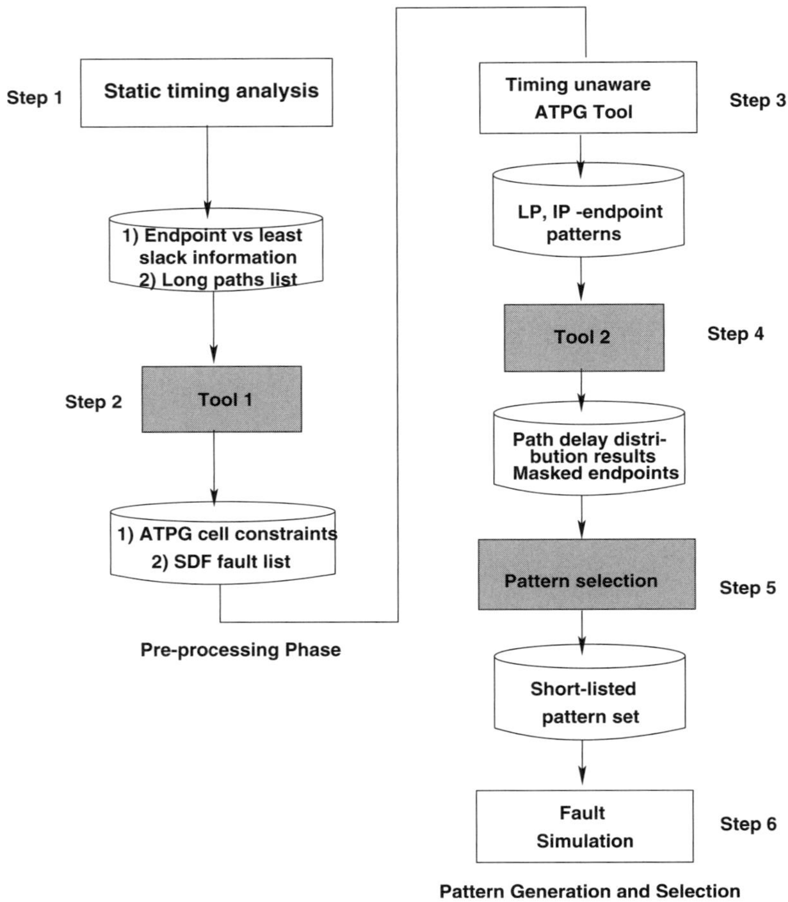

Fig. 7.7. Automation flow of the proposed pattern generation procedure.

Since the focus is to detect all small delay defects on long paths, all the paths in LP-region are extracted using static timing analysis (*Step 1*). These paths are converted to a fault list which will be used during the pattern generation process. This is performed by identifying all fault locations along each long path and a slow-to-rise and slow-to-fall fault is considered for each

Table 7.2. Number of endpoints with least slack in each path delay region [10].

Design	Total	LP	IP	SP	NP	GDFs	SDFs
s13207	626	10	138	453	25	15084	160
s15850	516	46	202	246	22	16178	230
s35932	1728	32	545	1149	2	44366	320
s38417	1564	208	735	600	21	44986	2120
s38584	1276	157	770	342	7	52874	2124

fault site. In case of multiple clock domains, the following path analysis and fault set extraction needs to be performed for each individual clock domain. Table 7.2 also shows the total number of gross delay faults (GDFs) on all paths, i.e. total number of transition faults and the small delay faults (SDFs) which lie only along the long paths. For example, for benchmark s38584, the static timing analysis tool gave 1491 long paths with a LP-cutoff of 30% cycle time. We, then extracted all the fault sites along the 1491 long paths to obtain 2124 transition faults, which forms our small delay fault list (see Table 7.2, last column).

After performing *Step 1*, the obtained data will be given to the developed Perl program (called *Tool 1*) which takes as input the worst slack information for each endpoint, along with design, clocking and cutoff limits for LP and SP region, and generates the atpg cell constraints for each path region (LP, IP and SP) in *Step 2*. As explained earlier, these cell constraints are required to ignore the value captured in the respective endpoints. However, an endpoint with a cell constraint can be loaded with a valid bit to activate and propagate a fault site. These cell constraints are used during the test pattern generation process. The tool also performs the extraction of transition faults (i.e. small delay faults (SDFs)) using long paths list.

7.5.2 Pattern Generation and Selection Phase

The pattern generation procedure (*Steps 3, 4, 5 and 6*) is performed for LP- and IP-endpoints only, as observe points one at a time, respectively. The IP-endpoints are used to augment the fault coverage in case the fault coverage achieved for small delay defects by observing only LP-endpoints is not sufficiently high. Inserting the IP-endpoints may not necessarily ensure detection of small delay defects, but makes the delay faults on long paths observable. Same holds true for SP-endpoints. If there are still some small delay faults left undetected in the fault list after using both LP- and IP-endpoints as observe points, SP-endpoints are used for observation. Due to large number of SP-endpoints, the detection of remaining small delay defects is highly ensured.

A commercial no-timing ATPG tool is used in the pattern generation procedure. Initially, path delay pattern generation is performed for the long paths. The generated path delay pattern set detects a subset of faults in the

SDF list which lie along the detectable paths. For example, for ISCAS'89 benchmark s38584, the path delay test pattern set for long paths provided 19.28% coverage of SDF transition fault list (2124 faults). This demonstrates the ineffectiveness of path delay fault test patterns for detecting small delay faults on long paths.

In (*Step 3*), for the remaining faults in the SDF fault list after fault grading path delay patterns, a 15-detect pattern generation with only LP-endpoints observable, followed by IP-endpoints only being observable is performed. The generated LP (IP) pattern sets are analyzed (*Step 4*), to measure the path delay distribution for the respective endpoints in each pattern set. The patterns are then re-ordered based on the percentage of active endpoints affected, as explained in Section 7.4. and a subset of the patterns with very high percentage of active endpoints is selected. Also, the endpoints affected by short paths are masked in each pattern, i.e. a delay less than 30% of the cycle time. This ensures that no short paths are exercised in the pattern set. Another software program (called *Tool 2*) have been developed, which performs the path delay distribution analysis, followed by masking of endpoints exercising short paths. It also re-orders the pattern set using the ATPG tool.

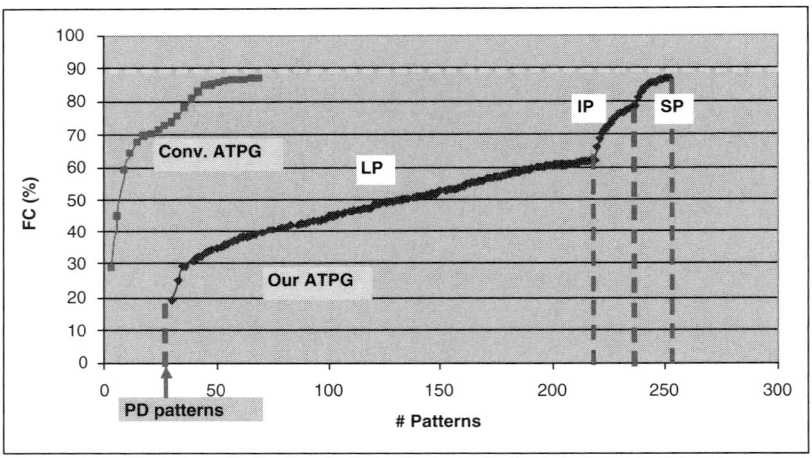

Fig. 7.8. Comparison between conventional and the presented pattern generation process for SDF fault list (s38584 benchmark) [10].

Since, the short paths are masked in the pattern set, the resulting pattern set is re-fault simulated (*Step 6*) to get the accurate coverage of SDF fault list. Figure 7.8 compares the fault coverage graphs of the proposed timing-based ATPG method and the conventional timing unaware ATPG. As shown, initially the path delay patterns (PD-patterns) are generated followed by LP-patterns, and IP-patterns based on LP- and IP-endpoints being observable, respectively. Note that, multiple-detect technique during LP-endpoints pattern generation may not cover all possible long paths. This may require a higher multiple-detect pattern generation and it will increase the run time. Hence, if it is possible to increase the frequency without any adverse effects on test power and IR-drop, IP-patterns can be generated for the remaining faults in the SDF fault list and higher frequency can be applied to detect them. After each step of fault detection, the detected faults are removed from the fault list. For example, after fault grading path delay patterns for transition fault model, the detected faults are removed from the fault list. Finally, after removing the detected faults by LP- and IP-patterns, conventional ATPG is performed using SP-endpoints. These faults in the SDF list covered by pattern generation using SP-endpoints are actually gross delay faults. The conventional no-timing ATPG generates less number of patterns because all the endpoints are observable and it can easily activate and propagate the fault through short paths. Therefore, the delay defect size required for such a pattern set will be very large.

The proposed pattern generation procedure, as shown in Figure 7.7, is not iterative. The final step of the procedure is *Step 6* where the short-listed pattern set is fault simulated to obtain the small delay fault coverage. Therefore, the complexity of the proposed procedure is low and the pattern generation time heavily depends on the size of the design and the efficiency of the existing ATPG and timing analysis tools. The two software programs developed in this work are quite fast when running on ISCAS'89 benchmark circuits as they identify small delay fault list and IP- and Lp-endpoints. It is worthy to mention that the proposed procedure can use any of the existing commercial ATPG and timing analysis tools.

Table 7.3 shows the ATPG results for the SDF fault list using conventional no-timing ATPG and the presented timing-based ATPG methods. Note that, the fault coverage numbers obtained from the conventional ATPG does not reflect the defect size coverage of each fault in contrast to the timing-based ATPG method. It is simply the gross delay defect size coverage (GDFC, shown in column 3), reported for the SDF fault list. As shown in Figure 7.8, after LP- and IP-patterns pattern generation are performed for SP-endpoints to get the final gross delay coverage of the timing-based ATPG technique. The small delay fault coverage (SDFC), obtained by this technique, is shown in column 6 using LP- and IP- patterns only. This coverage is excluding the additional coverage achieved using SP-patterns.

The number of patterns are higher for timing-based ATPG method due to limited number of endpoints being observable (to increase the probability of affected long paths) in this method. The best way to compare the timing-based ATPG method with the traditional timing unaware ATPGs is to observe the long and intermediate paths exercised. When using conventional ATPGs, neither timing analysis is required nor endpoint selection and pattern delay analysis are performed. Obviously, it is expected that the CPU time to be lower than our pattern generation procedure which performs all the above.

Among the benchmarks listed in the table, it is seen that s15850 provides lower small delay fault coverage (SDFC=51.74%). This issue was investigated and it was observed that this benchmark includes many hard-to-sensitize paths. As seen, even the gross delay fault coverage (GDFC) obtained by the ATPG tool was significantly lower (GDFC=73.04%) than other benchmark circuits.

Table 7.3. ATPG results [10].

Design			*Conv. ATPG*		*Timing-based ATPG*		
	SDFs	*GDFC* %	*Patt*	*Time [sec]*	*SDFC* %	*Patt*	*Time [sec]*
s13207	160	92.5	27	29	90.62	33	116
s15850	230	73.04	20	42	51.74	27	131
s35932	320	100	16	73	83.12	29	155
s38417	2120	99.43	63	114	98.87	174	534
s38584	2124	86.86	69	140	79.75	252	1768

The number of long paths affected using the pattern set generated using timing-based pattern generation procedure and commercial ATPG are analyzed. Figures 7.9 (a) and (b) show the comparison of long and intermediate paths exercised respectively for all benchmarks by the two methods. It can be noticed that the pattern set generated by the new technique affects higher number of long and intermediate paths in all benchmark circuits. This shows the effectiveness of the proposed technique in affecting more longer paths for small delay faults.

Due to the differences between the method shown in this chapter and other previously proposed methods it is not possible to fairly compare the obtained results. Most of the previous methods either increase the frequency to target small delay defects [15] [17] [21] or detect small delay defects on longest paths but not necessarily on long paths [14] [12]. However, the results are compared with those obtained from conventional ATPG tools that do not take the path length into account.

7.6 Summary

This chapter proposes new test pattern generation and pattern selection procedures to target small delay faults on long paths in deep-submicron designs. The technique divides the scan flip-flops into different categories based on the least slack. It then performs ATPG on each category to exercise more longer paths and is very efficient in detecting small delay defects. The experimental results showed that the proposed technique can detect a significant number of small delays through longer paths which otherwise would have escaped using

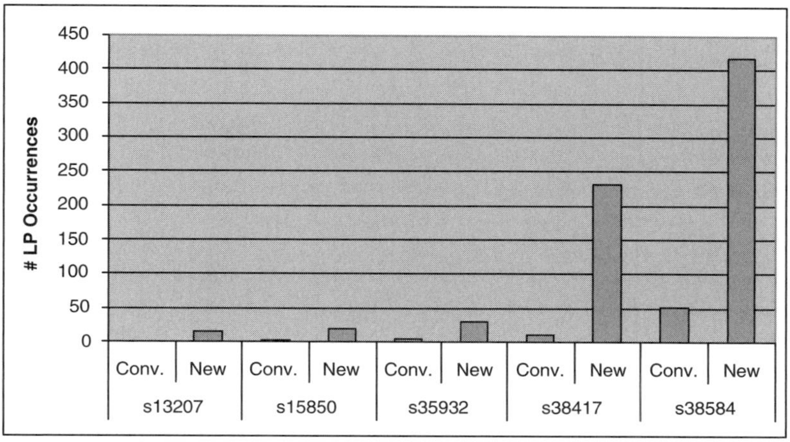

(a)

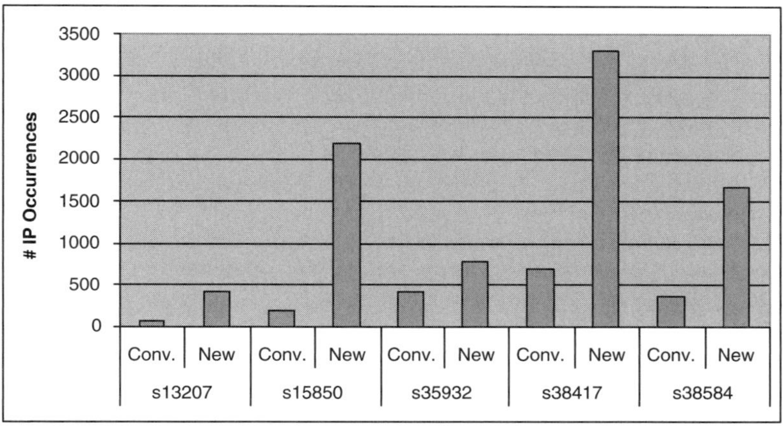

(b)

Fig. 7.9. Comparison of number of paths affected by conventional and the technique presented in this chapter, (a) long and (b) intermediate paths, for the same small delay fault coverage [10].

gross delay pattern set. As a result, this will increase the reliability of the designs.

References

1. X. Lin, R. Press, J. Rajski, P. Reuter, T. Rinderknecht, B. Swanson and N. Tamarapalli, "High-Frequency, At-Speed Scan Testing," *IEEE Design & Test of Computers*, pp. 17-25, Sep-Oct 2003.
2. V. Jayaram, J. Saxena and K. Butler, Scan-Based Transition-Fault Test Can Do Job, *EE Times*, Oct. 2003.
3. K. Cheng, "Transition Fault Testing for Sequential Circuits," *IEEE Transactions on Computer-Aided Design of Integrated Circuits and Systems*, vol. 12, no. 12, pp. 1971-1983, Dec. 1993.
4. T. M. Mak, A. Krstic, K. Cheng and L. Wang, "New challenges in delay testing of nanometer, multigigahertz designs," *IEEE Design & Test of Computers*, pp. 241-248, May-Jun 2004.
5. M. Bushnell and V. Agrawal, *Essentials of Electronics Testing*, Kluwer Publishers, 2000.
6. Cadence Inc., "http://www.cadence.com,", 2005.
7. Synopsys Inc., "User Manuals for SYNOPSYS Toolset Version 2004.06," Synopsys, Inc., 2004.
8. H. Hao and E.J. McCluskey, "Very-low-voltage testing for weak CMOS logic ICs," in Proc. *Int. Test Con. (ITC'93)*, pp. 275-284, 1993.
9. R. Foster, "Why Consider Screening, Burn-In, and 100-Percent Testing for Commercial Devices?," *IEEE Transactions on Manufacturing Technology*, vol. 5, no. 3, pp. 52-58, 1976.
10. N. Ahmed, M. Tehranipoor and V. Jayaram, "Timing-Based Delay Test for Screening Small Delay Defects," in Proc. *Design Automation Conference (DAC06)*, 2006
11. N. Ahmed, M. Tehranipoor and V. Jayaram, "A Novel Framework for Faster-than-at-Speed Delay Test Considering IR-Drop Effects," in Proc. *Int. Conf. on Computer-Aided Design (ICCAD'06)*, 2006.
12. X. Lin, et. al, "Timing-Aware ATPG for High Quality At-speed Testing of Small Delay Defects," in Proc. *IEEE Asian Test Symposium (ATS'06)*, 2006.
13. S. Kundu, P. Engelke, I. Polian and B. Baker, "On detection of Resistive Bridging Defects by Low-Temperature and Low-Voltage Testing," to appear in *IEEE Asian Test Symp. (ATS'05)*, 2005.
14. P. Gupta and M. S. Hsiao, "ALAPTF: A new transition fault model and the ATPG algorithm," in Proc. *Int. Test Conf. (ITC'04)*, pp. 1053-1060, 2004.
15. A. K. Majhi, V. D. Agrawal, J. Jacob, L. M. Patnaik, "Line coverage of path delay faults," *IEEE Trans. on Very Large Scale Integration (VLSI) Systems*, vol. 8, no. 5, pp. 610-614, 2000.
16. W. Qiu, J. Wang, D. M. H. Walker, D. Reddy, X. Lu, Z. Li, W. Shi and H. Balichandran, "K Longest Paths Per Gate (KLPG) Test Generation for Scan-Based Sequential Circuits," in Proc. *Int. Test Conf. (ITC'04)*, pp. 223-231, 2004.
17. B. Kruseman, A. K. Majhi, G. Gronthoud and S. Eichenberger, "On hazard-free patterns for fine-delay fault testing," in Proc. *Int. Test Conf. (ITC'04)*, pp. 213-222, 2004.

18. J. Saxena, K. M. Butler, V. B. Jayaram, N. V. Arvind, P. Sreeprakash and M. Hachingerr, "A Case Study of IR-Drop in Structured At-Speed Testing," in Proc. *Int. Test Conf. (ITC'03)*, pp. 1098-1104, 2003.
19. J. Rearick and R. Rodgers, "Calibrating Clock Stretch During AC Scan Testing," in Proc. *Int. Test Conf. (ITC'05)*, 2005.
20. B. Benware, C. Schuermyer, N. Tamarapalli, Kun-Han Tsai, S. Ranganathan, R. Madge, J. Rajski and P. Krishnamurthy, "Impact of multiple-detect test patterns on product quality," in Proc. *Int. Test Conf. (ITC'03)*, pp. 1031-1040, 2003.
21. B.N. Lee, L. C. Wang and M. S. Abadir, "Reducing pattern delay variations for screening frequency dependent defects," in Proc. *VLSI Test Symp. (VTS'05)*, pp. 153-160, 2005.

8

Faster-Than-At-Speed Test Considering IR-drop Effects

Interconnect defects such as weak resistive opens, shorts and bridges increases the path delay affected by a pattern during manufacturing test but not significant enough to cause a failure at functional frequency. Faster-than-at-speed tests have been proposed to detect such small delay defects. While these techniques increase the test frequency to reduce the positive slack of the path, they exacerbate the already well known issue of IR-drop during test. This may result in false identification of good chips to be faulty due to IR-drop rather than small delay defects. Although, it may be argued that such false failures can be easily identified by applying the faster-than-at-speed test pattern to a larger sample of good chips. A trivial solution to determine the maximum test frequency that a pattern can be applied would be to iteratively increase the applied frequency of the pattern on the tester until a good chip starts to fail. However, considering the test time impact and analysis required for a large test pattern set makes such a solution impractical. Also, it is impossible to apply each test pattern at an individual frequency either due to hardware limitations of the automatic test equipment (ATE) to generate higher frequencies or due to long synchronization times of on-chip clock generators (phase locked loops (PLLs)) affecting test time.

In this chapter, a new framework is presented for delay test pattern application to screen small delay defects generated using any commercial ATPG tool. Given a test pattern set, the technique groups the patterns into multiple sub-sets with close path delay distribution and determines an optimal test frequency considering both positive slack and performance degradation due to IR-drop effects. Since, the technique does not increase the test frequency to an extent that any paths exercised at the rated functional frequency may fail, it avoids any scan flip-flop masking. As most semiconductor companies currently deploy compression technologies to reduce test cost, scan cell masking is highly undesirable for pattern modification as it would imply pattern count increase and might result in pattern regeneration. Therefore, the solution is more practical as the test engineer can run the same pattern set without any changes to the test flow other than the at-speed test frequency [12].

8.1 Introduction

Transition fault model is widely practiced in industry to test delay-induced defects and it is a cost-effective alternative to functional pattern generation [1][2]. Traditionally, transition fault tests were generated assuming a fixed cycle time equivalent to the functional operating frequency for each clock domain. In general, designs have multiple clock domains and delay tests are generated/applied one clock domain at a time. These clock domains might be either synchronous or asynchronous in nature. Paths between two separate asynchronous domains are considered as false paths and hand-shaking techniques are used for data transfer between such clock domains. Also, paths between two separate synchronous domains are considered false as they might have different clock insertion delays which is the delay from the chip boundary to the flip-flops inside the design. Under the fixed cycle time assumption, a delay defect will be detected only when it causes a transition to reach an observe point (primary output or scan flip-flop) by more than the positive slack of the affected path. Slack of a path is a measure of how close a transition on the respective path meets the timing to an observable point, relative to the test cycle time.

A delay defect with defect size not large enough to cause a timing failure under the fixed cycle time notion is referred to as a small delay defect. A small delay defect might escape during test if is tested using a short path. While the same defect might be activated on a longer path during functional operation and it may cause a timing failure. Therefore, test coverage of small delay defects on long paths determines the quality of test patterns as these might cause immediate field failures. While, small delay defects on short paths might become a *reliability* issue as the defect might magnify during subsequent aging in the field and cause a failure of the device. Hence, it is important to detect such defects during manufacturing test using efficient techniques [3].

There is a growing industry concern for timing aware ATPG tools. However, adding such intelligence to the tool comes with the expense of higher pattern volume and longer CPU time as the tool has to perform extra processing to target each fault location through its longest path. Synopsys TetraMAX ™[4] pattern generation tool provides a solution to target the longest path through each fault site and utilizes node slack information in static timing analysis step during pattern generation. In [5], the authors report a 6% increase in affected long paths compared to conventional no-timing ATPG with a 2.5X increase in pattern count. Although, the technique improves the detection of small delay defects on long paths but short and intermediate paths still have enough slack to escape and requires faster-than-at-speed test application.

Encounter True-Time Delay Test Tool ™[6] uses actual design timing (Standard delay format (SDF)) information for ATPG. While it still uses efficient ATPG algorithms and pseudorandom data to achieve high coverage in fewer patterns, it uses back-annotated timing information (SDF) to apply them at faster-than-at-speed. It sets the transition test capture frequency based on the slack of the paths exercised and also includes the ability to test non-critical paths faster-than-at-speed for small delay defects. Since, the technique uses design timing information, the SDF must be calibrated with very high accuracy to correspond with the tester operating conditions at the worst process corner. The possible limitations of such techniques are: 1) the timing information must take into account the process variation effects, dynamic effects such as IR-drop and crosstalk and 2) pattern compression techniques are significantly effected by masking scan cells whose path delay exceeds the faster-than-at-speed clock period.

Various other techniques have been proposed in the past for improving the small delay defect screening quality of a pattern set. A number of these methods such as very-low-voltage (VLV) [7] and burn-in [8], modify the operating conditions of test environment and magnify the defect size, which escape at nominal conditions. However, in DSM designs, the effectiveness of VLV testing is reducing as I) the scaling of threshold voltage is not proportionate to supply voltage and II) issues like IR-drop and crosstalk are becoming more prominent and burn-in is associated with considerable high costs.

In [9], a new transition fault model, called As Late As Possible Transition Fault (ALAPTF) was proposed. The method tries to activate and propagate a transition fault at the target gate terminal through the least slack path possible. The ATPG method used is complex and will be more CPU intensive compared to a no-timing ATPG. In [10], the authors proposed a new ATPG tool to generate K longest paths per gate for transition fault test. The technique targets all the transition faults to find the longest path(s). A longest path does not reflect the detectable delay defect size. For example, if the least slack path of a gate is a short path then a small delay defect on such a gate output cannot be detected for the nominal frequency.

The technique proposed in [3] is based on detecting a smaller delay on a shorter path by increasing the frequency of operation. The method groups a conventional delay fault pattern set into multiple pattern sets which exercise almost equal-length paths. The different pattern sets are then applied at different frequencies to detect smaller delays. Due to increasing the frequency, the capture edge might occur in the hazard region for some of the observation points and requires additional steps to mask the respective endpoints to avoid false timing failures.

To enhance the effectiveness of screening frequency dependent defects, the authors in [11] proposed a pattern selection methodology to reduce the delay variation of the selected pattern set and higher frequency is used for pattern application. The method uses a multiple-detect transition fault pattern set and it uses statistical timing analysis techniques to reduce pattern delay variations. The above methods are limited by the highest possible frequency of operation which exacerbates the already well known issues of peak power during test and IR-drop. Power ($P \propto V^2 * F$) is directly proportional to the frequency of operation and for example, a two-fold increase in frequency increases the power proportionately.

8.1.1 Overview of the Faster-Than-At-Speed Test Technique

Most of the proposed techniques for screening small delay defect detection take advantage of applying the patterns at higher frequencies to reduce the positive slack of the paths. Increasing the frequency impacts the performance of the chip due to adverse IR-drop effects. In this chapter, the practical issues during faster-than-at-speed delay test application are presented. A case-study of a design will be presented to illustrate the increase in both peak and average IR-drop effects due to faster-than-at-speed pattern application. Increase in IR-drop directly relates to performance degradation due to effective voltage reduction reaching the gates in the circuit. Therefore, it is very important to consider the performance degradation due to IR-drop effects along with the positive slack when frequency is increased for small delay fault detection.

Also, a framework is presented for the application of transition fault patterns. The technique groups the transition fault test patterns generated using any commercial ATPG tool (here Synopsys Tetramax [4]) into different groups based on the maximum path delay affected in each pattern. We then perform worst IR-drop analysis and estimate the related performance degradation in each group based on the switching cycle average power (SCAP) and determine the maximum frequency of pattern group. This reduces the risk of any false identification of good chips to be faulty due to IR-drop effects rather than small delay defects.

8.2 Case Study: Design Implementation

In this section, the physical synthesis of the case study design is described. It is a 32-bit processor core [15] with the following design characteristics:

- 20K bytes of memory
- approx. 50K gates
- 124 IO pads (53 bi-directional)
- 4K scan flip-flops

Scan-based test insertion was performed using Synopsys DFT Compiler [4] with eight scan chains. During test mode, all the bi-directional pins are controlled in input mode to avoid any congestion problem. The memory modules are wrapped to provide controllability and observability of the outputs and inputs, respectively, during scan test. The at-speed test methodology was to implement functional launch-off-capture transition fault test with a slow speed scan enable signal. The physical synthesis was performed with Cadence SOC Encounter place and route tool [16] using $180nm$ standard cell and IO ($3.3V$) library [17]. The design was timing closed for an operating frequency of $100MHz$ at nominal operating voltage ($1.8V$) and temperature (25^oC) conditions. The scan shift path was closed for a lower frequency of $10MHz$.

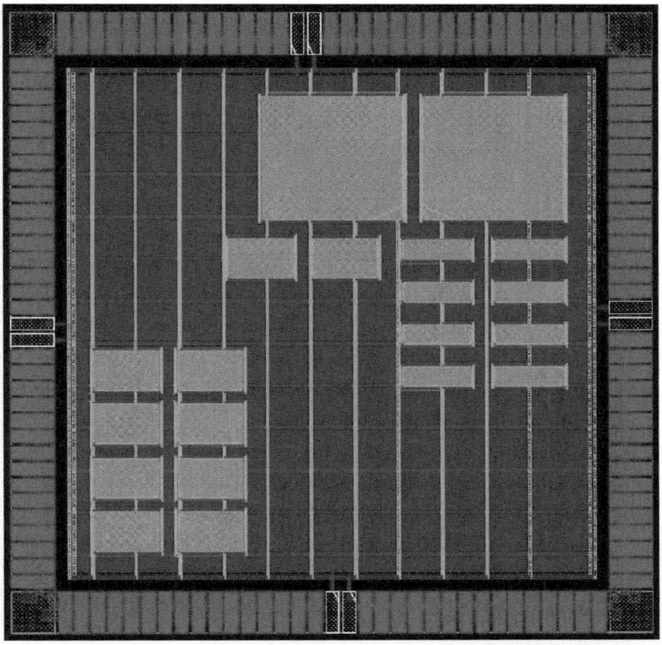

Fig. 8.1. Design floorplan with power/ground distribution network.

The power-planning for the design was performed assuming a net toggle probability of 20% during functional operation. Figure 8.1 shows the power/ground distribution network of the chip. Power and ground rings ($width = 20\mu m$) were created using higher-level routing layers (Metal5 and Metal6) which supply power around the standard cell core area. Also, power and ground rings ($width = 10\mu m$) were created around each memory module. Four power (VDD) and ground (VSS) pads each were inserted one on each side of the chip and connected to the respective rings with wires referred to as trunks. After creating the power rings, power and ground network was routed to the standard cells using horizontal and vertical stripes. The stripes ($width = 10\mu m$) were created using routing layers Metal4 and Metal5 with a distance of $200\mu m$ between adjacent stripes connecting the rings. After the power distribution stage, the design was timing-driven placed and routed along with clock-tree synthesis and scan cell ordering to minimize scan chain wirelength. The clock tree phase-delay (delay from clock pad to flip-flops) was $2.2 - 2.3ns$ with a rise/fall skew of $145ps$ and $120ps$ respectively.

In order to determine an estimate of IR-drop during functional operation, the design net parasitics (resistance and capacitance) were extracted using an extraction tool (Synopsys STAR-RCXT [4]). The average statistical IR-drop using vector-less approach was measured for both VDD and VSS nets considering 20% net toggle probability. The results showed 3.7% voltage drop in VDD network and a voltage bounce of 3.5% in the VSS network, which can be considered negligible. However, it will be illustrated in the following sections that the actual IR-drop during transition fault test pattern application is much higher compared to the measured statistical IR-drop due to high switching activity resulting in performance degradation of the circuit.

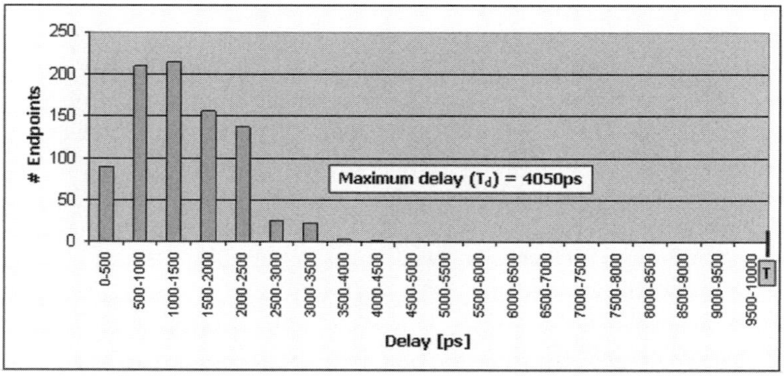

Fig. 8.2. Path delay distribution across all endpoints for a single launch-off-capture transition fault test pattern (*P1*).

8.3 Test Pattern Delay Analysis

IR-drop is directly related to the switching activity and the time frame in which it occurs. Therefore, it is important to understand the path delay distribution of a pattern to identify potential high IR-drop patterns. In this section, a detailed analysis of path delay distribution for a transition fault test pattern is presented including dynamic IR-drop effects, applied in two different cases: *Case 1:* at rated functional speed and *Case 2:* faster-than-at-speed to detect small delay defects.

The ATPG algorithms are based on zero-delay gate models and most of the existing commercial ATPG tools are timing un-aware for pattern generation process. The transition fault test patterns are generated based on the ease of finding an affected path, instead of a least slack path through the target gate. Figure 8.2 shows the path delay distribution of a transition fault test pattern (*P1*) across all endpoints, generated using a commercial ATPG tool (Synopsys TetraMax [4]). The pattern was simulated using the design timing information in IEEE standard delay format (SDF). The SDF file was generated by a delay calculator (Synopsys PrimeTimeSI [4]) using the parasitics extracted during physical synthesis at nominal operating conditions (25^oC and $1.8V$).

An observation point at the end of a path (primary output or scan flip-flop) is referred to as an *endpoint*. Since, the primary outputs are not observed due to insufficient timing accuracy of a low cost tester to strobe at functional speed [1][18], here, an *endpoint* refers only to a scan flip-flop. An endpoint which does not observe a transition, referred to as a *non-active* endpoint, is excluded in the above path delay distribution. It can be noticed that only a subset of endpoints observe transitions and for this particular pattern, approximately 20% endpoints. These are referred to as *active* endpoints.

The functional clock period ($f = 100MHz$) is represented by $T = 10,000ps$. In this particular pattern, the maximum path delay to an active endpoint was $T_d = 4050ps$. Also, it can be noticed that a majority of the active endpoints fall in the range of $1000 - 3000ps$. This indicates that only a delay defect size greater than half the clock cycle can be detected on paths terminating at such endpoints. Note that, the above pattern delay analysis was performed with timing information at nominal operating condition without taking IR-drop effects into account.

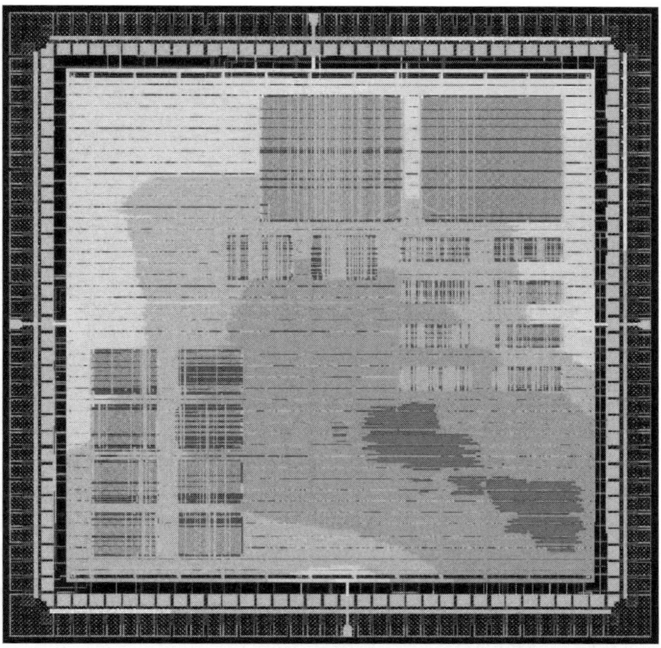

Fig. 8.3. IR-drop plot in VDD network for pattern *P1* at rated functional frequency f.

8.3.1 Dynamic IR-drop Analysis at Functional Speed

The IR-drop effects in a transition fault test pattern, *P1*, applied at rated functional frequency ($f = 100MHz$), is referred to as the *inherent IR-drop*. To measure the IR-drop of the pattern, the switching activity inside the circuit was captured in the standard value change dump (VCD) format during gate-level timing simulation. The switching activity information (VCD file) along with physical design and technology library information is used by SOC Encounter tool [16] to estimate the dynamic IR-drop of the pattern. Figures 8.3 shows the average inherent IR-drop plot on the power (VDD) network for pattern *P1*. The red region represents the area observing a voltage drop greater than 10% of VDD ($\Delta V \geq 0.18V$).

To measure the performance degradation due to dynamic IR-drop analysis, the average VDD and VSS voltage information of each instance in the design within the launch-to-capture window is stored. This information is then used to measure the cell delay degradation for each instance using the formulation: $ScaledCellDelay = CellDelay \times (1 + k\_volt * \Delta V)$, where *k_volt* is a factor specified in the vendor supplied technology library that accounts for non-linear delay scaling model and ΔV is the effective voltage decrease. Here, a value of 0.9 was used for *k_volt*, which means for a 5% effective cell voltage decrease ($\Delta V = 0.1V$), the cell delay increases by 9%. Figure 8.4 compares the path delay distribution of the same pattern (*P1*) across all endpoints considering inherent IR-drop effects and no IR-drop effects. It can be seen that the distribution curve has been shifted to the right hand side which indicates slow down (performance degradation) of the cells due to IR-drop effects. The maximum path delay to an active endpoint considering inherent IR-drop effects increased to $T_d = 5030ps$ (21% delay increase). However, there is still enough slack for small delay defects to escape.

In order to increase the small delay defect screening capability, based on the maximum path delay (T_d), the test clock timing can be adjusted in the range $T_d + \tau_{su} \leq T' < T$, where τ_{su} is the setup time of the scan flip-flop. Note that, for the purpose of clarity, the timing margin for process variation effects was not shown in the equation, which can be easily incorporated. The potential faster-than-at-speed clock period for this pattern are as shown in Figure 8.4. Increasing the frequency improves the small delay defect screening capability of the test pattern as the path delay affected by the pattern become relatively close to the clock period. However, applying faster-than-at-speed frequency increases the IR-drop which will impact the performance of the design. In this particular case, the faster-than-at-speed clock period was selected to be $T' = 7000ps$ ($f' \simeq 145MHz$), where $\tau_{su} = 200ps$ and enough timing margin was added for process variation and dynamic effects such as IR-drop and crosstalk.

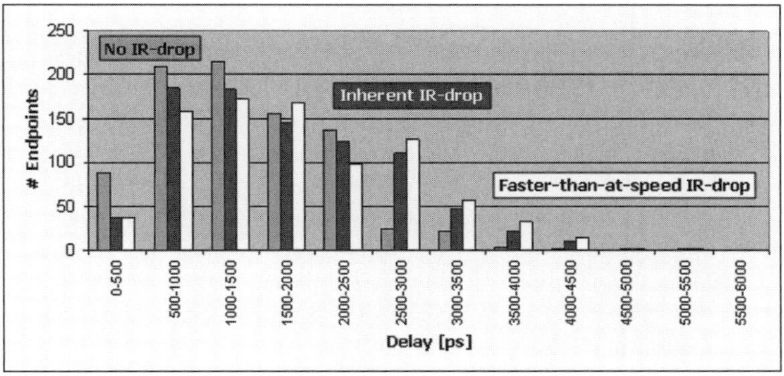

Fig. 8.4. Comparison of path delay distribution across all endpoints for pattern *P1* with no IR-drop effects and inherent IR-drop effects.

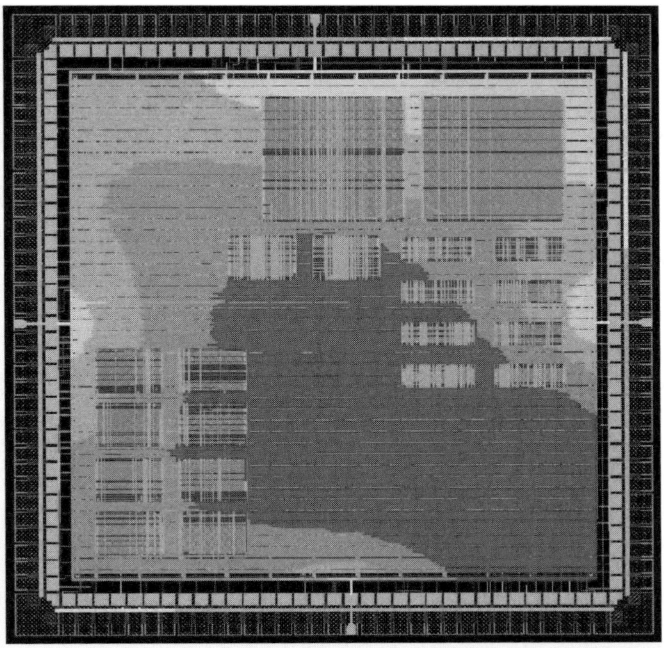

Fig. 8.5. IR-drop plot in VDD network for pattern *P1* at faster-than-at-speed frequency ($f' = 1.4 \times f$).

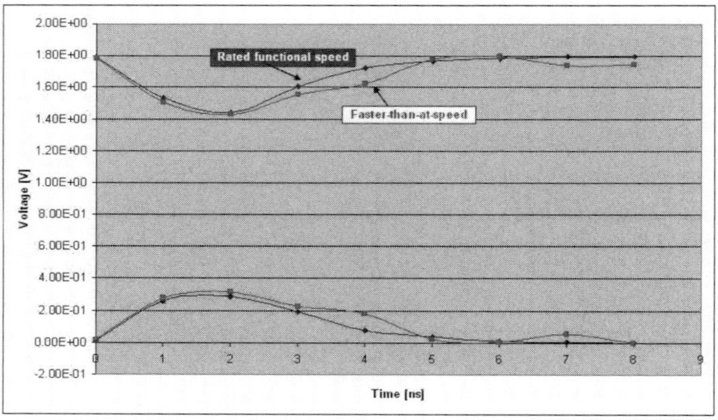

Fig. 8.6. IR-drop effects for rated functional speed and faster-than-at-speed of a transition fault test application.

8.3.2 Dynamic IR-drop Analysis at Faster-than-at-speed Test

The IR-drop in the VDD network for pattern *P1* applied at the new selected higher frequency to detect small delay defects is shown in Figure 8.5 and it is referred to as the *faster-than-at-speed IR-drop*. It can be noticed that the chip region observing voltage drop greater than 10% *VDD* has increased significantly. As the test frequency is increased, the IR-drop increases due to two factors: 1) higher data path switching speed and 2) occurence of negative clock edge switching activity in the clock network towards the early cycle period compared to the rated functional frequency. This results in increase of both peak and average IR-drop due to faster-than-at-speed pattern application as shown in Figure 8.6.

To measure the voltage curve, the launch-to-capture window was split into $1ns$ time frames and the IR-drop analysis tool reports the worst average IR-drop in each time frame. It can be noticed that the IR-drop effect is maximum in the beginning of the clock cycle due to high simultaneous switching activity and gradually decreases. The peak IR-drop for VDD increased from $0.28V$ to $0.31V$ (approximately by 10%) and the average IR-drop during the switching activity time frame window considering both VDD and VSS ($V_{VDD,IR-drop} + |V_{VSS,IR-drop}|$) increased from $0.26V$ to $0.35V$ (approximately 15%).

To measure the performance degradation, the pattern *P1* was simulated again with cell delay scaling technique based on each instance voltage obtained during the *faster-than-at-speed IR-drop* analysis. The path delay analysis showed some interesting results and it was found that some of the endpoints which previously observed transitions or glitches were static. This might be because some of the re-converging paths were slowed down due to IR-drop effects and the transitions/glitches were absorbed. Also, the transition fault test pattern generation is a non-robust technique and fault detection is dependent on the off-path delays unlike robust path-delay fault detection. However, the maximum path delay ($T_d = 5250ps$) was observed at the same endpoint and it increased by only 4% compared to inherent IR-drop. Further analysis showed that the longest path exercised by pattern *P1* was only partially affected by the *faster-than-at-speed* IR-drop which explains the slight increase in the maximum path delay. This may not likely occur in other patterns and may cause pattern failures due to IR-drop effects rather than small delay defects. Also, note that the faster-than-at-speed test frequency was selected based on the inherent IR-drop pattern delay analysis with enough timing margin (almost $2ns$) and this avoided the selection of even higher frequency leading to more adverse effects.

Figure 8.7 shows the path delay observed for five different endpoints in pattern *P1* for three different cases: Case1) no IR-drop effects, Case2) *inherent IR-drop* at functional frequency and Case3) *faster-than-at-speed IR-drop*. The performance degradation due to *faster-than-at-speed IR-drop* increased by up to 29% and 10% compared to no IR-drop effect and *inherent IR-drop*, respectively. For another pattern during this case study, the maximum pattern delay for faster-than-at-speed application increased from $3970ps$ (Case1) to $6050ps$ (Case3) which was slightly beyond the selected fast-than-at-speed clock period ($T' = 6000ps$) and resulted in a timing failure. This pattern would failure on tester due to IR-drop effects rather than a small delay defect. Therefore, it is very important to consider the IR-drop effects during faster-than-at-speed application along with the positive slack for detecting small delay defects.

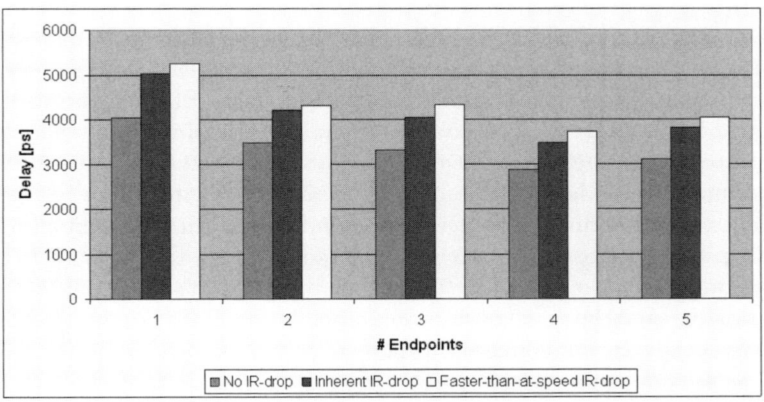

Fig. 8.7. Path delay observed for five different endpoints, after applying a transition delay pattern, in three cases: Case1) No IR-drop, 2) *inherent IR-drop* and 3) *faster-than-at-speed IR-drop*.

8.4 Pattern Generation Framework

As explained in the previous section, to improve the small delay defect screening capability of a test pattern, the timing can be adjusted such that the path delay affected by a pattern is relatively close to the clock period (*pattern slack* near zero). However, any such timing adjustment, i.e. test pattern application at a higher test frequency results in increase in IR-drop and related performance degradation needs to be considered. Therefore, the problem is to determine the maximum test frequency that a given transition fault test pattern can be applied without failing a good chip.

Each transition fault test pattern has varying delay distribution and switching activity and it is very difficult to perform detailed path delay analysis of each pattern as explained in Section 8.3. Also, even if such a detailed analysis was feasible, it may be difficult to apply each pattern at a different frequency. This might be either due to hardware limitations of the automatic test equipment (ATE) to generate multiple higher frequencies or due to the test time limitation because of long synchronization times required for on-chip clock generators (phase locked loops (PLLs)). Therefore, to simplify the problem, the test pattern set is grouped into a user-defined number of subsets with very close pattern delay distribution and then determine the maximum frequency for each group considering *faster-than-at-speed IR-drop* effects.

8.4.1 Pattern Grouping

In order to group the patterns with relatively close path delay distribution, the patterns were sorted in increasing order of *pattern slack*. The *pattern slack* is referred as the least slack (maximum path delay) across all the endpoints in the respective pattern. The launch-off-capture transition fault pattern set (1024 patterns) was generated using Synopsys Tetramax [4]. Figure 8.8 shows the maximum path delay affected in each pattern. Note that each pattern can effect various endpoints, but only the endpoint with the maximum delay (minimum slack) for each pattern is considered. The functional operating cycle time period is represented by $T{=}10ns$. It can be noticed that the affected paths in each of the pattern has considerable amount of slack for the small delay defects to escape during manufacturing test. This is because in most cases, the ATPG tools can find a shorter path to activate and propagate the transition fault effect.

Figure 8.9 shows the maximum path delay for each pattern in the resulting sorted pattern set. For faster-than-at-speed pattern application, the patterns with very close *pattern slack* distribution are grouped together (as shown in Figure 8.9). There are five groups $G1$ through $G5$ and the *pattern slack* range for each group was selected as $500ps$. Although, a more tight pattern slack range can be selected but it increases the number of groups and complexity as more processing will be required to perform IR-drop analysis for each group. Assuming a fixed faster-than-at-speed cycle time for each group, the first

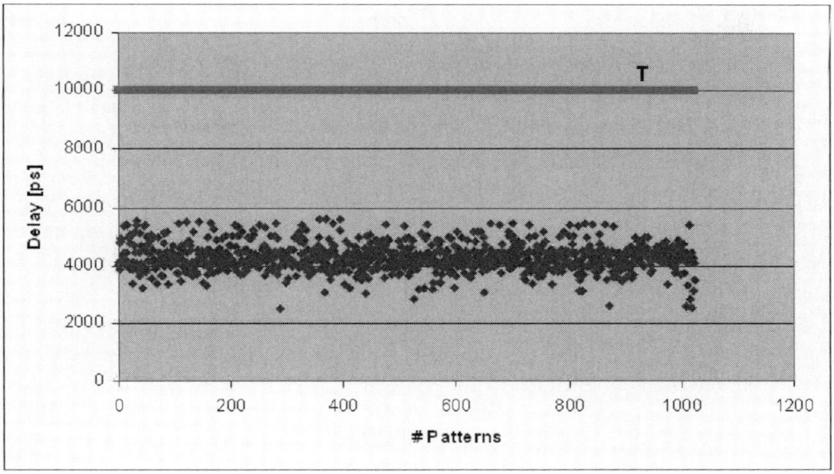

Fig. 8.8. Maximum path delay affected in each pattern of transition fault pattern set.

pattern in each group will have the least *pattern slack* and hence, it is used in determining the faster-than-at-speed test clock period for the respective group. If T'_{Gi} represents the new faster-than-at-speed clock period for a group then T'_{Gi} can be formulated as $T'_{Gi} \geq T_{Gi} + |\Delta T'_{Gi}| + \tau_{su}$, where T_{Gi} represents the maximum path delay for pattern group i without considering IR-drop effects, $\Delta T'_{Gi}$ represents the worst-case performance degradation due to *faster-than-at-speed IR-drop* for group Gi and τ_{su} is the setup time for the scan flip-flops.

8.4.2 Estimation of Performance Degradation

To determine the minimum test clock period (T'_{Gi}) for each pattern group, it is required to estimate the worst-case performance degradation ($\Delta T'_{Gi}$) due to IR-drop effects at the respective test frequency. Therefore, it is an iterative process to select a new faster-than-at-speed clock period (T'_{Gi}) and measure $\Delta T'_{Gi}$ for each pattern in the group until the performance degradation fails a pattern in the group. This is computationally very expensive and in order to reduce the complexity, two patterns *P1* and *P2* are selected in each group with the least slack and the highest switching cycle average power (SCAP) during the launch-to-capture window, respectively.

SCAP is an average power model and it will be discussed in detail in the next chapter. It provides a very good measure of the IR-drop as it considers both the switching activity and the time frame in which it occurs. Therefore, pattern *P2* will experience the highest IR-drop (ΔV) in the group for any applied test frequency. To measure the worst performance degradation, the

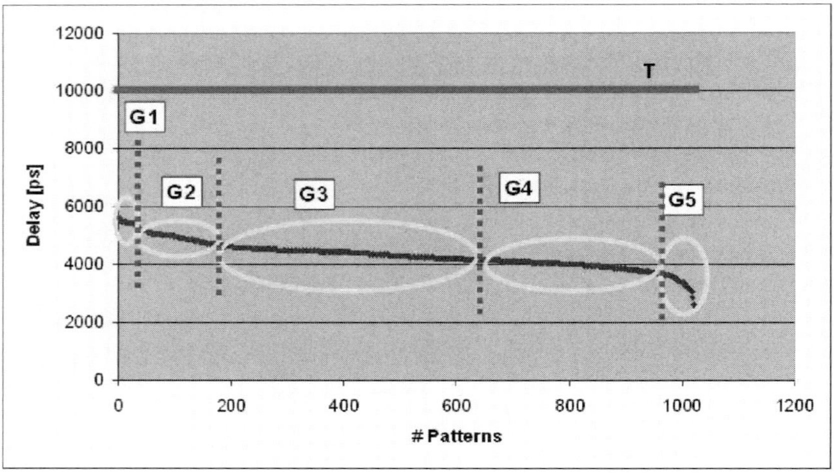

Fig. 8.9. Sorted transition fault pattern set based on maximum path delay.

IR-drop for pattern *P2* is then measured at T_{Gi}, i.e. the minimum possible test clock period without considering IR-drop effects. The corresponding worst cell delay degradation ($k\_volt * \Delta V$) if applied to pattern (*P1*) with the maximum path delay in the group would provide the worst possible path delay (T'_{Gi}) which can be used as the fastest at-speed clock period for the respective group. This procedure will provide enough timing margin for the entire pattern group considering faster-than-at-speed IR-drop effects.

The new faster-than-at-speed test pattern application framework is shown in Figure 8.10. The framework assumes that a delay pattern set is available, generated using any ATPG tool. Two programmable language interface (PLI) routines, *monitorDelay* and *calculateSCAP* have been developed. The PLI provides a standard interface to the internal data such as the nets switching inside the design during simulation. The PLIs can be plugged into any gate level simulator (here Synopsys VCS). The first PLI, *monitorDelay* monitors the switching activity across the functional pin (D) of each scan flip-flop and creates a pattern delay profile with the entire path delay distribution of each pattern. The maximum path delay of each pattern is then extracted and the patterns are grouped into subsets with relatively close path delay distribution.

The second PLI, *calculateSCAP* creates the patter power profile and measures the SCAP of each pattern in the VDD and VSS network. The capacitance of each gate instance is extracted from the RC parasitics file (*Standard parasitics exchange format (SPEF)*) generated with an extraction tool (Synopsys STAR-RCXT). It then uses the switching activity and parasitics information to measure the SCAP value ($(\Sigma C_i \times VDD^2)/STW$ [13] [14]) in both VDD and VSS network for each pattern during the launch-to-capture window, where *STW* is the *switching time-frame window*. It is the time span during

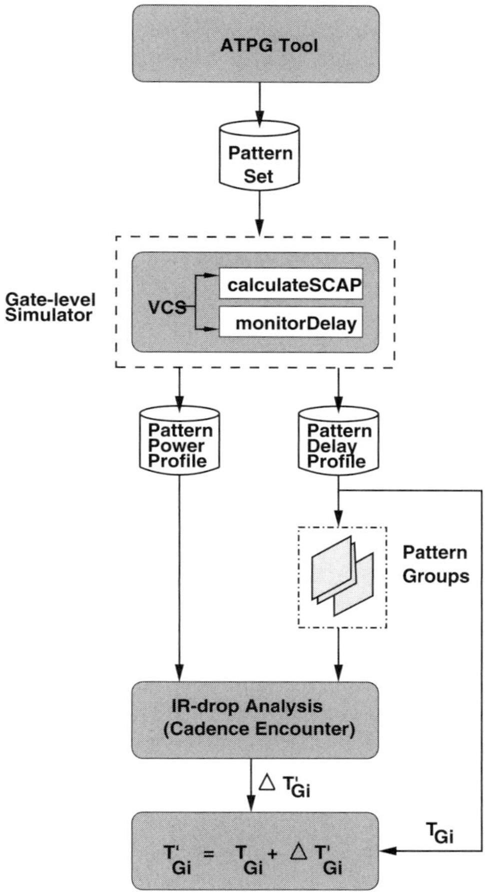

Fig. 8.10. New faster-than-at-speed test pattern application framework.

which the entire switching activity occurs between the two at-speed capture pulses. Figure 8.11 shows the SCAP measured in VDD network in each transition fault test pattern during the launch and capture window. The above procedure using PLI interface avoids large industry standard *value change dump (VCD)* file generation for estimation of switching power. However, the VCD file is still required for IR-drop analysis but it is performed only on one pattern in each group with the highest SCAP value. After determining the pattern in each group with the highest SCAP value, the entire net toggle activity for this particular pattern is captured in VCD file and the worst-case IR-drop analysis is performed. In the next step, the corresponding worst performance degradation ($\Delta T'_{Gi}$) is estimated and the faster-than-at-speed test period for each group is determined as explained in Section 8.4.2.

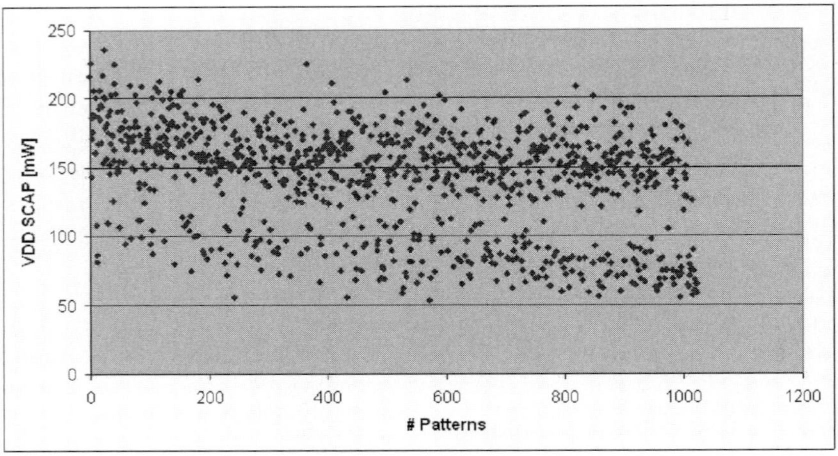

Fig. 8.11. Switching cycle average power (SCAP) in VDD network for each transition fault test pattern.

Table 8.1. Pattern grouping and worst IR-dtop results per patter group.

Group	# Patterns	Worst-delay (T_{Gi})	Worst Avg. (ΔVDD_{IRdrop})	Worst Avg. (ΔVSS_{IRdrop})
G1	54	5618	0.167	0.148
G2	135	5072	0.183	0.160
G3	468	4618	0.235	0.213
G4	321	4118	0.231	0.199
G5	46	3604	0.244	0.239

Table 8.2. Estimated faster-than-at-speed test clock results for different pattern groups.

Group	Worst-delay (T_{Gi})	Worst Performance degradation (ΔT_{Gi})[ps]	Faster-than-at-speed period ($T'_{Gi} = T_{Gi} + \Delta T_{Gi} + \tau_{su}$)
G1	5618	1752	7570
G2	5072	1718	6990
G3	4618	2047	6865
G4	4118	1750	6068
G5	3604	1722	5526

8.5 Experimental Results

The launch-off-capture transition fault pattern set (1024 patterns) was generated using Synopsys Tetramax [4]. As explained in the previous section, the framework divides the entire pattern set into multiple groups. The groups are divided based on the relative closeness of *pattern slack*. Here, the *pattern slack* range was selected as $500ps$ to generate a reasonable number of pattern groups (5 groups in this case). However, this is configurable and decreasing the *pattern slack* range increases the number of pattern groups.

Table 8.1 show the results obtained for each group $G1$ through $G5$ after sorting the pattern set based on maximum path delay of each pattern. The worst-delay (T_{Gi}) of a group is the maximum delay of a pattern in each group (Column 3). To estimate the worst case average IR-drop in each group, a pattern with the highest SCAP value was selected and an IR-drop analysis was performed for a clock period (T_{Gi}) as explained in Section 8.3. Columns 4 and 5 in Table 8.1 show the worst case IR-drop for VDD ($\Delta VDD_{IR-drop}$) and VSS ($\Delta VSS_{IR-drop}$), respectively.

In order to take into account the performance degradation effect due to *faster-than-at-speed IR-drop*, the design timing information at the respective effective voltage ($\Delta V = \Delta VDD_{IR-drop} + \Delta VSS_{IR-drop}$) needs to be generated for each group. For example, in case of group $G1$, the effective voltage is $VDD - \Delta V = 1.8 - (0.167 + 0.148) = 1.48V$. The design timing information at the new operating voltage condition can be generated by two methods. The standard cell library can be characterized at the new operating voltage and timing analysis tool can be used to generate the SDF file. The other method is to take the effective voltage reduction and apply cell delay degradation ($ScaledCellDelay = CellDelay \times (1 + k\_volt * \Delta V)$) to generate the new design timing information, where *k_volt* is a factor specified in the vendor supplied technology library that accounts for non-linear delay scaling model. Here, we used a value of 0.9 for *k_volt*, which means for a 5% effective cell voltage decrease ($\Delta V = 0.1V$), the cell delay increases by 9%. Hence, for group $G1$ with effective voltage reduction of $\Delta V = 0.32V$, due to IR-drop, each cell delay on the longest path will increase by 29% at the faster-than-at-speed test period (T'_{Gi}) compared to no IR-drop at functional clock period. The technique applies the voltage drop in thehighest SCAP pattern to the worst-case slack pattern andtheassumption is that the entireworst-slack path will experience the worst voltage drop.

After generating the design information, the pattern with the maximum delay in the group is simulated to obtain the worst performance degradation due to *faster-than-at-speed IR-drop* (column 3 in Table 8.2). Finally, the resulting worst-case performance degradation is used to determine the *faster-than-at-speed* clock period ($T'_{Gi} = T_{Gi} + \Delta T_{Gi} + \tau_{su}$) for the respective pattern group. Figure 8.12 shows the original rated functional period and *faster-than-at-speed* clock timing for each of the group considering IR-drop effects. It can be noticed that there is extra slack provided by the new technique between the maximum delay of the pattern group and the faster-than-at-speed clock period to take into account the performance degradation due to increase in IR-drop effects.

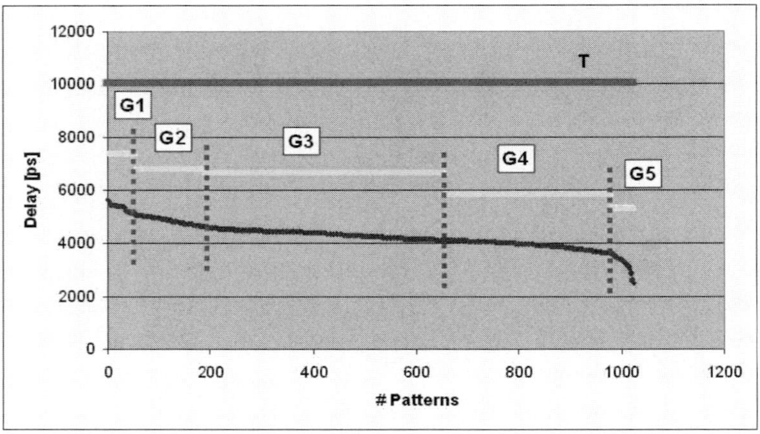

Fig. 8.12. Transition fault pattern groups with their respective faster-than-at-speed clock period.

8.6 Summary

In this chapter, a detailed analysis of faster-than-at-speed techniques utilized for small delay fault detection was presented. The analysis illustrated that the IR-drop is exacerbated during faster-than-at-speed pattern application (upto 16% compared to IR-drop at rated functional speed) and it is important to consider the performance degradation of the design due to increase in IR-drop effects. A new framework for applying transition fault test patterns at faster-than-at-speed was presented considering both the performance degradation due to adverse IR-drop effects and positive slack. The technique groups the pattern set based on their affected maximum delay and determines the worst case performance degradation for each pattern group. This avoids false identification of good chips to be fault due to IR-drop effects rather than small delay defects.

References

1. J. Saxena, K. M. Butler, J. Gatt, R. Raghuraman, S. P. Kumar, S. Basu, D. J. Campbell, J. Berech, "Scan-Based Transition Fault Testing - Implementa-

tion and Low Cost Test Challenges," in Proc. *International Test Conference (ITC'02)*, pp. 1120 - 1129, Oct. 2002.
2. X. Lin, R. Press, J. Rajski, P. Reuter, T. Rinderknecht, B. Swanson and N. Tamarapalli, "High-Frequency, At-Speed Scan Testing," *IEEE Design & Test of Computers*, pp. 17-25, Sep-Oct 2003.
3. B. Kruseman, A. K. Majhi, G. Gronthoud and S. Eichenberger, "On hazard-free patterns for fine-delay fault testing," in Proc. *Int. Test Conf. (ITC'04)*, pp. 213-222, 2004.
4. Synopsys Inc., "User Manuals for SYNOPSYS Toolset Version 2007.03," Synopsys, Inc., 2007.
5. S. Allampally, V. Prasanth, R. K. Tiwari and S. Krishnamurthi, "Small delay defect testing in DSM technologies -Needs and Considerations," in *Synopsys Users Group (SNUG'07) Conference*, India, 2007.
6. Cadence Inc., "Encounter True-time TestATPG", http://www.cadence.com, 2006.
7. H. Hao and E.J. McCluskey, "Very-low-voltage testing for weak CMOS logic ICs," in Proc. *Int. Test Con. (ITC'93)*, pp. 275-284, 1993.
8. R. Foster, "Why Consider Screening, Burn-In, and 100-Percent Testing for Commercial Devices?," *IEEE Transactions on Manufacturing Technology*, vol. 5, no. 3, pp. 52-58, 1976.
9. P. Gupta and M. S. Hsiao, "ALAPTF: A new transition fault model and the ATPG algorithm," in Proc. *Int. Test Conf. (ITC'04)*, pp. 1053-1060, 2004.
10. W. Qiu, J. Wang, D. M. H. Walker, D. Reddy, X. Lu, Z. Li, W. Shi and H. Balichandran, "K Longest Paths Per Gate (KLPG) Test Generation for Scan-Based Sequential Circuits," in Proc. *Int. Test Conf. (ITC'04)*, pp. 223-231, 2004.
11. B.N. Lee, L. C. Wang and M. S. Abadir, "Reducing pattern delay variations for screening frequency dependent defects," in Proc. *VLSI Test Symp. (VTS'05)*, pp. 153-160, 2005.
12. N. Ahmed, M. Tehranipoor and V. Jayaram, "A Novel Framework for Faster-than-at-Speed Delay Test Considering IR-Drop Effects," in Proc. *Int. Conf. on Computer-Aided Design (ICCAD'06)*, pp. 198-203, 2006.
13. N. Ahmed, M. Tehranipoor and V. Jayaram, "Supply Voltage Noise Aware ATPG for Transition Delay Faults," in Proc. *IEEE VLSI Test Symposium (VTS'07)*, 2007.
14. N. Ahmed, M. Tehranipoor and V. Jayaram, "Transition Delay Fault Test Pattern Generation Considering Supply Voltage Noise in a SOC Design," in Proc. *Design Automation Conference (DAC07)*, 2007.
15. Gaisler Research, http://gaisler.com, 2006.
16. Cadence Inc., "User Manuals for Cadence Encounter Tool set Version 2004.10," Cadence, Inc., 2004.
17. http://crete.cadence.com, 0.18μm standard cell GSCLib library version 2.0, Cadence, Inc., 2005.
18. N. Ahmed, C.P. Ravikumar, M. Tehranipoor and J. Plusquellic, "At-Speed Transition Fault Testing With Low Speed Scan Enable," in Proc. *IEEE VLSI Test Symp. (VTS'05)*, pp. 42-47, 2005.
19. B. Benware, C. Schuermyer, N. Tamarapalli, Kun-Han Tsai, S. Ranganathan, R. Madge, J. Rajski and P. Krishnamurthy, "Impact of multiple-detect test patterns on product quality," in Proc. *Int. Test Conf. (ITC'03)*, pp. 1031-1040, 2003.

9

IR-drop Tolerant At-speed Test Pattern Generation

Due to shrinking technology, increasing functional frequency and density, and reduced noise margins with supply voltage scaling, the sensitivity of designs to supply voltage noise is increasing. The supply noise is much larger during at-speed delay test compared to normal circuit operation since large number of transitions occur within a short time frame. Existing commercial ATPG tools do not consider the excessive supply noise that might occur in the design during test pattern generation. This chapter presents two case studies of an ITC'99 benchmark and a SOC design to show detailed IR-drop analysis, measurement and its effects on design performance during at-speed test. Next, a method is presented to measure the average power of at-speed test patterns, referred to as switching cycle average power (SCAP). Using SCAP model provides a cost-effective solution to identify patterns with high IR-drop and avoids expensive dynamic IR-drop analysis. A new practical pattern generation procedure is presented to generate supply noise tolerant delay test patterns using existing capabilities in commercial ATPG tools. The procedure will be implemented on two large designs. The results demonstrate that the new patterns, while slightly larger, will minimize the supply noise effects on path delay.

9.1 Introduction

Shrinking technology along with increase in design density and frequency have posed serious design and test challenges. One important issue of testing today's nanometer high-speed designs is the increasing number of timing-related defects and another issue is power supply noise. Among existing delay test methods, transition delay fault test is widely practiced in industry to target delay-induced defects and is considered as a cost-effective alternative to functional pattern generation [1] [2]. The at-speed launch and capture in addition to large number of switchings in the circuit during transition delay fault testing can cause excessive peak power and result in large IR-drop.

IR-drop effects have become more significant in recent years and needs to be efficiently taken into consideration, as it poses design, test and reliability challenges for the chip manufacturers/foundries. This situation has grown more complicated with reducing supply voltage and the limitation of further reduction of threshold voltage. The reduced voltage difference between the VDD and VSS pins of a standard cell reduces the cells operating performance and may result in chip performance reduction if the cell is on a critical path. The IR-drop also reduces the cell's noise immunity and in some cases may lead to functional failures [3] [4].

In order to simplify the pattern generation process, traditionally ATPGs consider zero delay gate model and target as many faults per pattern as possible in order to reduce the test pattern volume. In other words, operating and manufacturing conditions (e.g. supply noise, temperature, process variations, etc.) are ignored during ATPG. Patterns generated using such ATPGs may cause large number of transitions in the circuit which may not necessarily occur during functional operation. As a result, a design that may *not* have a delay fault may fail a delay test pattern due to excessive IR-drop related effects. Therefore, new pattern generation methods are required to generate test patterns that reliably distinguish between good and bad chips, i.e. the test patterns should not generate excessive supply noise in the design under test. This issue would be even more problematic when testing system-on-a-chip (SOC) designs where different on-chip blocks generate different IR-drops and in some cases the blocks are tested in parallel to reduce test time. The power consumption must be taken into account [5] [6] and the IR-drop performance degradation effects need to be considered during ATPG.

Launch-off-shift [7], launch-off-capture [8], and enhanced scan [9] are three major scan-based techniques proposed for transition delay fault testing. In all the three methods, a pattern pair (V1, V2) is applied to target delay faults but with different launch mechanisms. Pattern V2 for launch-off-shift, launch-off-capture, and enhanced scan is generated using last shift, functional response, and arbitrary using ATPG, respectively. Various techniques have also been proposed to improve the quality of at-speed test by increasing fault coverage and reducing pattern count, avoiding functionally untestable faults, or reducing scan enable design effort [10] [11] [12].

Several approaches have been proposed for power supply noise analysis and estimation in recent years. Some closed-form equations are derived in [4] to calculate simultaneous switching noise. Estimation of ground bounce, caused by the switching in internal circuitry for deep-submicron circuits, using a scaling model is discussed in [13]. Reference [14] proposes a simulated switching circuit model to estimate PSN which includes IR voltage drop and ΔI noise based on an integrated package-level and chip-level power bus mode. Modeling of PSN on distributed on-chip power networks is described in [15]. ATE and neural network are used to find the patterns generating maximum instantaneous current [16]. The test scheduling for SOC designs considering power consumption is discussed in literature [5] [6] to minimize test time while ensuring the SOC test power is lower than functional power threshold.

The issue of overkill during delay test is addressed in [17] and a vector-based approach for power supply noise analysis during test compaction is proposed. A power supply noise model is developed and used during test compaction. The procedure may become slow for large designs since all the patterns are generated without *random-fill* and the power supply noise needs to be estimated in every compaction loop. The method proposed in [18] verifies test vectors for IR-drop failures and identifies failing vectors. The method estimates the average current drawn from power rails and compares it against a pre-defined threshold set by designer. A pattern generation technique is proposed in [19] by building current/voltage libraries to maximize the power supply noise along targeted paths and cause longer propagation delays for the nodes along the paths. The computation complexity of the pattern generation procedure is high since it targets one pattern at a time. Authors in [20][21] propose a low-capture power, X-filling pattern generation method. The proposed method reduces the switching activity in scan flip-flops for launch-off-capture patterns.

9.1.1 Overview of the IR-drop Tolerant Pattern Generation Method

In this chapter, a new method is presented to measure the average power during at-speed test (during fast launch-to-capture cycle), referred to as *switching cycle average power (SCAP)*. The method considers both length of the paths affected by each pattern and number of transitions occurred during the switching time frame window as opposed to calculating switching power for entire clock cycle in statistical approach. A pattern generation procedure taking supply voltage noise into account is presented ensuring that the supply noise will always be lower than defined threshold. Note that, in this chapter, the shift IR-drop is not addressed as lower frequencies are used during test pattern shift and is of less concern for test engineers compared to at-speed launch and capture IR-drop. The pattern generation procedure uses existing commercial ATPG tools and can be easily adopted in current DFT flows. The results show that the new pattern set generated for two large designs using the IR-drop tolerant pattern generation procedure significantly reduces the IR-drop and minimizes the performance degradation.

9.2 Case Study 1: ITC'99 Benchmark b19

In this section, the physical design implementation and statistical IR-drop analysis are described. A detailed dynamic IR-drop analysis for two types of patterns is presented with different path delay distribution. Also, a new power model is explained to measure the average power of at-speed test patterns which takes both the switching activity and the pattern path delay distribution into account.

9.2.1 Physical Design Implementation

The physical design synthesis for the ITC'99 benchmark *b*19 was performed using Cadence SOC Encounter place and route tool [27]. The design contains almost 219K gates, 51 IO pads, and about 6,642 flip-flops. Scan-based test insertion was performed using Synopsys DFT Compiler [26] with eight scan chains and a slow speed scan enable is used for launch-off-capture transition fault test. During physical design, the design is timing closed for an operating frequency of 142MHz at nominal operating voltage ($1.8V$) and temperature (25^oC) conditions. A slow scan shift speed of 10MHz was used. It is implemented in $180nm$ standard cell library [28]. Note that at this point of experiments, no decoupling capacitances were inserted.

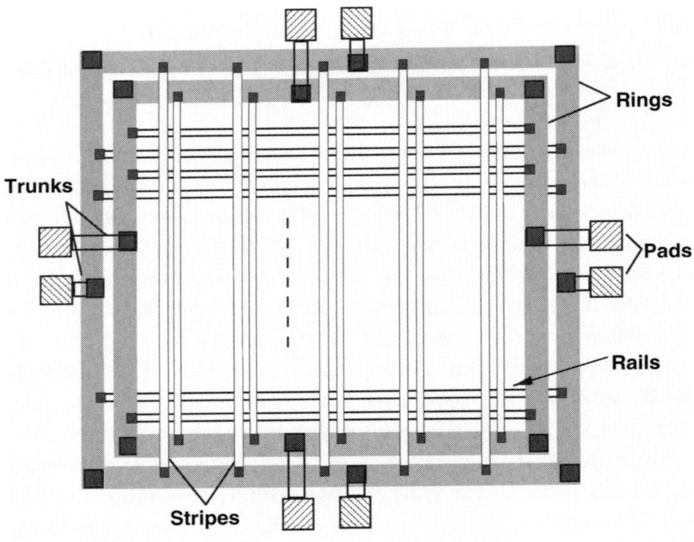

Fig. 9.1. Power/Ground Distribution Network [23].

The power-planning for the design was performed assuming a net toggle probability of 20% during functional operation. Figure 9.1 shows the power/ground distribution network of the chip. Power rings ($width = 20\mu m$) were created using higher-level routing layers (Metal5 and Metal6) and carry power around the standard cell core area. Four power (VDD) and ground (VSS) pads each were inserted and connected to the respective rings with wires referred to as trunks. After creating the power rings, power and ground is routed to the standard cells using stripes and rails. The stripes ($width = 10\mu m$) were created using routing layer Metal4 and a distance of $100\mu m$ between adjacent stripes connecting power rings. The design was then placed and routed along with clock-tree synthesis and scan cell ordering to minimize scan chain wirelength.

9.2.2 Statistical IR-drop Analysis

In order to determine an estimate of functional IR-drop, the design net parasitics (resistance and capacitance) were extracted using Synopsys STAR-RCXT [26] extraction tool. The average statistical IR-drop using vector-less approach was measured for both VDD and VSS nets considering 20% net toggle probability during functional operation. The results showed 2.8% voltage drop in VDD and a voltage bounce of 4.5% for the VSS net, which can be considered negligible. However, such an analysis provides an underestimation of both average and peak IR-drop even during functional operation. This is because the tool considers the probability of net toggle activity over the entire cycle period (vector-less approach). However, to measure IR-drop more accurately during test a vector-based IR-drop method must be devised to consider both the average time frame window for each pattern during which the entire switching occurs and the simultaneous switching.

To measure the average IR-drop experienced by the transitions, it is important to estimate the average switching time frame. The time span during which all the transitions occur is referred to as the *switching time frame window (STW)*. For a transition fault pattern, the maximum path length affected determines this time frame. Note that for different test vectors, the longest path exercised will be different. From previous experiments on the same design during transition fault test patterns, it has been observed that an *average switching time frame window* is close to half the clock cycle period which is mainly because the ATPG tools tend to detect delay faults through short paths. This shows that the actual average functional power surge observed during an average switching time frame for a pattern is almost twice of the measured value during one cycle period. This observation shows that if N flip-flops or nets toggle during one capture event, the same N flip-flops will toggle irrespective of the frequency. The effect on IR-drop will now be more pronounced if the toggle happens in a smaller window, thus tying IR-drop effects to the path delays affected by test patterns.

Table 9.1 shows the statistically measured average power consumption for the entire cycle period *(Case1)* and average switching time frame *(Case2)* for half cycle period. It also shows the average IR-drop reported for the two cases measured using Cadence SOC Encounter tool. It can be noticed that the average IR-drop is almost doubled when the switching time frame window is reduced to half of the cycle period. Although, this might appear over pessimistic but it provides a good estimate of the IR-drop which the design will experience during functional operation. Also, *Case2* provides an average *power threshold* that can be used to identify high toggle activity transition fault test patterns at a later stage.

Table 9.1. Statistical functional IR-drop analysis results for ITC'99 benchmark (*b19*) [23].

	Avg. Switching Power [mW]	*Avg. IR-drop [V]*	
		VDD	VSS
Case1 (Full cycle period)	96.3	0.05	0.084
Case2 (Half cycle period)	190.6	0.11	0.162

9.2.3 Dynamic IR-drop Analysis

The actual IR-drop during transition fault test patterns is much higher compared to statistical IR-drop due to very high switching activity and smaller switching time frame. In this section, analysis of IR-drop effects is presented for two types of transition fault test patterns (*P1 and P2*) with almost the same switching activity but with different switching time frame windows. The maximum path delay for patterns *P1* and *P2* are $t_{d(P_1)} = 4854ps$ and $t_{d(P_2)} = 2313ps$, respectively and the clock period being $T = 7000ps$. This implies that pattern *P2* exercises more number of short paths (that is, larger number of simultaneous switching) which increases the probability of higher IR-drop.

Figure 9.2 shows the clock timing waveform used for dynamic IR-drop analysis. The clock insertion delay is the time taken by the clock signal from the chip periphery to all the registers in the design and it is represented as t_i. For clarification, the clock edge uncertainty is not shown which is also included in t_i. The entire clock network switching activity for each clock edge occurs in this time frame. Since, the clock switching power forms a major component in the total power drawn, it is also included in the power measurement. If the switching time frame window is slightly less than half the cycle period (see $t_{d(P_2)}$ in Figure 9.2), the negative clock edge switching activity is also considered. This is reasonable as the clock network already starts to switch close to the end of the switching window ($t_{d(P_2)}$), assuming clock uncertainty and process variations. In general, the switching time frame window for dynamic IR-drop analysis is measured using the following formulation:

$$STW_i = \begin{cases} t_i + T/2 & \text{if } t_{d(P_i)} \leq T/2 \\ t_i + t_{d(P_i)} & \text{if } t_{d(P_i)} > T/2 \end{cases}$$

To measure the IR-drop of the pattern, the switching activity inside the circuit was captured in the standard value change dump (VCD) format during

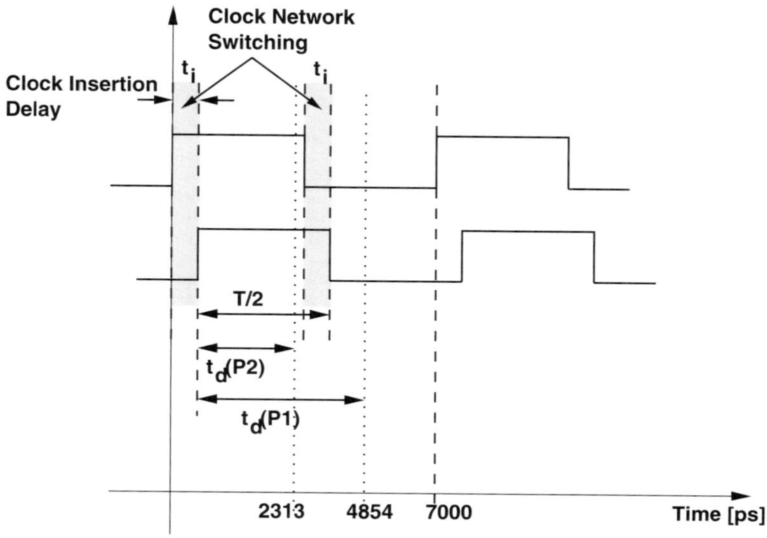

Fig. 9.2. Clock timing waveform used for dynamic IR-drop analysis [23].

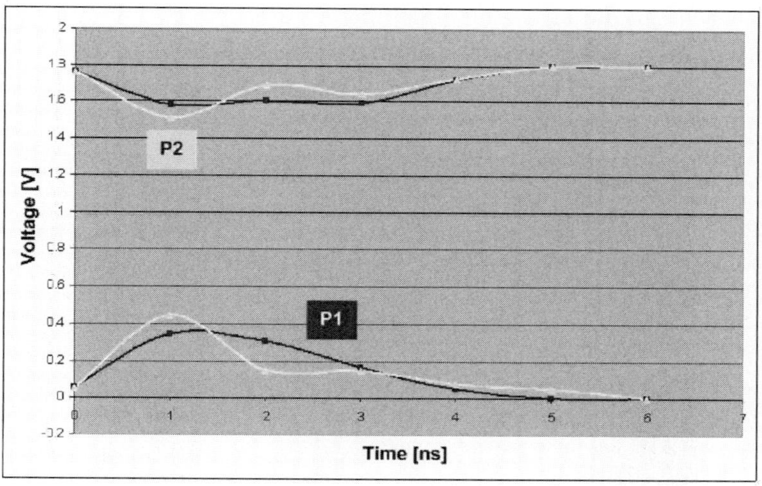

Fig. 9.3. IR-drop effects on VDD and VSS during pattern *P1* and *P2* application within 7 *ns* capture window [23].

gate-level timing simulation. The timing information of the gates and the extracted parasitic interconnect delay information was back-annotated using the standard delay format (SDF) file. The switching activity information (VCD file) along with physical design and technology library information is used by SOC Encounter tool [27] to estimate the dynamic IR-drop of the pattern. Figure 9.3 shows the VDD (VSS) voltage waveforms during the at-speed launch and capture cycles for pattern *P1* and *P2*. To measure the IR-drop, the launch-to-capture window $(7ns) + t_i$ was split into $1ns$ time frames and average IR-drop was measured in each time frame. It can be noticed that the effect of IR-drop is maximum in the beginning of the clock cycle due to high simultaneous switching activity and gradually decreases. Also, the effect of IR-drop is maximum in pattern *P2* as high switching activity occurs in a smaller switching time frame window.

Figures 9.4 and 9.5 show the average IR-drop plots on the ground (VSS) network for patterns *P1* and *P2* during their respective switching time frame windows. The IR-drop plots were obtained from the Cadence SOC Encounter tool [27] measured across the respective switching time frame for each pattern. Note that, for pattern P2, the IR-drop in a large portion of the chip increases which results in reduced effective voltage difference between the VDD and VSS ports observed by each gate in that region. This might result in higher performance degradation or functional failure of the circuit due to excessive noise.

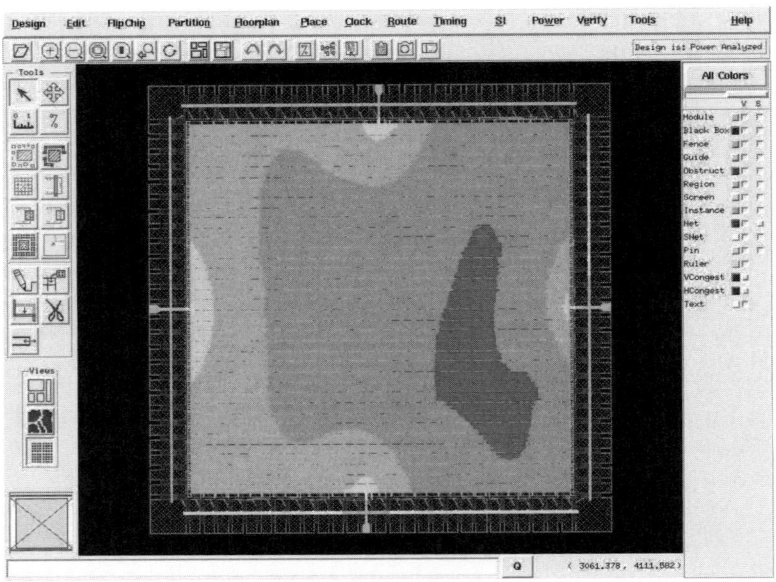

Fig. 9.4. IR-drop plot in VSS net for pattern *P1* [23].

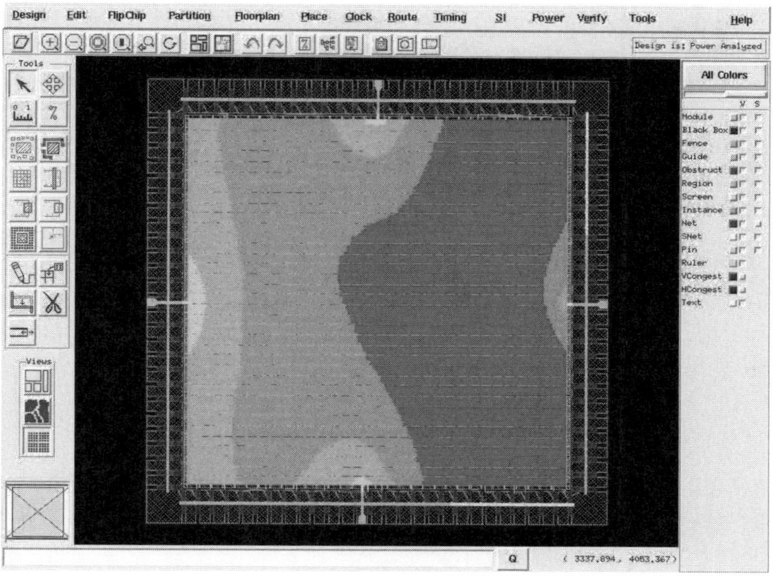

Fig. 9.5. IR-drop plot in VSS net for pattern *P2* [23].

9.2.4 Average Power Model

As it was explained in the previous sections, the IR-drop depends on the total switching capacitance and the time frame window during which it occurs. Since, dynamic IR-drop analysis for each delay test pattern is prohibitively expensive, a model is required to identify test patterns which have a high probability of failure due to IR-drop effects, during their application. The *cycle average power (CAP)* [25] is defined as the average power consumed during a single tester cycle. However, it does not factor in the varying time frame window of the entire switching activity for each pattern. Therefore, a pattern with relatively lesser switching activity but with a very short switching time frame window will not be considered as a potential pattern of IR-drop failure by the CAP power model. Therefore, a new term referred to as *switching cycle average power (SCAP)* is defined which is the average power consumed by the test pattern during the time frame of the entire switching activity (STW). CAP and SCAP are calculated by:

$$CAP = (\Sigma C_i \times VDD^2)/T$$
$$SCAP = (\Sigma C_i \times VDD^2)/STW$$

where C_i is the output gate capacitance of gate G_i. Table 9.2 shows the comparison of average power and IR-drop analysis of pattern *P2* using the CAP and SCAP model. It can be noticed that the power surge during the switching time frame window (SCAP) is $1.3X$ higher compared to the cycle average power. Also, the average IR-drop using CAP model ($0.16V$) reported is within the V_{min} operating conditions for which the IR-drop effect will not be of concern. In general, during sign-off, the design is made sure to work under V_{min} and V_{max} operating conditions. However, with the SCAP model, the average IR-drop experience by the design on VSS network ($0.216V$) during the switching interval exceeds it by 34 %.

Table 9.2. Average dynamic power/IR-drop analysis results of a pattern for CAP and SCAP model [23].

	Avg. Switching Power [mW]	*Avg. IR-drop [V]*	
		VDD	VSS
CAP	163	0.120	0.161
SCAP	211	0.136	0.216

9.2.5 Pattern Generation Framework

As explained in Section 9.2, the switching cycle average power provides a more practical measure to identify patterns with very high IR-drop effects bypassing the expensive dynamic IR-drop analysis per pattern. Since, the transition fault pattern set has varying path delays and switching activity, the pattern generation problem can be divided into two sub-problems: *A*) to measure the switching cycle average power (SCAP) for each pattern and *B*) to generate a new pattern set ensuring that the IR-drop will remain under a pre-defined threshold. Both will be discussed in the following sub-sections.

SCAP Calculator

To determine the SCAP of each pattern in the transition fault pattern set, the following information is needed: 1) the gates switching inside the circuit, 2) output capacitance of each gate and 3) the switching time frame window. Simulation-based techniques can be used to capture the switching activity information in the standard *value change dump (VCD)* format. But, this technique is sufficient only to analyze a very small number of patterns due to the extremely large size of VCD files for large designs.

To overcome this problem, programming language interface (PLI) routines are used during gate-level verilog simulation. The PLI provides a standard interface to the internal data representation of the design during simulation. Figure 9.6 shows the SCAP calculation flow. The capacitance per each gate instance is extracted from the RC parasitics file (*Standard parasitics exchange format (SPEF)*) generated using Synopsys STAR-RCXT extraction tool. A

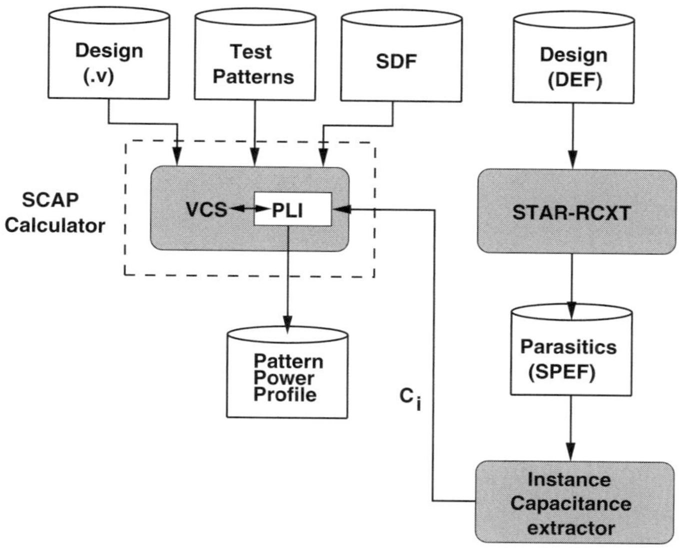

Fig. 9.6. SCAP calculator using Synopsys VCS simulator [23].

PLI have been developed which can be plugged into Synopsys VCS gate level simulator which acts as SCAP calculator during simulation. It reports the SCAP value for VDD and VSS power network for each pattern during the launch-to-capture window in the launch-off-capture transition fault pattern set. The above procedure using PLI interface avoids the VCD file generation for estimation of switching power.

Figure 9.7 shows the SCAP value for VDD and VSS network in each transition fault test pattern during the launch and capture functional cycles of the launch-off-capture pattern. It can be noticed that the initial patterns have higher switching cycle average power and then very gradually reduces. This is because the initial patterns detect most of the transition faults (exercise more paths) and the later patterns target the hard-to-detect faults.

Pattern Generation

In the first phase, the launch-off-capture transition fault pattern set (2630 patterns) were generated using Synopsys TetraMax [26] in the conventional manner with random fill of don't-care bits during ATPG. Figure 9.8 shows the new pattern generation framework which uses existing commercial ATPG tools. In the pre-processing phase, statistical IR-drop analysis is performed to set a threshold for the SCAP value. Also, during pattern generation, fault list manipulation techniques are applied to reduce switching activity which will be explained in detail during the SOC case study in Section 9.3.

The generated patterns are simulated using a gate-level simulator and the SCAP value for each pattern is measured as explained in Section 9.2.5.

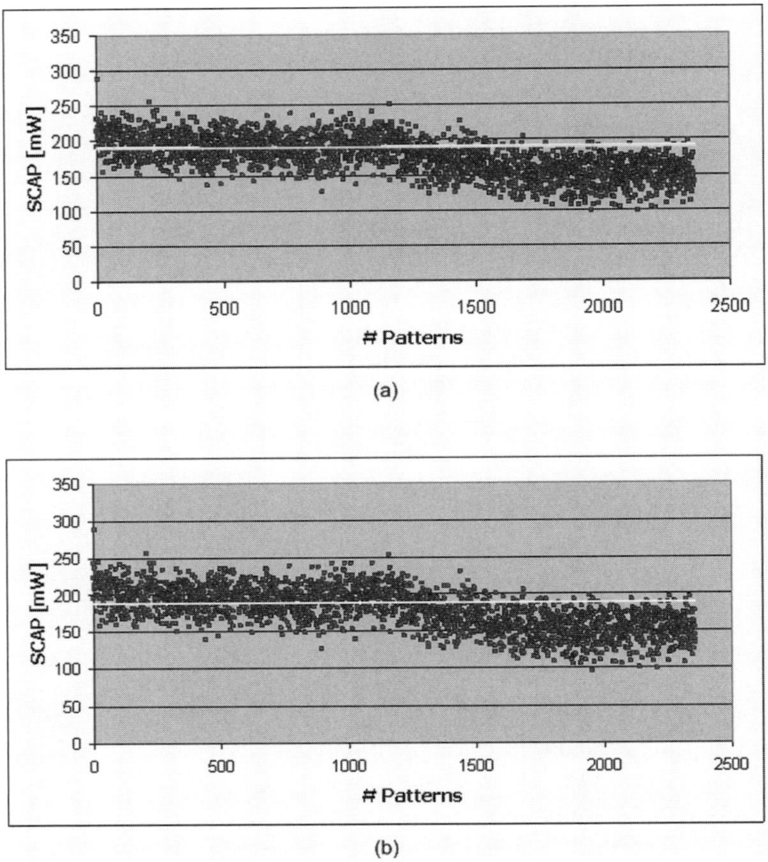

Fig. 9.7. Switching cycle average power (SCAP) measured for each transition fault test pattern (benchmark b19) (a) VDD network and (b) VSS network [23].

After obtaining the power pattern profile, it is important to set a threshold to shortlist the patterns with very high SCAP value which relates to a lower tolerance to IR-drop effects. For this, the average functional switching power estimated using statistical-based approach is used as explained in Section 9.2.2 for the SCAP threshold as shown in Table 9.1. A 20% toggle activity over an average switching time frame of half the clock cycle period is assumed. This number is usually defined by designer during power network synthesis. The remaining patterns are called *IR-drop tolerant patterns.*

Based on the functional operation SCAP threshold for both the VDD and VSS network, the short-listed patterns are fault simulated using the Synopsys TetraMax tool to obtain the list of extra faults detected on top of the remaining IR-drop tolerant patterns. In the next phase of pattern generation, new patterns are generated for these set of faults with *adjacent-fill* or *fill-X* ($X = 0$ or 1) options. Adjacent fill causes don't-care scan cells to be filled with

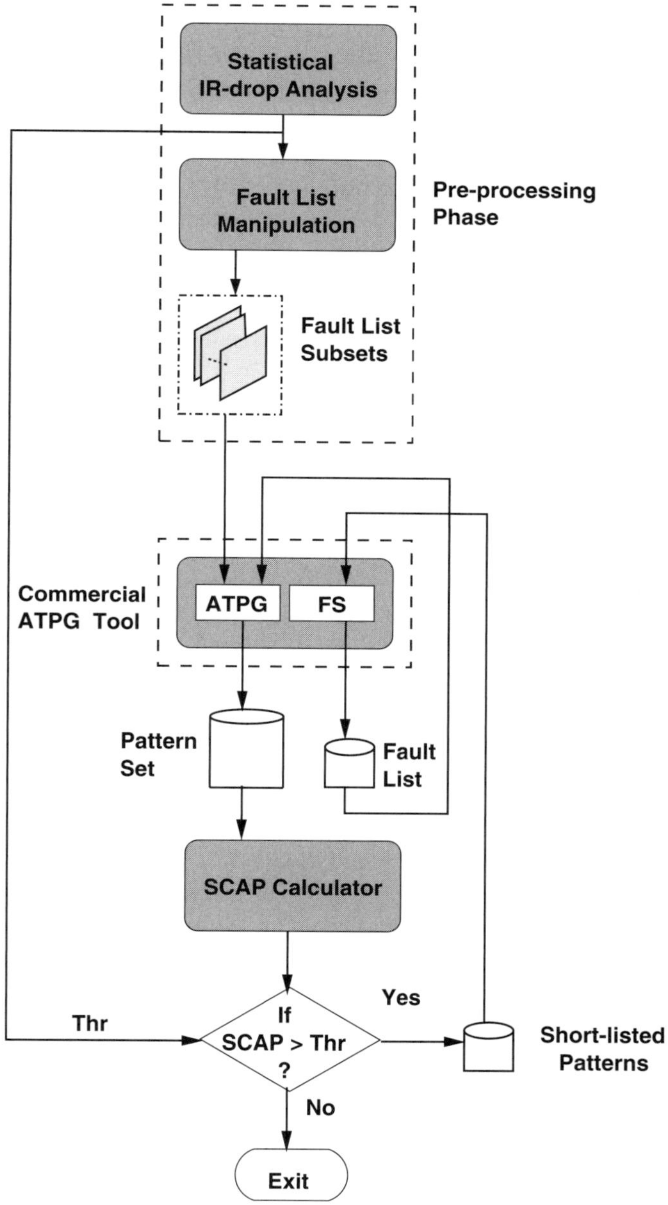

Fig. 9.8. New pattern generation framework [23].

the value of the first downstream scan cell with a defined/care value (i.e. it creates runs of 0's and 1's in the test pattern). This option in the ATPG tool is useful to minimize power usage as it significantly increases the correlation between the initialization and launch patterns. Although, this reduced signal switching comes at the expense of slightly higher pattern count.

9.2.6 Experimental Results

The launch-off-capture transition fault test patterns (2360 patterns) were generated using conventional random fill method. As shown in Figure 9.7, the SCAP threshold was set to $190.6mW$ based on the average power measured in Table 9.1 for 20% toggle activity during functional operation for a switching time frame window of half the clock cycle. Based on this threshold, approximately 860 patterns were short listed with high SCAP value compared to the threshold. In the new ATPG, first the IR-drop tolerant patterns are fault simulated followed by the short listed patterns and it was observed that they contribute almost 10% of test coverage to the total coverage over the IR-drop tolerant test patterns.

In the next step, the patterns are generated using the don't-care fill options built in Synopsys TetraMax for low power patterns. Note that other previously proposed low power pattern generation techniques could also be used. The patterns were generated with three different fill options: *Case1: fill-0*, causes all don't-care scan cells to be filled with 0's, *Case2: fill-1*, causes all don't-care scan cells to be filled with 1's and *Case3: fill-adjacent*, causes don't-care scan cells to be filled with the value of the first adjacent scan cell with a defined/care value. *Case3* is mostly useful to minimize power usage during scan shifting by reducing signal switching at the expense of higher pattern count. However, in the experiments the goal is to reduce the switching activity between the launch and capture window of the launch-off-capture patterns.

Figure 9.9 shows the switching cycle average power for the additional patterns generated in the three different cases. The number of patterns generated were 939, 957 and 900 in *Case1*, *Case2* and *Case3*, respectively. It can be noticed that *fill-0* provided the best results with almost all the patterns below the threshold. Figure 9.10 shows the test coverage curve for the two pattern generation methods. It can be seen that the new technique generates approximately 957 patterns (*fill-0*) for the 10% coverage of the short-listed patterns, which is a very slight increase in the number of patterns (approximately 97 additional patterns). Therefore, the increase in test time is not very significant with almost all of the patterns in the new pattern set within the threshold limit.

9.3 Case Study 2: Cadence SOC Design 'Turbo-Eagle'

In this case study, experimention is done with an industrial-strength SOC design (Turbo-Eagle). Table 9.3 shows the design characteristics. It is a dual-

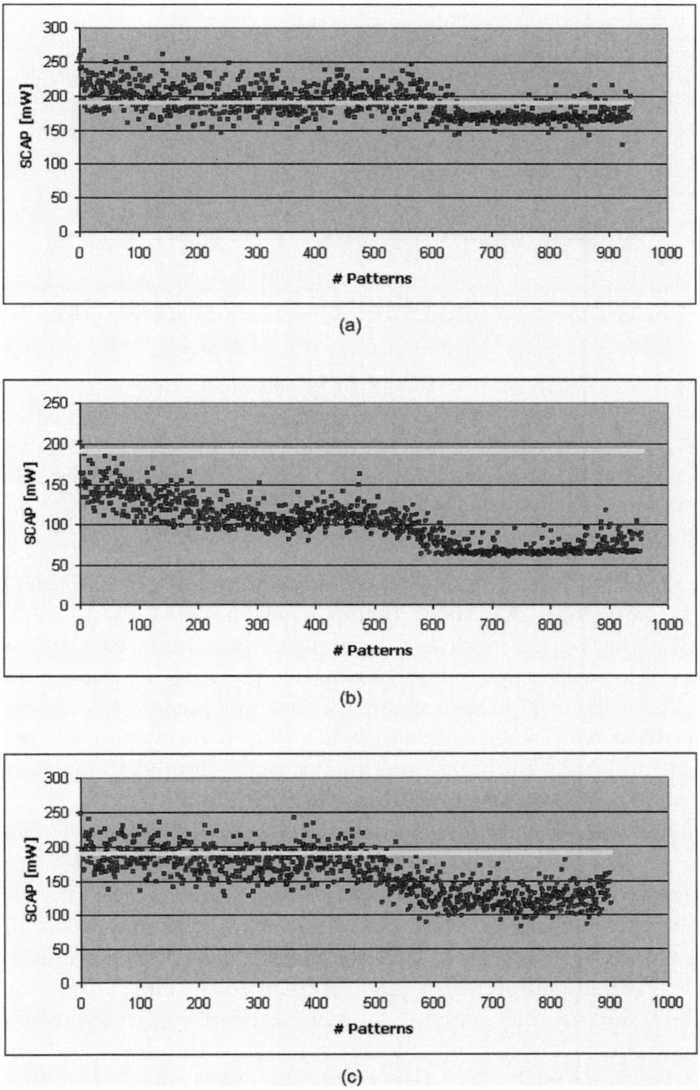

Fig. 9.9. Switching cycle average power (SCAP) in VDD network for the low switching activity test patterns generated in three cases (benchmark b19): (a) *fill-1*, (b) *fill-0* and (c) *fill-adjacent* [23].

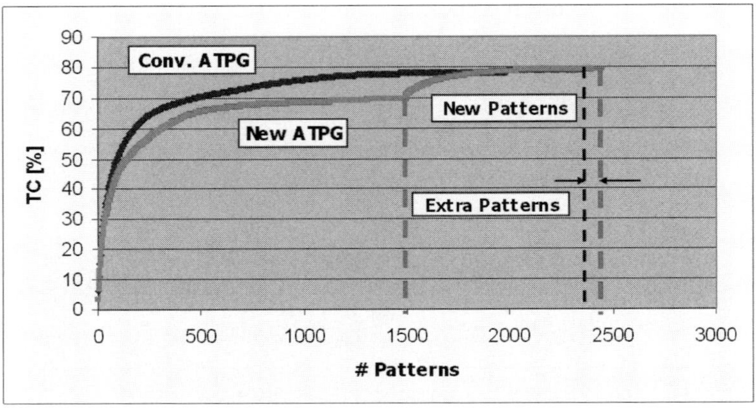

Fig. 9.10. Test coverage curves for conventional ATPG and the new supply aware ATPG results for benchmark b19 [23].

processor SOC [28] and contains a host of peripherals like USB, HDD, VGA and DMA controllers. All these blocks are connected by the AMBA bus from ARM. The design-for-test was implemented hierarchically using full-scan methodology (Synopsys DFT Compiler [26]) with approximately $23K$ scan cells stitched into 16 scan chains. There are six internal clock domains and 22 negative edge scan cells which are placed on a separate scan chain. During test mode, the bi-directional pins were configured to operate in input mode to avoid any congestion problems.

The physical design implementation was performed using Cadence SOC Encounter place and route tool [27]. The maximum frequency of the design is $100MHz$ for the master processor. However, a slow scan shift frequency of 10MHz was used during test pattern application. The design was implemented in $180nm$ standard cell library [28] and timing closed at nominal operating voltage ($1.8V$) and temperature (25^oC) conditions. The design contains 37 power (VDD) and ground (VSS) pads each inserted uniformly around the entire chip periphery. The placement and routing of the design was performed hierarchically along with clock-tree synthesis and scan cell ordering to minimize scan chain wirelength. Finally, the empty spaces in the design were inserted with filler cells and metal fill was performed to increase the density of the metal layers, which makes the topology of the layers more uniform. Figure 9.11 shows the entire chip floorplan with six major sub-blocks *B1* through *B6*.

Table 9.3. Design Characteristics [24].

Clock Domains	6
Scan Chains	16
Total Scan Flops	23420
Negative Edge Scan Flops	22
Transition Delay Faults	9651568

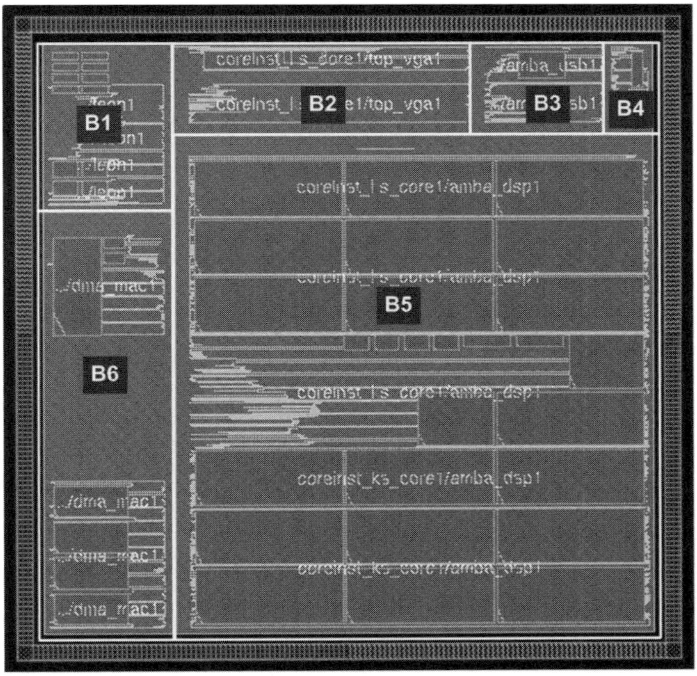

Fig. 9.11. SOC floorplan [24].

9.3.1 Test Strategy using Statistical IR-drop Analysis

The IR-drop experienced depends on a number of design specific factors such as the power and ground network, number of VDD/VSS pads and their placement, package type and the placement and size of decoupling capacitance. It is very difficult to take all these factors into account during test pattern generation step. However, the physical design characteristics cannot be ignored and simply reducing the switching activity does not necessarily reduce IR-drop effects. Therefore, the results of the vector-less statistical IR-drop analysis step performed by the physical design engineer for design validation and sign-off are utilized. This comprehensively captures all the design specific parameters and there is no extra effort by the test engineer.

Although, statistical IR-drop analysis is a very simple technique but it gets complicated to obtain a good IR-drop estimate with several clock domains and blocks in a SOC design. For a given functional toggle activity, the objective is to identify design regions which consume more power and are likely to observe higher IR-drop. Later, it is taken into consideration during test pattern generation to avoid such hot spot regions. For a SOC design, the process is simplified by concentrating at block level rather than dividing the power grid into smaller regions. Also, this is reasonable as different blocks in a SOC might have independent power and voltage domains.

Transition fault test patterns are mostly generated per clock domain. A clock domain with a very high number of controlled scan flip-flops is likely to create more switching activity. Therefore, firstly the number of scan flip-flops are split in each individual clock domain and identify the dominant clock domains. A clock domain with very high number of controlled scan flip-flops is referred to as the dominant clock domain. Table 9.4 shows the number of scan flip-flops in each of the six clock domains. It can be noticed that *clkA* clock domain is the dominant clock domain with approximately $17K$ scan flip-flops. After identifying the dominant clock domains, in the next step, the blocks consuming high power in each dominant clock domain are identified (in this design only for *clkA* domain) in order to reduce the switching activity in them during test pattern generation.

Table 9.4. Clock Domain Analysis [24].

Clock Domain	*#Scan Cells*	*Frequency [MHz]*	*Blocks Covered*
clkA	17966	50	B1 - B6
clkB	1165	100	B1
clkC	1673	50	B3
clkD	724	25	B6
clkE	1560	25	B6
clkF	142	25	B2

Apart from the design parameters, there are two pattern dependent parameters, which determine the IR-drop effects: 1) switching activity and 2) time frame in which it occurs. However, to estimate the average functional power and IR-drop statistically (vector-less approach), in general, designers assume a 20% net toggle activity occurring during one functional clock period. But here, the analysis is performed for a greater toggle activity rate (30%) as later this average switching power threshold will be used to identify delay test patterns with high IR-drop. The reason for such a pessimistic analysis is because the switching activity during test is far greater and non-uniform than during functional operation and the average switching power threshold limit determines the number of patterns generated. The lower the threshold set implies less transition faults detected by each patten and greater number of delay test patterns generated during ATPG process.

The statistical IR-drop estimation tools assume uniform toggle activity over the entire design region and measures average IR-drop and power over a full cycle period. This might never occur during real time or test pattern application and not only non-uniform distribution of switching activity and varying switching time frame window is observed but also most of the switchings occur during the early clock cycle period. Therefore, the IR-drop analysis considered over the entire cycle underestimates the average IR-drop during functional operation. The time span during which all the transitions occur is referred to as the *switching time frame window (STW)*. For a transition fault pattern, the maximum path length affected determines this timing window. Note that for different test vectors, the longest path exercised will be different. From previous experiments on ITC'99 benchmark design (*b19*) during transition fault test patterns [22], an *average switching time frame window* close to half the clock period has been observed. This shows that the actual average power surge observed during an average switching time frame for a pattern is almost twice of the measured value during one full cycle period.

Table 9.5. Statistical functional IR-drop analysis results for each block in SOC [24].

	Case1 (Full cycle period)			*Case2 (Half cycle period)*		
	Avg. Switching Power [mW]	*Worst Avg. IR-drop [V]*		*Avg. Switching Power [mW]*	*Worst Avg. IR-drop [V]*	
		VDD	VSS		VDD	VSS
B1	20.8	0.033	0.033	30.6	0.034	0.034
B2	34.5	0.035	0.036	87.2	0.043	0.044
B3	12.9	0.028	0.028	17.6	0.029	0.029
B4	4.8	0.019	0.019	9.3	0.020	0.020
B5	108.6	0.076	0.076	204.9	0.119	0.120
B6	63.8	0.045	0.045	114.6	0.051	0.050
Chip	265.2	0.077	0.077	404.5	0.126	0.125

Table 9.5 shows the statistical IR-drop analysis results for the entire cycle period *(Case1)* and average switching time frame *(Case2)*. Also, it shows the average power consumption reported for the two cases measured using Cadence SOC Encounter tool. Two important observations can be derived from this table: 1) the average switching power in all blocks is almost doubled when the switching time frame window is halved, but not the worst average IR-drop. It is because most of the blocks *B1*, *B2*, *B3*, *B4* and *B6* are smaller and closer to the chip periphery and therefore, the logic in these blocks observe relatively lower IR-drop even though the switching time frame window is reduced. 2)

Blocks *B5* and *B6* consume most of the switching power and observes high IR-drop when the switching time frame window was reduced. This shows that more focus is needed on reducing the switching activity in block *B5* and *B6* during Pattern generation to avoid IR-drop problems. Although, the above analysis (*Case2*) might appear over pessimistic but it provides a good estimate of the average IR-drop and identifies the blocks in the design which will experience higher IR-drop during both functional and delay test pattern application. Also, *Case2* provides an average functional power threshold that can be used to identify high toggle activity transition fault test patterns during final pattern validation step.

9.3.2 Switching Cycle Average Power (SCAP) Model

CAP and SCAP are calculated by the same equations shown in previous section:

$$CAP = (\Sigma C_i \times VDD^2)/T$$

$$SCAP = (\Sigma C_i \times VDD^2)/STW$$

where C_i is the output gate capacitance of gate G_i. Table 9.6 compares the average power and IR-drop analysis results over the entire chip for a delay test pattern exercising the dominant clock domain *clkA* using the CAP and SCAP model. The STW for this pattern and the clock period were $8.34ns$ and $20ns$, respectively. It can be noticed that the power surge using SCAP model is more than $2X$ compared to CAP model. Also, the worst average IR-drop on *VDD* and *VSS* reported using CAP model (0.128 V and 0.134 V respectively) were within the V_{min} $(0.9 \times VDD = 0.18V)$ operating conditions for which the IR-drop effect will not be of concern. In general, during sign-off, the design is made sure to work under V_{min} and V_{max} operating conditions. However, for the same pattern with the SCAP model, the worst average IR-drop experienced by the design on *VDD* and *VSS* network is doubled. This shows the importance of STW and therefore, SCAP is a better model to identify high power dissipating patterns.

Table 9.6. Average dynamic power/IR-drop analysis results of a pattern for CAP and SCAP model [24].

	Avg. Switching Power [mW]		*Worst Avg. IR-drop [V]*	
	VDD	VSS	VDD	VSS
CAP	206.3	224.4	0.128	0.134
SCAP	457.5	451.6	0.272	0.275

9.3.3 Fault List Manipulation and Pattern Generation

In this section, the issue of fault list manipulation is briefly discussed which is important for the IR-drop tolerant pattern generation procedure. The switching activity of test patterns is very high compared to functional operation due to two main reasons: 1) the ATPG tool tries to reduce the test pattern volume by increasing the number of faults detected by each pattern and 2) random-fill of scan cells (don't-care bits) which do not have any effect on the fault activation or propagation. The initial set of patterns generated by the ATPG tool are highly compacted and detect a very large set of faults, also referred to as easy-to-detect faults which require very less activation and propagation effort. While the subsequent generated patterns observe a very high percentage of don't-care bits as they target hard-to-detect faults and they are randomly filled by the ATPG tool to increase fortuitous detection of un-modeled faults. Therefore, to reduce the switching activity of test patterns, not only the number of faults targeted by each pattern is to be limited but also the don't-care bits need to be filled carefully.

Figure 9.12 shows the fault coverage curve and the care-bits per pattern. The launch-off-capture transition fault test patterns were generated for clock domain *clkA* with an option to fill the scan cells not effecting the transition launch and propagation at the target fault sites with don't-care value (*'X'*) in each pattern. This reduces the dynamic pattern compaction capability of the tool and it generates more number of patterns (9100 patterns) compared to random-fill (5846 patterns). It can be observed that the percentage of care-bits in the initial patterns starts at approx. 18% and decreases sharply. For most of the patterns the care-bit percentage is less than 5%, reaching less than 1% for the last patterns. This shows that a large portion of scan flip-flops are don't-cares and randomly filled by the ATPG tool, thereby increasing the switching activity inside the circuit.

The ATPG tool tries to detect most active (un-detected) faults per pattern and if faults in all blocks are targeted simultaneously, the initial set of generated patterns will observe very high switching activity. Since, there is no option in ATPG tools to limit the maximum number of faults targeted by a pattern, a fault list manipulation technique was applied. The workaround is to provide the ATPG tool with target faults only in a subset of blocks. For example, faults in blocks *B1*, *B2*, *B3* and *B4* can be targeted simultaneously as they observed the least IR-drop (see Table 9.5). In this case, the generated patterns will have very high don't-care bits in the remaining blocks *B5* and *B6* and appropriate fill options can be applied, thereby reducing power dissipation in them. Similarly, faults only in block *B5* and *B6* are targeted respectively, to reduce the switching activity. In other words, during pattern generation, the ATPG tool observes only a limited set of active faults at any point of time. However, this will increase the number of patterns but this procedure is only applied for the dominant clock domains (*clkA*).

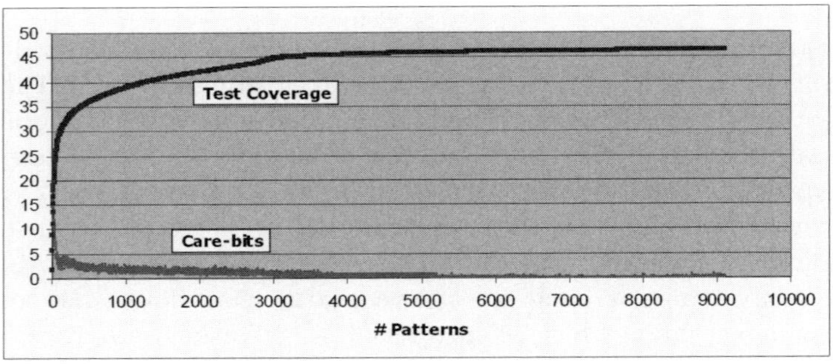

Fig. 9.12. Test coverage and care-bits per pattern for conventional ATPG [24].

9.3.4 Experimental Results

Launch-off-capture transition fault test patterns were generated per clock domain with Synopsys TetraMax [26] tool. During ATPG, the primary inputs were held constant and primary outputs not measured due to low-cost tester speed limitations. Two separate pattern sets were generated, one using conventional random-fill for don't-care scan cells and the new pattern generation procedure using fault list manipulation technique described in Section 9.2.5.

Figure 9.13 shows the test coverage curves for the two methods. For the dominant clock domain *clkA* which was the major concern, 5846 delay test patterns were generated using conventional method. It can be noticed that in the new technique for *clkA* domain, the ATPG process was divided into three steps: *Step1:* In this step the ATPG tool was provided with fault list for blocks *B1* through *B4* only, *Step2:* fault set in block *B6* alone targeted and *Step3:* fault set in block *B5* only is targeted. Finally, the patterns generated in these three steps were fault simulated and top-off patterns were generated for any un-detected faults between the blocks. In each step, *fill-0* option was used for don't-care cells. Different don't-care fill options, such as *fill-0(1)* have been tried which causes all don't-care scan cells to be filled with 0(1)'s and *fill-adjacent* that fills don't-care scan cells with the value of the first adjacent scan cell with a defined/care value. But, here only the results of *fill-0* option are presented which provided the best results. For *clkA*, the new technique generated slightly higher number of patterns (644 extra patterns) compared to conventional random-fill pattern set. For the remaining clock domains, the ATPG is similar in both the methods.

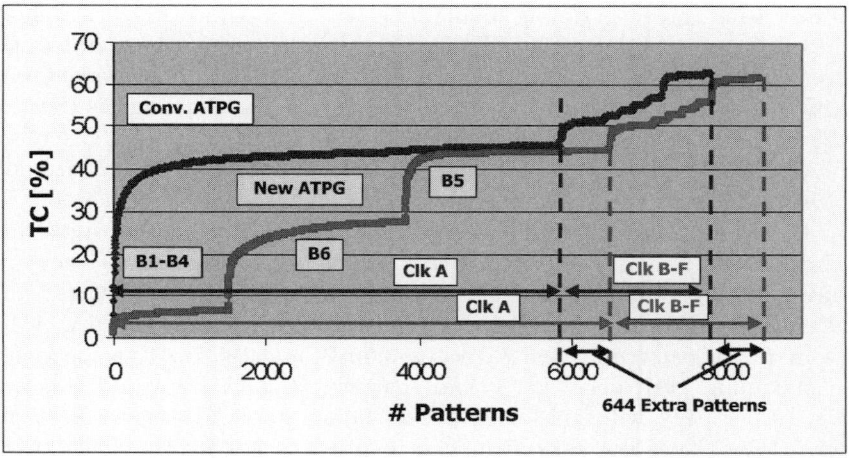

Fig. 9.13. Test coverage curves for conventional ATPG and the new pattern generation procedure [24].

SCAP-based Pattern Validation

In this section, validation of the test patterns using SCAP model is presented. The block level average switching power thresholds derived using statistical IR-drop analysis technique (explained in Section 9.2.2) was used as a measure to identify patterns with high SCAP value resulting in high IR-drop.

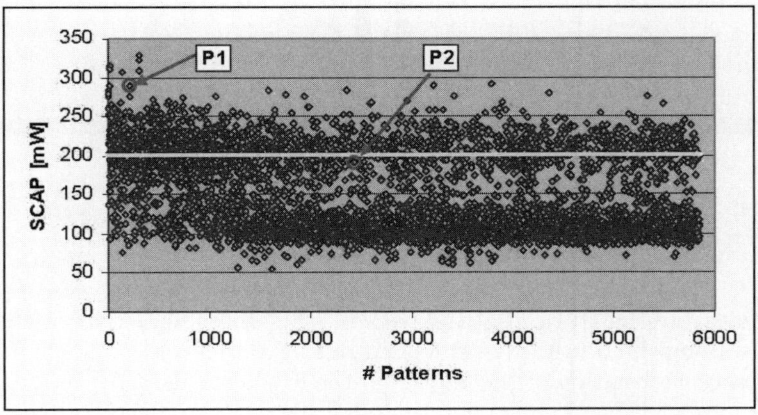

Fig. 9.14. SCAP measured in block *B5* for conventional transition fault pattern set in *clkA* domain [24].

The SCAP value for each pattern was measured at both block-level and top-level using gate level timing simulations with Synopsys VCS [26] simulator and PLI. Figure 9.14 shows the measured SCAP value on VDD network for the conventional test patterns (*clkA* domain) in block *B5* which was observed to be the most power consuming block during statistical IR-drop analysis (Section 9.2.2). It can be seen that a large number of patterns (2253 patterns) exercised a SCAP value greater than the average functional switching power threshold ($204mW$) determined during statistical IR-drop analysis for block *B5*. To show the relationship between the measured SCAP value and corresponding IR-drop in the design, two patterns *P1* and *P2* were selected. As shown in Figure 9.14, pattern *P1* experiences very high SCAP whereas the SCAP value for *P2* is close to the threshold limit. The measured VDD network SCAP value in block *B5* for pattern *P1* and *P2* was $283.5mW$ and $190.7mW$, respectively. The maximum path delay (STW) for pattern *P1* and *P2* were $7.8ns$ and $8.6ns$, respectively with the clock period being $20ns$. Therefore, STW for these patterns does not vary significantly and this implies that pattern *P1* observes very high switching activity compared to *P2*.

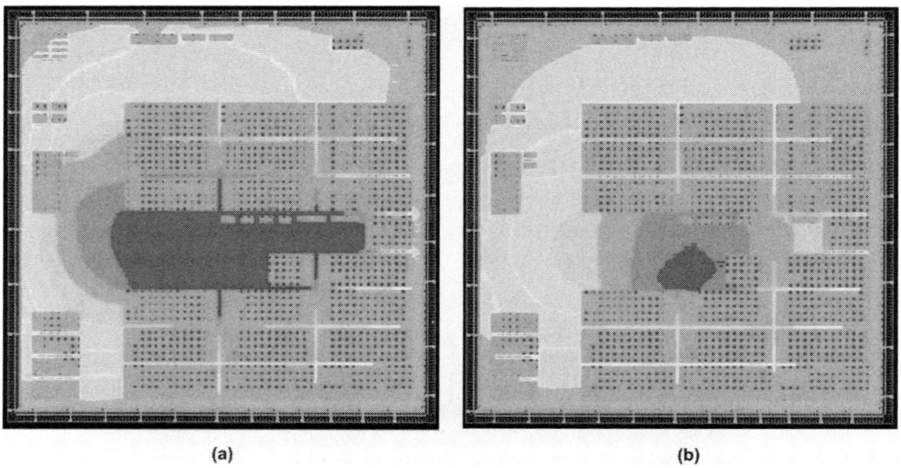

Fig. 9.15. VDD IR-drop plots using SCAP model for two patterns (a) *P1* and (b) *P2* [24].

To measure the IR-drop of the pattern, the switching activity inside the design was captured in the VCD format during gate-level timing simulation. The timing information of the gates and the extracted parasitic interconnect delay information was back-annotated using the standard delay format (SDF) file. The switching activity information (VCD file) along with physical design and technology library information is used by SOC Encounter tool [27] to estimate the dynamic IR-drop of the pattern. Figure 9.15 shows the average IR-drop plots using SCAP model on the VDD network for both of these patterns. The red region shows the portion of the design where the voltage drop is greater than 10% VDD (0.18V). The worst average IR-drop was reported to be 0.28V and 0.19V for pattern *P1* and *P2*, respectively. This shows that the average functional power estimated using statistical IR-drop analysis using SCAP model is a good measure to identify patterns with high IR-drop effects. Also, SCAP calculator provides a cost-effective solution than using expensive dynamic IR-drop analysis for each pattern.

In the next step, SCAP is measured for the new test pattern set generated. Figure 9.16 shows the measured SCAP value in the VDD network of block *B5* for the new test patterns in *clkA* domain only (6490 patterns). It can be seen that the initial patterns (up to approximately 4000 patterns) have very low and nearly the same SCAP value. This is because these patterns target faults in other blocks and *fill-0* option maintains block *B5* in a quiet state with very less switching activity. Also, notice that there is a sudden disturbance and a high switching activity is observed in the later patterns when the ATPG tool targets the faults in block *B5*. This shows that the ATPG tool applies greedy algorithm to target as many faults in block *B5* even with *fill-0* option and it is unaware of the power consumption. However, the number of patterns above the SCAP threshold are extremely low (approximately 57 patterns) using the new pattern generation technique compared to random-fill at an expense of approximately 8% increase in test pattern count.

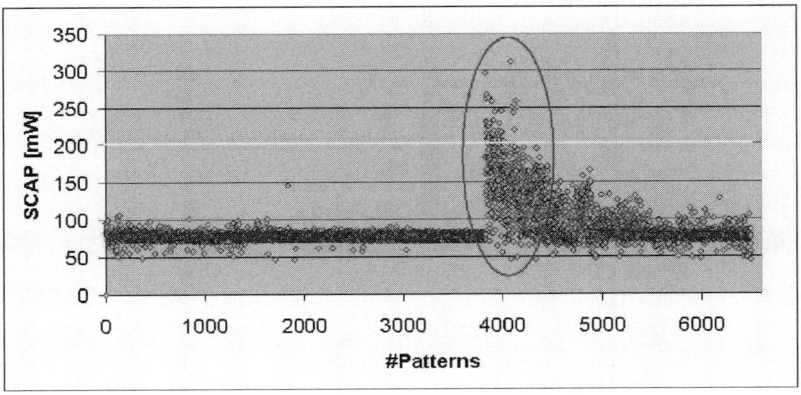

Fig. 9.16. SCAP measured in block *B5* for the new test pattern set [24].

Gate-level Simulation with IR-drop Effects

To further validate some of the delay test patterns exercising long paths, we wanted to perform pattern simulation including IR-drop effects. However, this requires transistor-level simulation with the power/ground network parasitics which is not a feasible solution for a large design. In general, presently during test pattern sign-off, the patterns are simulated at the best and worst-case corners. This is either over optimistic or pessimistic as the corner conditions are applied to all the portions of the design which is not the case as seen in Figure 9.15. In order to take advantage of relatively faster gate-level simulation but still take IR-drop effects into account, another PLI was developed which can be plugged into any gate level simulator (here Synopsys VCS). It modifies the cell delays during a pattern simulation based on the voltage of every instance reported during dynamic IR-drop analysis using SOC Encounter tool for the respective test pattern. The cell delay degradation is calculated by the following formulation:

$$ScaledCellDelay = Delay \times (1 + k\_volt * \Delta V)$$

where *k_volt* is a factor that accounts for non-linear delay scaling model and it is specified in the vendor supplied technology library. Here, a value of 0.9 for *k_volt* was used, which means for a 5% cell voltage decrease ($\Delta V = 0.1V$), the cell delay increases by 9%.

Figure 9.17 shows the delay observed at each endpoint of the design for a single test pattern in two cases: *Case1:* no IR-drop effect and *Case2:* scaled cell delays due to IR-drop effects. The pattern was selected with most faults being tested in block *B5* (circled region in Figure 9.16) but with the SCAP value below the threshold limit. An observation point at the end of a path (scan flip-flop) is referred to as an *endpoint*. An endpoint which does not observe a transition, referred to as *non-active*, is represented with zero delay. It can be seen that the delay observed by a certain number of scan flip-flops (*Region 1*) has increased (upto 30% in some cases). This is because the gates in the input logic cone of these scan flip-flops observed high IR-drop and the delay scaling factor was high for them. Also, it was noticed that these flip-flops were in the higher IR-drop region in block *B5*. Also, the delay observed by some endpoints has reduced *(Region 2)*. This is because the path delay observed at each endpoint was measured based on the reference clock signal reaching the respective endpoint. The clock reaches different endpoints at different times due to clock skew and cell delay scaling due to IR-drop effects. Therefore, if the clock signal reaching the capture flip-flop is delayed relative to the clock signal of launch flip-flop due to IR-drop effects, then the path delay measured at the capture flip-flop decreases.

This kind of pattern simulation is more practical and gives a better insight of the endpoints and paths effected by IR-drop in each pattern. However, the above analysis is still very computationally expensive and it requires two simulations, one to generate VCD file for dynamic IR-drop analysis and then use the reported instance voltages for the next simulation including cell delay degradation effects. Hence, it is preferable to apply this technique for only a limited set of patterns for more analysis or to debug any pattern which is identified to fail due to IR-drop effects.

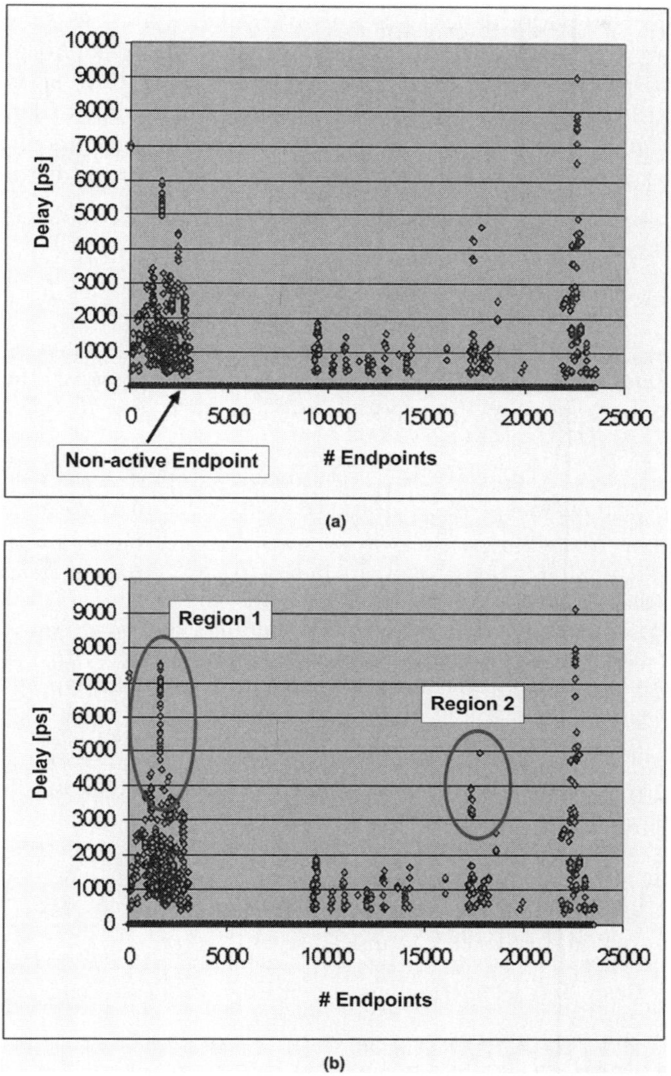

Fig. 9.17. Path delay variation of a test pattern in two cases: (a) no IR-drop effects and (b) scaled cell delays due to IR-drop effects [24].

9.4 Summary

In this chapter, a new model called SCAP has been presented to measure the impact of a transition delay fault test pattern on supply noise and its negative effect on performance. The presented model and pattern generation procedure were implemented on two large designs. The comparison between the new pattern set and and the one generated using conventional ATPG shows that the supply noise significantly decreases in a cost of slight increase in pattern count. The presented procedure uses existing commercial ATPGs and wrapper is added around them to generate new IR-drop tolerant pattern set.

References

1. J. Saxena, K. M. Butler, J. Gatt, R. Raghuraman, S. P. Kumar, S. Basu, D. J. Campbell, J. Berech, "Scan-Based Transition Fault Testing - Implementation and Low Cost Test Challenges," in Proc. *International Test Conference (ITC'02)*, pp. 1120 - 1129, Oct. 2002.
2. X. Lin, R. Press, J. Rajski, P. Reuter, T. Rinderknecht, B. Swanson and N. Tamarapalli, "High-Frequency, At-Speed Scan Testing," *IEEE Design & Test of Computers*, pp. 17-25, Sep-Oct 2003.
3. M. Nourani, M. tehranipoor and N. Ahmed, "Pattern Generation and Estimation for Power Supply Noise Analysis," in proc. *VLSI Test Symposium (VTS'05)*, pp. 439-444, 2005.
4. R. Senthinatharr and J. L. Prince, *Simultaneous Switching Noise of CMOS Devices and Systems*, Kluwer Academic Publishers, 1994.
5. K. Chakrabarty, "Test scheduling for core-based systems using mixed-integer linearprogramming," in proc. *IEEE Trans. on Computer-Aided DEsign of Integrated Circuits and Systems*, Vol. 19, No. 10, 2000, pp. 1163-1174.
6. Y. Huang, et. al, "Resource Allocation and Test Scheduling for Concurrent Test of Core-Based SOC Design," in proc. *IEEE Asian Test Symposium (ATS'01)*, pp. 265-270, 2001.
7. J. Savir, "Skewed-Load Transition Test: Part I, Calculus," in Proc. *Int. Test Conf. (ITC'92)*, pp. 705-713, 1992.
8. J. Savir and S. Patil, "On Broad-Side Delay Test," in Proc. *VLSI Test Symp. (VTS'94)*, pp. 284-290, 1994.
9. B. Dervisoglu and G. Stong, "Design for Testability: Using Scanpath Techniques for Path-Delay Test and Measurement," in Proc. *Int. Test Conf. (ITC'91)*, pp. 365-374, 1991.
10. X. Liu and M. Hsiao, "Constrained ATPG for Broadside Transition Testing," in Proc. *IEEE Int. Symp. on Defect and Fault Tolerance in VLSI Systems (DFT'03)*, pp. 175-182, 2003.
11. Z. Zhang, S. M. Reddy, and I. Pomeranz, On Generating Pseudo-Functional Delay Fault Tests for Scan Designs, in proc. *IEEE Intl. Symposium on Defect and Fault Tolerance in VLSI Systems (DFT'05)*, pp. 398405, 2005.
12. N. Ahmed, C.P. Ravikumar, M. Tehranipoor and J. Plusquellic, "At-Speed Transition Fault Testing With Low Speed Scan Enable," in proc. *IEEE VLSI Test Symposium (VTS'05)*, pp. 42-47, 2005

13. Y.-S. Chang, S. K. Gupta and M. A. Breuer, "Analysis of Ground Bounce in Deep Sub-Micron Circuits," in Proc. *IEEE VLSI Test Symposium (VTS'97)*, pp. 110-116, 1997.
14. H. H. Chen and D. D. Ling, " A Power Supply Noise Analysis Methodology for Deep Submicron VLSI Chip Design," in Proc. *CM/IEEE Design Automation Conference (DAC'97)*, pp. 638-643, 1997.
15. L. Zheng, B. Li, and H. Tenhunen, "Efficient and Accurate Modeling of Power Supply Noise on Distributed On-Chip Power Networks," in Proc. *Int. Symposium on Circuits and Systems (ISCAS'00)*, pp. 513-516, 2000.
16. E. Liau and D. Landsiedel, "Automatic Worst Case Pattern Generation Using Neural Networks & Genetic Algorithm for Estimation of Switching Noise on Power Supply Lines in CMOS Circuits," in Proc. *European Test Workshop (ETW'03)*, pp. 105 -110, 2003.
17. J. Wang, Z. Yue, X. Lu, W. Qiu, W. Shi, D. M. H. Walker, "A Vector-based Approach for Power Supply Noise Analysis in Test Compaction," in Proc. *Int. Test Conf. (ITC'05)*, 2005.
18. A. Kokrady and C.P. Ravikumar, "Static Verification of Test Vectors for IR-drop Failure," in Proc. *Int. Conf. on Computer-Aided Design (ICCAD'03)*, pp. 760-764, 2003.
19. A. Krstic, Y. Jiang and K. Cheng, "Pattern Generation for Delay Testing and Dynamic Timing Analysis Considering Power-Supply Noise Effects," in *IEEE Transactions on CAD*, vol. 20, nNo. 3, pp. 416-425 March 2001.
20. X. Wen, Y. Yamashita, S. Kajihara, L. T. Wang, K. K. Saluja and K. Kinoshita, "On low-capture-power test generation for scan testing," in proc. *VLSI Test Symposium (VTS'05)*, pp. 265-270, 2005.
21. X. Wen, S. Kajihara, K. Miyase, T. Suzuki, K. K. Saluja, L. T. Wang, K. S. Abdel-Hafez and K. Kinoshita, "A New ATPG Method for Efficient Capture Power Reduction During Scan Testing," in proc. *VLSI Test Symposium (VTS'06)*, 2006.
22. N. Ahmed, M. Tehranipoor and V. Jayaram, "A Novel Framework for Faster-than-at-Speed Delay Test Considering IR-Drop Effects," to appear in *Int. Conf. on Computer-Aided Design (ICCAD'06)*, 2006.
23. N. Ahmed, M. Tehranipoor and V. Jayaram, "Supply Voltage Noise Aware ATPG for Transition Delay Faults," in Proc. *IEEE VLSI Test Symposium (VTS'07)*, 2007.
24. N. Ahmed, M. Tehranipoor and V. Jayaram, "Transition Delay Fault Test Pattern Generation Considering Supply Voltage Noise in a SOC Design," in Proc. *Design Automation Conference (DAC07)*, 2007.
25. J. Saxena, K. M. Butler, V. Jayaram, S. Kundu, N. V. Arvind, P. Sreeprakash and M. Hachinger, "A Case Study of IR-Drop in Structured At-Speed Testing," in Proc. *International Test Conference (ITC'03)*, pp. 1098 - 1104, Oct. 2003.
26. Synopsys Inc., "User Manuals for SYNOPSYS Toolset Version 2005.09," Synopsys, Inc., 2005.
27. Cadence Inc., "User Manuals for Cadence Encounter Tool set Version 2004.10," Cadence, Inc., 2004.
28. http://crete.cadence.com, 0.18μm standard cell GSCLib library version 2.0, Cadence, Inc., 2005.

10

Pattern Generation for Power Supply Noise Analysis

This chapter presents an automatic pattern generation methodology to stimulate the maximum power supply noise in deep submicron CMOS circuits. This information can benefit both the design and failure anaysis teams. The generated test patterns can also be used for targeting supply noise effects during fabrication test. The design team can use this information to further analyze the power/ground network for driving maximum current to the circuit without affecting the circuit performance.

The ATPG-based approach, presented in this chapter, first generates the required patterns to cover $0 \rightarrow 1$ and $1 \rightarrow 0$ transitions on each node of internal circuitry. Then, a greedy heuristic is applied to find the worst-case (maximum) instantaneous current and stimulate maximum switching activity inside the circuit. The quality of these patterns were verified by SPICE simulation. Experimental results show that the pattern pair generated by this approach produces a tight lower bound on the maximum power supply noise.

10.1 Introduction

Power supply noise (PSN) due to switching current has become an important factor for deep submicron (DSM) designs. This noise effect is becoming more detrimental as VLSI technology scales. As the number of interconnect layers and gate density increases, the switching activity increases which lead to increase in current density and voltage drop along the power supply net. Increasing the frequency and decreasing the rise/fall transition time in today's designs cause more simultaneous switching activity within a small time interval and increase the instantaneous currents. The power supply noise reduces the actual voltage level reaching a device, which increases the signal delay and results in signal integrity loss and performance degradation. It may also cause logic errors, degradation in switching speed and hence timing errors.

PSN includes the inductive ΔI noise ($L * \frac{dI}{dt}$) and IR voltage drop. The former is derived from the distributed RLC model of on-chip power lines and the latter is caused by the switching inside the circuit as well as input and output buffers. Applying input patterns to a CMOS circuit creates the signal switching and causes the switching currents. To activate the switching in a circuit, a pair of patterns is required to be applied to the inputs of the circuit. Assuming there are n number of inputs, $2^n * 2^n = 2^{2n}$ number of pattern pairs are required for an exhaustive search to find the pair of patterns that generate the maximum PSN. Due to process variation effects on clock, glitches can also significantly contribute to the total power supply noise. Therefore, to maximize PSN while considering variations, a set of pattern pairs may be required.

In any case, applying all possible patterns to a circuit to find such pairs is possible only for the circuits with very small number of inputs. New techniques are needed to estimate power supply noise efficiently and find the pattern(s) that generate the maximum PSN in reasonable amount of time.

A straightforward approach to determine the worst case power supply noise is to simulate all possible input pattern combinations and identify the vector pair(s) that induce the maximum switching noise. Since it is impossible to simulate the design for exhaustive pattern set due to excessive number of patterns using SPICE, there are two general solutions for the problem. The first solution is used to develop a fast simulator with reasonable accuracy and apply statistical approaches like Monte Carlo and Genetic Algorithm (GA) to iteratively select a pattern set and search for the patterns that cause the maximum switching noise. The second solution is to generate deterministic set of patterns based on heuristics with reasonably high quality to stimulate maximum power supply noise.

Several approaches have been proposed for power supply noise analysis and estimation in recent years. Some closed-form equations are derived in [12] to calculate simultaneous switching noise. Estimation of the ground bounce, caused by the switching in internal circuitry for deep-submicron circuits, using a scaling model is discussed in [2]. Reference [3] proposes a simulated switching circuit model to estimate PSN which includes IR voltage drop and ΔI noise based on an integrated package-level and chip-level power bus mode. Modeling of PSN on distributed on-chip power networks is described in [4].

ATE and neural network are used to find the patterns generating maximum instantaneous current [5]. The neural network is used to learn the behavior of chip power consumption and changes due to different input patterns applied. Several genetic algorithms for finding pattern(s) that stimulate the worst cases are proposed in [6-10]. In [7], the standard cells in the technology library are pre-characterized with SPICE to derive the delay and switching current waveform characteristics and a event-driven simulator along with a delay lookup table is used to perform timing analysis of switching events. A combination of Monte Carlo and genetic algorithm is employed to search for the worst case input vector pair(s) that induce the maximum switching noise.

The current waveform of the entire design is not a direct superposition of the individual block current waveforms when RC power/ground network is considered. The wire/substrate capacitances provide some of the current drawn and help in reducing the instantaneous current surge. The authors in [6][9] tackle this problem. Current/voltage waveform libraries for each cell in a library are derived using SPICE. A current waveform simulator is used to simulate a small set of patterns derived iteratively using a genetic algorithm. Finding the maximum voltage drop in the power bus of digital VLSI circuits using a genetic algorithm is discussed in [10]. In this work, the fitness value for different input vector pairs is the worst-case voltage drop at a specified node in the power bus.

10.1.1 Overview of the Method

In previous test pattern generation methods, impact of noise on the transient characteristics is not taken into consideration during the initial generation of current/voltage waveform libraries. Hence, the estimated noise level may not be accurate. On average 10% overestimate in noise voltage was reported compared to SPICE for $0.25\mu m$ technology [6]. This estimation error may increase as the technology scales down. In this chapter, a new pattern generation algorithm that targets power supply noise is presented. The methodology employs common Automatic Test Pattern Generation (ATPG) technique applied to conventional stuck-at faults. With the aid of an ATPG, the technique quickly and accurately identifies the transient characteristics of gates for a given pattern and its relationship with PSN. The pattern generation process is independent of the physical layout information and preprocessing of library cells and guarantees a tight lower bound for maximum PSN.

10.2 Power Supply Noise (PSN) Model

In general, PSN includes two components: inductive ΔI noise and power net IR voltage drop and is given by $PSN = L * \frac{dI}{dt} + IR$. The inductive ΔI noise $(L * \frac{dI}{dt})$ depends on the rate of change of the instantaneous current, while the IR voltage drop is caused by the instantaneous current through the resistive power and ground network. The inductance is mainly due to package lead and wire/substrate parasitics.

Simultaneous switching of a large number of gates often induces a very large current spike on the power/ground lines in a short time interval. With low-k copper (Cu) interconnects being used in deep-submicron designs, the resistance of the wires is drastically reduced. This will generate considerable inductive noise $L * \frac{dI}{dt}$ even though the inductance L can be relatively small. In order to create maximum switching noise, it is important to analyze the characteristics of the switching current waveform.

The switching current waveform of each gate is determined by the propagation delay (t_d) and its drive capacity (I_{max}). Empirical evidence shows that all switching currents last for approximately $3t_d$ and the peak drive current

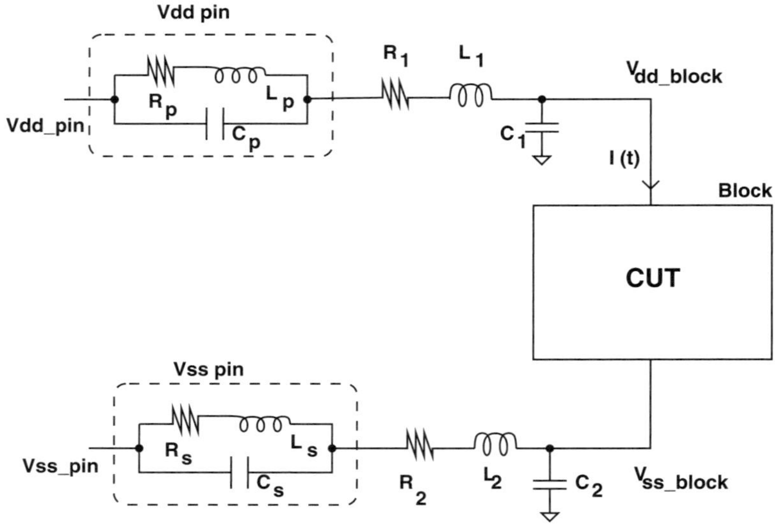

Fig. 10.1. The circuit model and power supply noise measurement [13].

may slightly change for different capacitive load. The propagation delay is directly related to the fanout of the gate. Therefore, a gate with a smaller fanout has less propagation delay and hence a shorter current waveform duration, i.e the rate of change of current $\frac{dI}{dt}$ is higher. Hence, it induces greater inductive noise.

To illustrate the effect of fanout on power supply noise, a simple experimentation was performed. Consider a block consisting of 10 2-input NAND gates which switch simultaneously in two different cases. In $Case1$, each gate has a fanout of 3 minimum-sized inverters while in $Case2$, each gate has a fanout of 2 minimum sized inverters. The circuit model used for power/ground pin and power/ground network is shown in Figure 10.1. Each V_{dd} and V_{ss} pin is modeled as an RLC circuit. The pin parasitics are R_p, L_p and C_p for V_{dd} pin and R_s, L_s and C_s for V_{ss} pin. The power/ground network is essentially modeled as a lumped RLC network.

The simulations are performed on the circuit implemented in $0.13\mu m$ technology. Figure 10.2 shows the SPICE simulation results for the block current waveforms in the two different cases. The variation of peak current value in the two cases is insignificant. The duration of the switching current waveform in $Case1$ is greater than in $Case2$ due to large propagation delay which is proportional to fanout. Figure 10.3 shows the corresponding rate of current change. The rate of change of switching current $\frac{dI}{dt}$ is greater in $Case2$ and hence induces greater inductive noise.

Figure 10.4 shows the corresponding PSN waveforms. The power supply noise was computed from the transient voltage waveforms on the power/ground lines as:

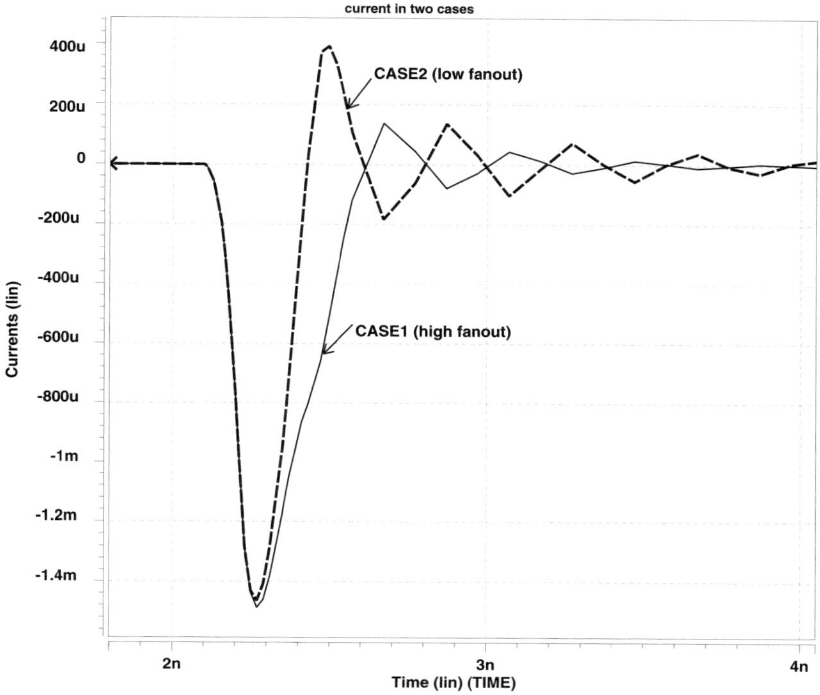

Fig. 10.2. Current $I(t)$ of the entire block in two different cases [13].

$$V_{PSN}(t) = \{V_{dd\_pin} - V_{ss\_pin}\} - \{V_{dd\_block}(t) - V_{ss\_block}(t)\}$$

where V_{dd_pin} (V_{ss_pin}) is the input supply (ground) voltages to the package lead (Here, 1.2 V and 0 V respectively), $V_{dd\_block}(t)$ and $V_{ss\_block}(t)$ are the transient voltage waveforms on the power and ground network, respectively. It is clear that noise induced in $Case2$ is greater than in $Case1$ even though the peak current occurs in $Case1$. This confirms that maximum switching current does not necessarily generate maximum switching noise.

Based on these analytical and empirical observations, as a main guideline to generate maximum PSN, more preference would be given to patterns that cause more switching in gates with smaller fanouts. More formally, suppose a circuit G has n gates $g_1, g_2, \cdots, g_n$ with fanout values of $f_1, f_2, \cdots, f_n$ corresponding to those gates. Let $g_i(V_1)$ and $g_i(V_2)$ be the output of gate g_i for two input patterns V_1 and V_2, respectively. $s_i^{V_1 \rightarrow V_2} = |g_i(V_1) - g_i(V_2)|$ will be a binary variable indicating if gate g_i has a transition in its output when pattern pair (V_1, V_2) is applied. According to the above guideline, to maximize PSN, pattern (V_1, V_2) pairs need to be selected such that $\sum_{i=1}^{n} s_i^{V_1 \rightarrow V_2}$ is maximized while simultaneously minimizing $\sum_{i=1}^{n} s_i^{V_1 \rightarrow V_2} \cdot f_i$.

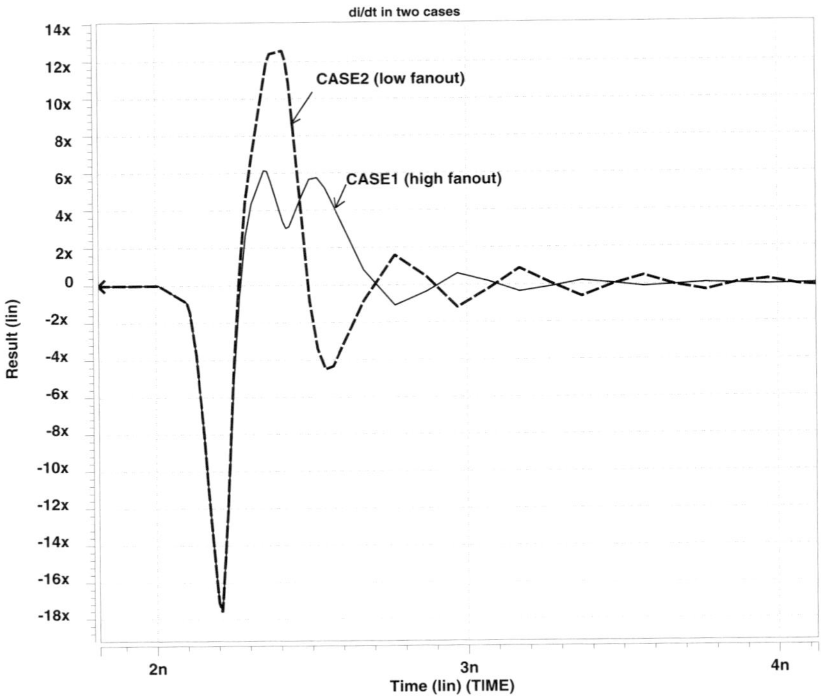

Fig. 10.3. Rate of change ($\frac{dI}{dt}$) of block current in two different cases [13].

10.3 Pattern Generation

PSN depends on the switching activity inside the circuit and thus is highly input pattern dependent. The spatio-temporal correlation among signals determines the level of switching activity inside a circuit. To stimulate the power supply noise, a pattern pair, i.e. (V_1, V_2), is required to be applied to the circuit under test. In order to stimulate the worst case (maximum PSN), it is necessary to maximize both the rate of change of current $\frac{dI}{dt}$ and also the peak current drawn. As explained in Section 10.2, the inductive noise dominates the resistive noise. More specifically, the rate of current change can be increased by stimulating simultaneous switching in large number of gates with low fanout in a circuit.

10.3.1 Timing of Switching Events

The propagation delay of a gate depends on many factors such as fanout load, input rise/fall time and drive strength. Due to difference in propagation delay of the gates, a change in primary input (PI) will trigger a sequence

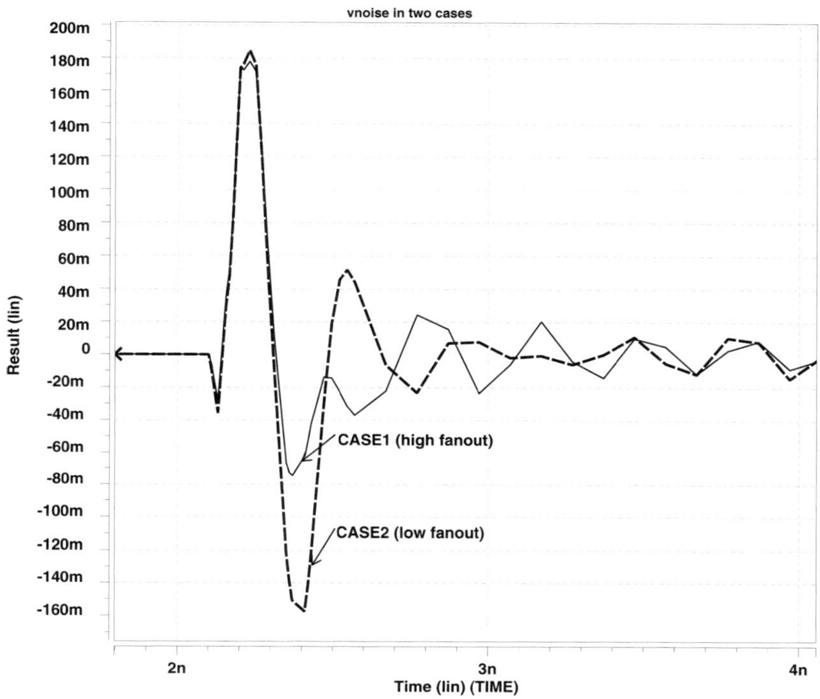

Fig. 10.4. PSN induced in two different cases [13].

of switching events in the gates that are directly or indirectly connected to it. Since the switching activity inside the circuit determines the switching noise, it is important to find the time intervals where maximum simultaneous switchings occur. To determine the simultaneous switching activity within a clock cycle T, the clock cycle is broken down into N small time frames. Each time frame has a duration of $\frac{T}{N}$ and N is chosen based on the required resolution.

A random pattern pair for ISCAS'85 benchmark circuit ($c432$) was simulated that generated high PSN, using PowerMill tool in Synopsys which is an event driven transistor-level simulator [11]. Figure 10.5 shows the number of switchings over a period of time when the pattern pair (V_1, V_2) is applied. The horizontal axis plots the time intervals ($\frac{T}{N} = 10ps$) from the time the second vector (V_2) is applied . Maximum simultaneous switching activity occurs at the beginning of the simulation time frame and small peaks occur later in the simulation period. Simulation results for all ISCAS'85 benchmarks confirm that the maximum simultaneous switching activity occurs in the early period of the simulation cyle.

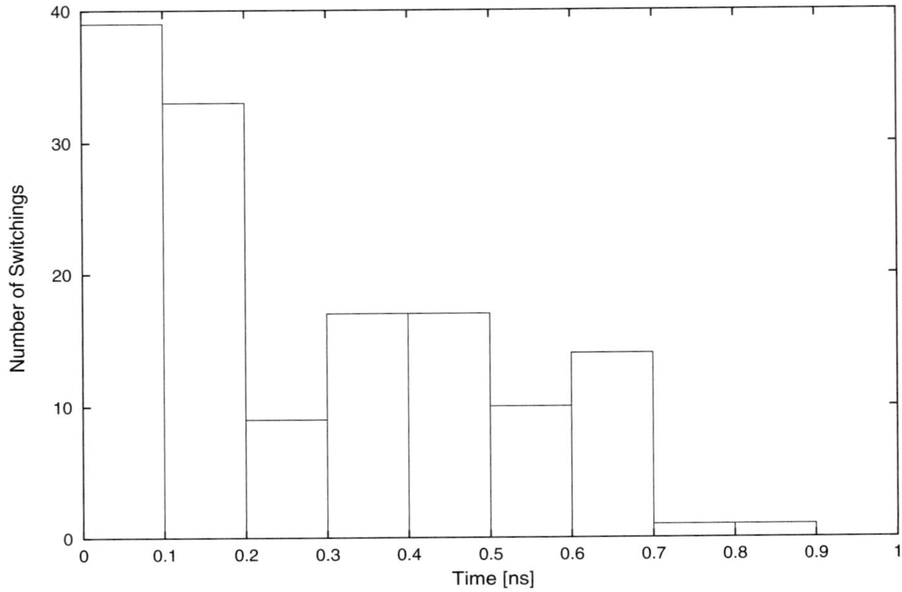

Fig. 10.5. Switching activity vs time for circuit c432 [13].

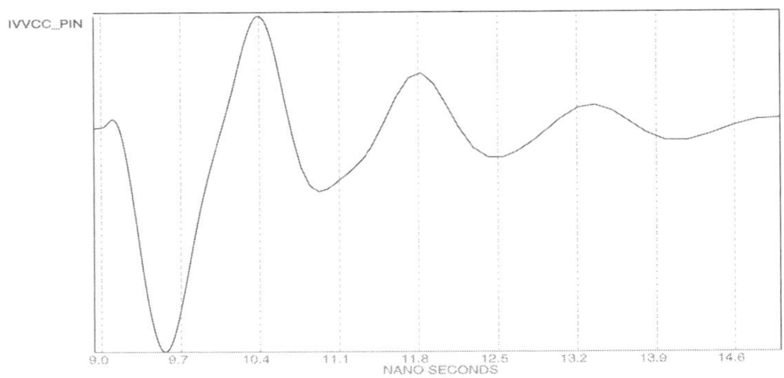

Fig. 10.6. Current drawn from the V_{dd} supply voltage [13].

Figure 10.6 shows the corresponding SPICE simulation current waveform for the pattern applied to circuit c432. It shows that the current drawn from the power supply is maximum at the early stage of the clock cycle and decreases later on. The first peak in the current waveform is due to the initial maximum simultaneous switching activity. Therefore, to generate the steepest maximum current (maximum noise) the number of switchings in the low-fanout gates of the circuit needs to be increased during the early period of the cycle.

10.3.2 Preprocessing Phase

For each time frame T, a subset of active gates in a circuit G will be chosen and the pattern generation works according to the following three guidelines:

1. *Gate Fanout:* Sort gates in increasing order of their fanouts.

2. *Gate Level:* Within each group, formed in the previous step, sort them according to the *level* that they are positioned in. A level of a gate is the distance of the gate from the primary inputs (PI). When back-traced, a gate close to the PI's has more number of don't-cares (X's) in the input pattern than a gate far away from the PI's. Hence, choosing a node with more number of X's, i.e. a gate in lower level, leaves us with more choices of assigning transitions on the other nodes and increases the chance of generating maximum switching activity in the time frame.

3. *Gate Transition:* Both types of transitions ($0 \rightarrow 1$ and $1 \rightarrow 0$) are tested in each iteration. Depending on the topology of the circuit, the location of the gate and the way that it affects others, one of these transitions may have a better chance in maximizing PSN.

In the next subsection, it is shown how the algorithm uses these guidelines and the conventional model of stuck-at-fault (*saf*) and ATPG process to justify a transition at each gate and find vector pairs that maximize PSN.

10.4 Algorithm

The pattern generation algorithm is shown in Figure 10.7. Given a design, in the preprocessing phase the target time frames with the likelihood of having switching activity are obtained. An ATPG algorithm independent of the simulation method is then used to find target time frames.

10.4.1 Pseudocode

For each target time frames, the corresponding set of gates (G) that switch in this time frame are extracted (line 03). The gates are sorted (lines 04-05) using the criteria explained in the Section 10.3. In the pattern generation process (lines 08-21), transitions are assigned and justified to gates from the sorted list of gates. To justify a transition at a node, an ATPG mechanism is applied (i.e. TetraMax [12]) originally used for stuck-at fault testing. The algorithm obtains patterns to justify a value '0' at a node, (viewed as a stuck-at-1 ($sa1$) fault) and a '1' at a node, (viewed as a stuck-at-0 ($sa0$) fault). When both patterns exist (line 11-15), a ($0 \rightarrow 1$) or ($1 \rightarrow 0$) transition can be generated at the output of a selected gate.

The ATPG process generates the pattern pair (V_1, V_2) based on zero-delay model. Now, a power simulator (i.e. PowerMill [11]) is used to accurately measure PSN (lines 13-14). PowerMill is a variable delay event-driven simulator and it takes into account the hazards and glitches caused due to difference in the gate propagation delays during the PSN measurement. The result of PSN measurement for this pattern pair is compared to the maximum found so far (PSN_{max}) and the worst case scenario of power supply noise is saved. The vector pair set (V_{max}) are also updated accordingly (line 19). Note that the ATPG based pattern generation is technology independent and does not require any pre-processing of library cells. On the other hand, the power estimator is used to evaluate the patterns based on the library/technology. Instead of finding one pair of patterns, the procedure can be slightly modified to find and report all pattern pairs that create noise in a given range, e.g. $[PSN_{max}, PSN_{max} + \Delta]$.

10.4.2 Example

For purpose of illustration consider Figure 10.8 showing generic test pattern generation process applied to a small example circuit. The stuck-at fault patterns are generated by back-tracing the node towards the primary inputs and are listed in Figure 10.8(b). In conventional stuck-at fault test generation, the observation points are the primary outputs. However, in the our method the target node itself is considered to be the observation point as it needs to be justified to a particular value. Therefore, it requires only back-tracking from the target node and no forward tracing.

Initially, the input vector pair (V_1, V_2) is assumed to be all unknown (X) values. The gates are sorted in increasing order of fanout as shown in Figure 10.8(c). An untried node with the lowest fanout is selected from the sorted list and a transition is assigned to it. For example, in the first iteration, node f is selected and a $0 \rightarrow 1$ transition is assigned to it. The $0 \rightarrow 1$ transition assignment can be viewed as a ($sa1, sa0$) fault pair at the node. Since f is the first node in the list and a $0 \rightarrow 1$ transition is selected, therefore V_1 and

```
For each target time frame T
{
  G ← gates switch in time frame T
  Sort gates in G in increasing order of their fanouts
  Sort equal-fanout gates in G in increasing order of their levels
  PSN_max = 0
  V_max = {}
  for (i = 1, ··· , |G_i|)
  {
    Perform ATPG using TetraMax to get sa0/sa1 patterns for g_i ∈ G
    if (both patterns V_1 and V_2 exists)
    {
      Try V_1 → V_2; if successful run PowerMill to compute PSN_1
      Try V_2 → V_1; if successful run PowerMill to compute PSN_2
    }
    if (successful and PSN_1 and/or PSN_2 are computed)
    {
      PSN_max ← MAX{PSN_1, PSN_2, PSN_max}
      V_max ← Update input vector pair(s) accordingly
    }
  }
}
Return PSN_max and vector pair set V_max creating it.
```

Fig. 10.7. Deterministic test pattern generation procedure [13].

V_2 patterns are equal to $sa1$ and $sa0$ patterns for node *f*, respectively. Note carefully, there is no conflict for the first chosen node.

A conflict occurs when there is a mismatch in the comparison of the respective stuck-at fault patterns with the input vector pair. If the patterns match then the input pattern pair is updated by replacing the corresponding 'X' values with known justified values in the stuck-at fault patterns. For example, after justifying a $0 \rightarrow 1$ transition at node *f*, the input vector pair becomes $(V_1, V_2) = (1X1XX, 0XXXX)$. The updated input pattern after each gate transition justification is shown in the last column of Figure 10.8(c). The same procedure is repeated for the next node i and a $1 \rightarrow 0$ transition is justified. The input pattern pair becomes $(V_1, V_2) = (1X1X0, 0X0X1)$ after a $1 \rightarrow 0$ transition is justified on gate i. When there is a mismatch, then the assigned transition cannot be justified and thus the opposite transition is tried. For node *z1*, when a $0 \rightarrow 1$ transition is tried to be justified, a conflict occurs. In case of a conflict, an opposite transition, i.e. $1 \rightarrow 0$ transition is tried. If both transition assignments fail, the node is skipped and the next node in the list is tried. The process is repeated for all the nodes in the list. After processing the entire list, any leftover X's in the generated pattern input pair (V_1, V_2) will be changed to create transitions because they might still induce more glitches

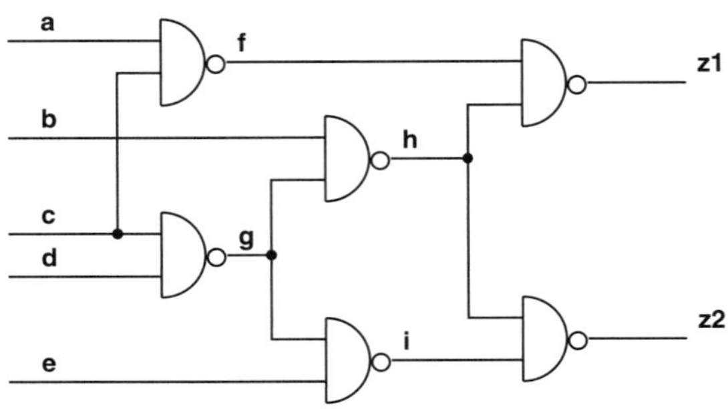

(a)

Primary Input	f		i		z1		z2		g		h	
	sa0	sa1	sa0	sa1	sa0	sa1	sa0	sa1	sa0	sa1	sa0	sa1
a	0	1	X	X	1	0	X	X	X	X	X	X
b	X	X	X	X	X	0	X	0	X	X	0	1
c	X	1	X	0	1	X	0	X	0	1	X	0
d	X	X	X	X	X	X	X	X	X	1	X	X
e	X	X	0	1	X	X	1	0	X	X	X	X

(b)

Sorted Gates	Transition Selected	Backtracing Result	(sa0, sa1) patterns	(V1 , V2)	
				(XXXXX,XXXXX)	Initial pair
f	0->1	✓	(0XXXX,1X1XX)	(1X1XX,0XXXX)	
i	1->0	✓	(XXXX0,XX0X1)	(1X1X0,0X0X1)	
z1	0->1 1->0	conflict ✓	(1X1XX,00XXX)	(1X1X0,000X1)	
z2	1->0 0->1	conflict ✓	(XX0X1,X0XX0)	(101X0,000X1)	
g	0->1	✓	(XX0XX,XX11X)	(10110,000X1)	
h	1->0 0->1	conflict conflict	(X0XXX,X10XX)		
				(10110,00001)	Final pair

(c)

Fig. 10.8. ATPG process for a small example circuit [13].

and cause more power supply noise in the circuit. Based on this guideline, 'X' in $V_2 = 000X1$ will be replaced by '0' and the final pattern generated for the example shown in Figure 10.8(a) is $(V_1, V_2) = (10110, 00001)$.

10.5 Experimental Results

Several experiments were performed on ISCAS'85 benchmark circuits implemented in $0.25\mu m$ technology. The V_{dd} pin characteristic values used in these simulations are $R_p = R_s = 0.3\Omega$, $L_p = L_s = 8nH$ and $C_p = C_s = 4pF$. These typical values are chosen from the TSMC $0.25\mu m$ library application notes. The effective resistance and capacitance values in the power/ground network are estimated based on the parasitic values per unit length. The resistance and capacitance per unit length used for the power/ground lines are $r = 0.04\Omega/\mu m$ and $c = 10aF/\mu m$, respectively. The primary input's rise time is set to 100ps.

To show the quality of patterns generated by the our technique exhaustive simulation were performed for three small benchmark circuits. The presented algorithm generated the pattern pairs that cause maximum power supply noise compared to exhaustive pattern simulation results in much shorter CPU time. For benchmark circuit c17, it took 181 *sec* to perform the exhaustive simulation while it takes less than a second for the same vector pair to be generated by the proposed method. Table 10.1 shows the results of exhaustive simulation and compares the run times. In all three cases, the proposed method generates the worst case power supply noise test patterns, identical to those found by SPICE, in very short time.

The power supply noise is calculated from the transient voltage waveforms on the power/ground lines as [7]:

$$V_{PSN}(t) = V_{dd\_pin} - V_{dd\_block}(t) - V_{ss\_block}(t)$$

where V_{dd_pin} is the input supply voltage to the package lead (2.5 Volt in this case) $V_{dd\_block}(t)$ and $V_{ss\_block}(t)$ are the transient voltage waveforms on the power and ground network, respectively. When $V_{noise}(t)$ is positive, the effective supply voltage is less than the nominal supply voltage V_{dd_pin}. Table 10.2 shows the peak noise voltages at near end (node closest to the power/ground pins) and far end (node farthest from power/ground pins). As expected (see Section 10.2), the far end noise is more severe due to larger effective resistive parasitics experienced by the blocks close to the far end. The main advantage of the ATPG based method is its short runtime. For example, SPICE takes 12 minutes to simulate one input vector pair for circuit *c432*, while it takes 179 sec to generate and simulate 500 patterns for maximum power supply noise by the new method.

Table 10.1. Comparing exhaustive simulation and the presented method [13].

Circuit	*# PI's*	*# Gates*	*Peak Noise [V]*	*CPU Time [sec]*	
				SPICE	*Our Method*
c17	5	6	0.42	181	0.5
cm42	4	18	0.58	354	2.0
cm138	6	15	0.61	2682	3.5

Table 10.2. Experimental results for various ISCAS85 benchmark circuits [13].

Circuit	*# PI's*	*# Gates*	*Peak Noise (near end) [V]*	*Peak Noise (far end) [V]*	*CPU Time [sec]*
c432	36	160	0.76	0.86	179
c499	41	202	0.41	0.52	170
c880	60	357	0.81	0.99	246
c1355	41	514	0.52	0.64	332
c1908	33	880	0.73	0.87	386
c3540	50	1667	0.62	0.75	444
c5315	178	2290	0.82	0.99	568
c6288	32	2416	0.89	1.06	636

10.6 Summary

An automatic pattern generation mechanism to stimulate the maximum power supply noise has been presented in this chapter. The basic strategy is to maximize the switching activities of those gates in the first few levels of the circuit that have lower fanouts. The methodology uses conventional ATPG and power simulators to evaluate a gate-level circuit and finds patterns that cause maximum switching activity and thus maximum instantaneous current. The quality of the patterns have been verified using SPICE simulation. In all cases, the proposed method finds the same (or comparable) patterns while its running time is 2 to 3 order of magnitude faster than that of SPICE.

References

1. R. Senthinatharr and J. L. Prince, *Simultaneous Switching Noise of CMOS Devices and Systems*, Kluwer Academic Publishers, 1994.
2. Y. Chang, S. Gupta and M. Breuer, "Analysis of Ground Bounce in Deep Sub-Micron Circuits," in Proc. *VLSI Test Symp. (VTS'97)*, pp. 110-116, 1997.
3. H. Chen and D. Ling, "Power Supply Noise Analysis Methodology for Deep-Submicron VLSI Design," in Proc. *Design Automation Conf. (DAC'97)*, pp. 638-643, 1997.

4. L. Zheng, B. Li, and H. Tenhunen, "Efficient and Accurate Modeling of Power Supply Noise on Distributed On-Chip Power Networks," in Proc. *Int. Symposium on Circuits and Systems (ISCAS'00)*, pp. 513-516, 2000.
5. E. Liau and D. Landsiedel, "Automatic Worst Case Pattern Generation Using Neural Networks & Genetic Algorithm for Estimation of Switching Noise on Power Supply Lines in CMOS Circuits," in Proc. *European Test Workshop (ETW'03)*, pp. 105 -110, 2003.
6. Y. Jiang, K. Cheng and A. Deng, "Estimation of Maximum Power Supply Noise for Deep Sub-Micron Designs," in Proc. *Int. Symp. on Low Power Electronics and Design (ISLPED'98))*, pp. 233-238, 1998.
7. S. Zhao, K. Roy and C. Koh, "Estimation of Inductive and Resistive Switching Noise on Power Supply Network in Deep Sub-micron CMOS Circuits," in Proc. *Int. Conf. on Computer Design (ICCD'00)*, pp. 65-72, 2000.
8. S. Zhao and K. Roy, "Estimation of Switching Noise on Power Supply Lines in Deep Sub-micron CMOS Circuits," in Proc. *Thirteenth Int. Conf. on VLSI Design,*, pp. 168-173, 2000.
9. Y. Jiang, K. Cheng and A. Krstic, "Estimation of Maximum Power and Instantaneous Current Using a Genetic Algorithm," in Proc. *Custom Integrated Circuits Conf. (CICC'97)*, pp. 135-138, 1997.
10. G. Bai, S. Bobba and I. Haji, "Maximum Power Supply Noise Estimation in VLSI Circuits Using Multimodal Genetic Algorithms," in Proc. *Int. Conf. on Electronics, Circuits and Systems (ICECS'01)*, vol. 3, pp.1437 -1440, 2001.
11. *Synopsys Inc., Power Mill Reference Manual*, 2003.
12. *Synopsys Inc., TetraMAX Reference Manual*, 2003.
13. M. Nourani, M. tehranipoor and N. Ahmed, "Pattern Generation and Estimation for Power Supply Noise Analysis," in proc. *VLSI Test Symposium (VTS'05)*, pp. 439-444, 2005.

11

Delay Fault Testing in Presence of Maximum Crosstalk

High speed interconnects have been contributing to a majority of the delay present in modern sub-micron technologies. As the trend towards nanoscale continues, the effects from this delay will only worsen. Although it is possible to compensate for this with design tools, the limitations of testing tools are beginning to surface since parasitic coupling capacitance is not directly addressed by testing tools. A chip passing a manufacturing test with a specific pattern set only suggests that it will pass under the specific operating and stimulus conditions in which the patterns were applied on the tester. However, in the field, the surrounding paths around the critical paths may experience significantly different switching activity causing it to fail in the field. This chapter presents a structural test pattern generation procedure that magnifies the effect of parasitic crosstalk effects on critical paths. The pattern generation procedure considers the physical design and transition direction without simulation to increase the delay on the critical path. This work intends to minimize the escape ratio and improve in the field reliability. There are few modern testing tools that account for timing, but these products are not fully aware of the timing violations that may occur due to signal integrity degradation in modern technologies. This leads to silicon failures and escape.

11.1 Technology Scaling Effect on Crosstalk

Smaller process technologies along with high performance demands have posed tremendous design and test challenges with signal integrity being one among many [1]. The number of silicon failures and escapes caused by signal integrity problems is on the rise because existing design tools and test methodologies cannot fully address nanotechnology issues effectively. In nanotechnology designs, timing is dominated by interconnect-dependent RC delay, cross coupling and via resistance. As a result, for a successful tapeout and reliability, crosstalk issues must be resolved during design and must be tested for after manufacturing.

Traditionally, structural patterns are used to test DSPs, microprocessors and complex SoCs, but functional patterns may also be used to test and debug hard-to-detect, timing related defects. The need to magnify the impact of these defects during test becomes increasingly important to increase the probability of detection and reliability. For instance, the crosstalk effects can cause timing problems on a targeted critical path either by slowing it down or speeding it up. If the crosstalk effects are not carefully considered during manufacturing test, a chip may pass structural or functional patterns, but a field failure may still occur. Other sources of delay variation such as supply noise due to excessive IR-drop can cause either a timing or logic failure. Therefore, designers must take crosstalk and IR-drop effects into account during design and test engineers must generate efficient patterns to weed out the defective chips. To overcome these issues, common techniques such as line spacing and higher metal wire widths for power/ground wires are used. However, over conservative design processes may result in slower chips because of increased length of critical paths or large die area which negatively have impact on yield. Overly conservative designing a chip may not allow designers to fully take advantage of all potentials.

Normal functional patterns may not be able to effectively maximize the crosstalk effects on a chip when targeting critical paths. In addition to testing the critical paths, they must also model other functional use conditions in the rest of the design to effectively detect such hard-to-detect defects. However, when considering all sources of delay variation (crosstalk, supply noise, process variation, temperature, etc.), generating such efficient functional patterns will be a challenging task and can be prohibitively expensive. As a result, new automatic test pattern generation (ATPG) techniques are required to maximize such effects while still guaranteeing high fault coverage and low pattern count. Same needs can be justified for transition delay fault testing in presence of supply voltage noise, temperature and process variation [18]. Such structural test patterns must also be used for validating the design before sign off. And finally, in addition to design validation and defect screening, these new patterns must also assist in diagnosis and failure analysis to identify the root cause of the failure.

The larger aspect ratio and reduced wire spacing in nanotechnology create more coupling noise. This increases signal integrity problems significantly, magnifying the need for efficiency when generating patterns for test and validation. Recently, this issue is being taken very seriously in industry when applying delay fault test patterns since it requires two at-speed launch and capture clocks and generates a large number of transitions in the circuit. The increased process variation of nanotechnology has also become a serious challenge, as it adds another uncertainty for circuit performance and yield. Chemical mechanical polishing can result in as much as 15-20% variation in the thickness of metal and interlayer dielectric layers, a variation that greatly affects the coupling capacitances. Incorporating process-variation capability into layout parasitic extraction tools will be critical for accurate coupling

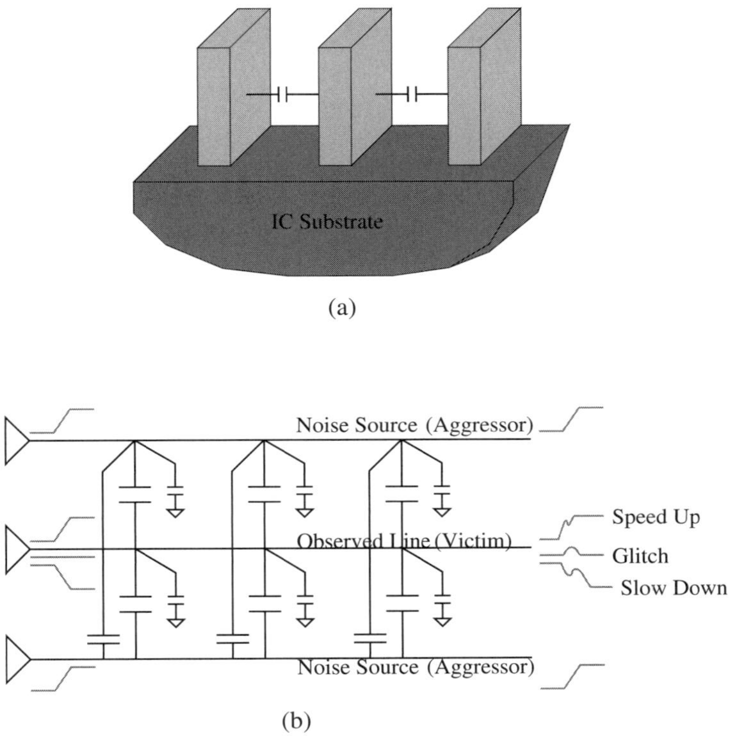

Fig. 11.1. (a) Sidewall capacitance effects increase with shrinking feature sizes and (b) Interconnect C only model (for simplicity R is not shown).

capacitance and inductance extraction, efficient critical path selection, and consequently the downstream test pattern generation.

One reason for the urgency in dealing with these issues is that mask sets that used to cost few hundred thousands of dollars will cost about a few millions of dollars for 90nm, 65nm and future technologies. It is required to validate the designs in presence of crosstalk to ensure correct functionality and performance. On the other hand, these effects are considered as timing related and hard-to-detect defects, i.e. a chip may pass the test but fail in the field (aka escape). So to avoid very expensive mistakes, design and test engineers must work together, understand these changes, and adapt their tools and methodologies accordingly. The design must be accurately validated during a post-layout step in presence of all delay sources using efficient test patterns and the fabricated chip must be tested to verify performance and reliability and increase the first silicon success.

Path delay fault tests are currently generated without taking any of such issues into consideration. When the test pattern is generated for a target critical path, the unspecified bits are filled in different ways, e.g. 0-fill, 1-fill, random-fill or adjacent fill. These methods may not cause maximum coupling effects on critical paths under test. Since current ATPG methods are timing unaware, the possible transitions after filling don't-care bits and location of transitions are not considered. New pattern generation procedures are required to consider transitions on all neighboring nets without extensive simulation. Also, new pattern generation methods are needed to maximize the coupling effects on critical paths.

Several techniques have been proposed to deal with crosstalk and signal integrity issues during verification and test. Crosstalk verification with interconnect process variation is discussed in [6]. The authors in [7] present fault modeling (called maximum aggressor (MA)) and simulation for crosstalk on SOC interconnects. In MA model, all aggressors switch in the same direction. The investigation has shown that the MA model cannot always ensure highest crosstalk effects [4] [11] especially when mutual inductance effects are considered.

Some researchers have proposed using on-chip sensors or glitch/delay detectors [2] [3] to detect noise and delay violations. The drawback of such techniques is that the sensors must be tuned and very accurate and adding one sensor per interconnect is prohibitively expensive. Due to their high sensitivity to voltage and timing, process variation can also negatively impact their operation. The authors in [4] [5] utilize the boundary scan cells to generate test patterns to detect noise and delay violations on a system chip and observe the responses which are then scanned out using boundary scan shift procedure. Authors in [8] propose validation and test generation for crosstalk-induced delay and noise for SOC interconnects. Most of the proposed techniques, mentioned above, target only buses or interconnects between cores in a SOC rather than critical paths.

Pattern generation to induce delay due to crosstalk was presented in [9] but only generates patterns for a single aggressor affecting a target path. The procedure proposed in [10] considers a genetic algorithm based approach when inducing crosstalk into delay test patterns. There has also been a proposed academic ATPG that considers crosstalk and transition arrival times during pattern generation [15] [16].

Power supply noise estimation, pattern generation and its impact on delay fault testing are discussed in [12] [13]. Supply voltage noise is considered a serious issue during both transition and path delay fault testing. Todays commercial ATPGs target as many transition delay faults as possible with one pattern to reduce pattern count, therefore cause a large number of simultaneous switchings in the circuit that can cause large IR-drop. Same holds true for path delay fault testing where the noise and delay tolerance are significantly lower since the target fault site is a critical path.

11.1.1 Overview of the Method

The focus of this chapter is on development of an effective structural delay test pattern generation procedure to magnify crosstalk effects across targeted paths. By considering parasitic information; actual location of wires and segments; and transition behavior, a pattern generation procedure is developed to accurately target such timing-related defects. The physical layout information is transported into extraction and timing analysis tools which are used to identify neighboring nets affecting targeted critical paths (i.e. hindering performance). The neighboring aggressor nets information is then used during pattern generation process to generate appropriate stimuli to maximize crosstalk effects. In the future, process variation, voltage fluctuation and temperature effects on timing of the target paths in presence of maximum crosstalk effects will be taken into account during test pattern generation. The technique presented in this chapter aims at minimizing the escape ratio and increase the reliability of the chips used in the field.

11.2 Preliminary Analysis: Proximity and Transition Direction

Signal integrity can be significantly affected by crosstalk. An aggressor line will induce a current upon a victim line based on the voltage change in the aggressor and the coupled capacitance between the two lines. The aggressor voltage rate of change is dependent on the rise and fall time. The coupling capacitance is proportional to the density of interconnects around the critical path. However, two additional factors affect the signal integrity of a path, namely transition direction of aggressors and timing of transitions on aggressors. Each of these components have a direct effect on coupling and propagation delay, but this work focuses on transition direction due to the difficulty in correctly predicting actual circuit timing without extensive simulation. Identifying the timing of transitions on aggressor and victim lines requires an extensive statistical analysis since various sources of delay, such as process variations, supply noise, etc, can have significant impact on the signal arrival time on each net.

As technologies continue to shrink beyond the ultra-deep submicron level, the parasitic coupling capacitance will continue to grow worse and will play a greater role not only in the design of a chip but also testing. Figure 11.1 shows sidewall capacitance between nearby wires in addition to interconnect C model (R is not shown for the sake of simplicity). Since the height of the wires remains almost the same as technology shrinks and the fact that the length of interconnects are becoming longer, the cross coupling effects are becoming increasingly dominant when compared to substrate capacitances. Due to reduced wire spacing, it is expected that the coupling effects will increase. The switching activity on the aggressor lines may either speed-up or slow-down a transition signal on the victim line. If the aggressor lines switch in the same direction as victim, they cause a speed-up on the signal on victim line. The slow down on victim line occurs if the aggressor lines switch in opposite direction of victim line. The effect of aggressor lines transitions on a quiescent victim line appears as glitches which can cause logic failure [3] [7].

Design and test engineers must assess, analyze and deal with these problems before signing off on a tapeout and after fabrication during manufacturing test (mainly during delay testing). Note that since the direction of transitions and transition arrival time on aggressor lines directly impact the signal delay on victim line then novel pattern generation procedures are required to maximize such effects. In other words, the total number of switchings around critical paths do not ensure maximum crosstalk.

11.2.1 Victim/Aggressor Proximity

The width and pitch of the interconnects have continued to scale as feature sizes have diminished. However, in order to keep wire resistance low, the height (thickness) of the interconnects have not been scaling down at the same rate. Along with the fact that interconnect lengths are also growing, parallel interconnects have become prime locations for large parasitic capacitance.

The coupling between two interconnects can be modeled using the spacing between the adjacent sidewalls, the length of the overlapping interconnects, and the height. Since the height of the interconnects cannot be altered, either the length of the parallel overlap must be minimized or the distance between the two interconnects increased to minimize parasitic capacitance.

11.2.2 Victim/Aggressor Transition Direction

The directions of the aggressor net transitions with respect to the victim net can have a significant effect on the propagation delay. Assuming transitions that occur on coupled nets occur at the same time, the signal on the victim can experience either speed-up or slow-down. When transitioning in the same direction as the victim, aggressors potentially speed-up the transition proportionally to the amount of coupling between the two lines. Similarly, when transitioning in the opposite direction of the victim, the aggressor will slow-down the signal and increase delay [14] [7].

Three minimally spaced interconnects are simulated at 180nm technology node of approximately $1mm$ in length (relatively long wires) with the middle interconnect considered as the victim in each simulation as shown in Figure 11.2. For the Case 1, only the victim had a rising transition with the surrounding nets held to a steady-state zero. Case 2 placed a simultaneous rise on all three nets. In the final simulation, Case 3, a rising transition is applied on the victim net with both of the aggressor nets falling.

Figure 11.3 demonstrates the affect each of the rising and falling transitions of the aggressors have on the victim. The rising transition on the far

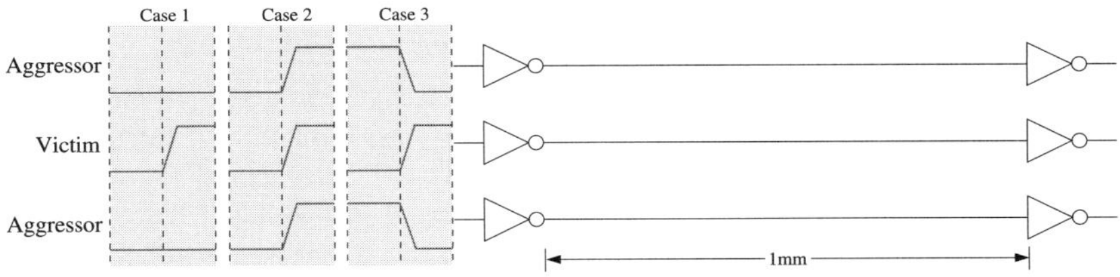

Fig. 11.2. Simulation setup to observe the effects of simultaneous switching on a victim line.

left represents the transition at the source of the interconnect. Case 2 is immediately to the right of the input transition. Case 1 is to the right of Case 2 and Case 3 is the far right transition. As can be seen from the figure, using Case 1 as a reference, Case 2 causes speed-up in the signal and Case 3 causes slow-down.

Varying combinations of rising and falling transitions around the victim net are possible to cause varying results. An additional observation made when applying a transition to one aggressor in the same direction as the victim and applying a transition in the opposite direction on the second aggressor, the victim line still experienced speed-up; although to a lesser degree than if all three were transitioning in the same direction. The intuitive response would be the transitions on the two aggressor nets would nullify each other, but since the victim is also transitioning, it also contributes to the overall speed-up for the particular segment of interconnects that were simulated.

11.3 Inducing Coupling Effects on Critical Paths

Characterization of coupling effects on signal integrity have previously required SPICE simulations or static crosstalk analysis. Although these methods may be effective for design, SPICE simulations are extremely time consuming while static analysis does not provide an accurate worst-case scenario for critical path timing. However, an approach is presented in this chapter that takes advantage of current static timing analysis tools and integrate them with an current pattern generation tools to generate a pattern to maximize crosstalk effects on critical paths.

11.3.1 Path Segmentation and Coupling

The approach divides the victim (critical) path into segments based on the gates that lie in the path. For example, the path in Figure 11.4 contains four gates and five segments between the source and the destination. It is possible to segment the path further by dividing at each bend and metal layer, however this would require interpreting the layout and unnecessarily complicate the

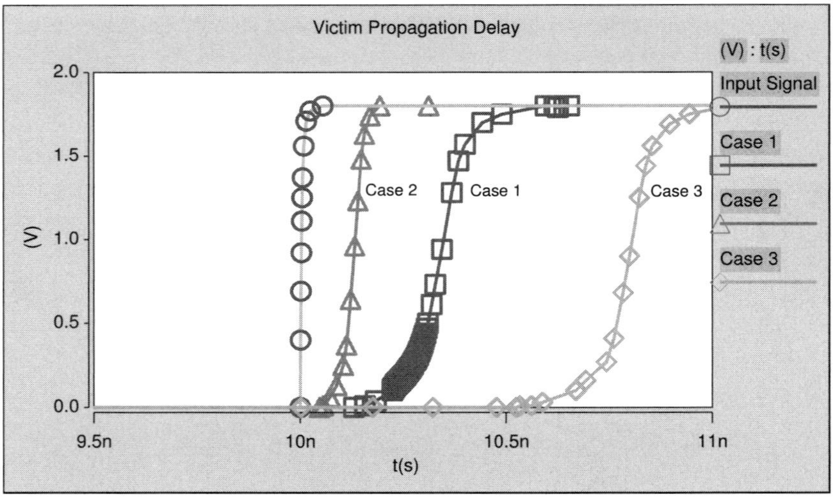

Fig. 11.3. Simulation results of applying a rising transition on an interconnect with the two neighboring interconnects with no transitions, transitioning in the same direction, and transitioning in the opposite direction.

problem since this flow does not use timing analysis tools that represent such a fine granularity of detail.

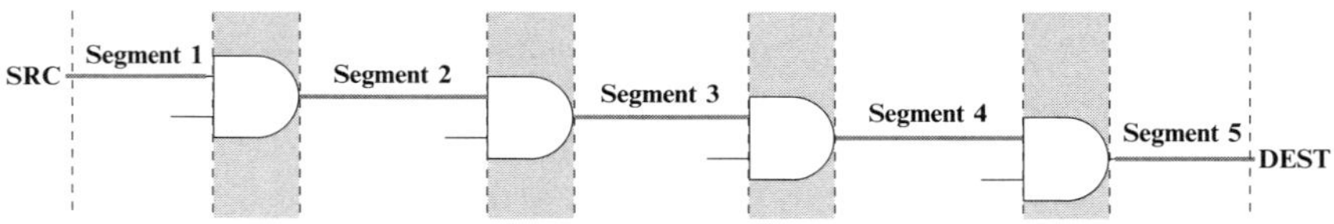

Fig. 11.4. A victim path divided into five segments based on the interconnects between gates. Since timing analysis tools do not consider physical layout aside from back-annotation of parasitic effects, further division is unnecessary.

Each segment of the victim path will have a multitude of coupling capacitance with neighboring interconnects. It is during this phase that the physical layout is taken into consideration. As two interconnects are routed closer and for a longer distance, the coupling capacitance value between the two will increase. So, these physical characteristics will be accurately modeled in the coupling capacitance and can be used to identify *neighboring nets*.

As shown in Figure 11.5, after routing, the victim path may have many aggressor paths that are minimally spaced but only parallel for short distances while having a few aggressors that are parallel to the segment for a significant distance but several pitches away. The relative coupling of the aggressors with

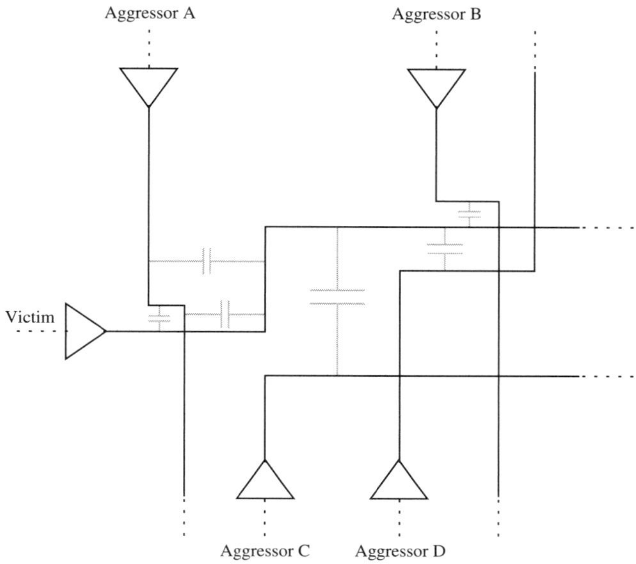

Fig. 11.5. The coupling capacitance effects on a victim line from four aggressors. The amount of coupling is represented in the size of the capacitor, which are larger for longer distances of parallel interconnects. For simplicity, coupling between aggressors are not shown.

the victim is shown by the size of the capacitor. The coupling value between the Victim and Aggressor C will reflect the pitch and parallel distance. This will be larger than the coupling between the Victim and Aggressor A, but less than the coupling between the Victim and Aggressor D. So, even though A is closer to the Victim than C, it is parallel for a very short distance, but since the pitch between D and the Victim is smaller, the coupling will be larger even though the parallel distance is shorter. Even though Aggressor A is physically closer than either Aggressors C or D, it may not be considered a neighboring net if the coupling value does not reach a *minimum threshold*.

A minimum coupling threshold is used to reduce the complexity during analysis by filtering some of the neighboring nets that have almost no effect on the victim path. This will eliminate nets that may be near each other but are routed perpendicularly.

11.3.2 Inducing Coupling Effects

As was shown in Figure 11.3, the direction of transitions on neighboring nets with respect to a segment of the critical path can play a significant role on the propagation delay. However, since a majority of the segments on the critical path are significantly shorter (for 180nm) than the length of the example provided in the Section 11.2, creating transitions that effect one segment may not induce enough noise to significantly effect the delay. So, transitions on neighboring nets of all the segments must be maximized in order to observe any significant change in the delay of the path.

The remainder of this chapter focuses on inducing crosstalk effects to increase propagation delay by maximizing opposite transitions on neighboring nets, but the approach can also be utilized to reduce propagation delay by forcing transitions in the same direction on neighboring nets. There are a variety of options available to drive an opposite direction transition on the neighboring path. The first option is to deterministically generate a pattern that tests the critical path while concurrently justifying the pattern for transitions on neighboring nets [9]. A second option would be to approach this problem with a genetic algorithm, which requires iterative simulations [10]. Our approach combines two established delay testing techniques, path delay fault (PDF) and transition delay fault (TDF) pattern generation, to maximize transitions on the neighboring aggressors.

The PDF patterns are used to establish a base pattern that will ensure testing of the critical paths. Then for each critical path, a list of transition delay faults containing only the neighboring nets for that particular critical path will be used during TDF pattern generation. To ensure the critical path is testable with this new TDF pattern, the PDF pattern is used as a constraint during pattern generation. The direction of the transition of a segment caused by the PDF pattern will determine whether a slow-to-rise (str) or a slow-to-fall (stf) TDF is added to the fault list.

Using the same circuit as before, Figure 11.6 shows the transition created by the PDF pattern in red on the Victim segment. For this example, stf faults are added to the fault list and used for pattern generation. The result of the pattern generated causes Aggressors A and C to remain quiescent for the duration of the test, but causes transitions on Aggressors B and D. Since a stf on Aggressor B was in the fault list, an opposite transition from the critical path segment was successfully generated. However, for Aggressor D, due to a combination of constraints from the PDF and inducing transitions at other fault sites, a rising transition is generated leaving the stf fault undetectable. So, the total number of detectable transition faults relies on the constraints imposed by using the PDF pattern and prior faults that have already been detected.

11.4 Pattern Generation Flow with Neighboring Crosstalk Sensitization

The presented pattern generation process involves three major steps as shown in the flow diagram in Figure 11.7. The three steps consist of parasitic extraction of the physical design; identification of the critical path and segmentation;

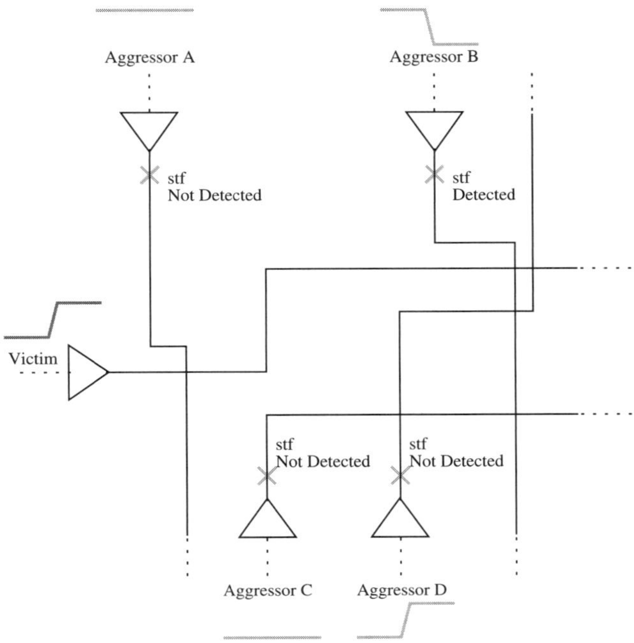

Fig. 11.6. A path delay fault pattern that successfully detects a slow-to-rise fault on the critical path is used as constraints when generating a single pattern that will also detect slow-to-fall transition faults on neighboring aggressors.

and pattern generation, which combines both PDF and TDF pattern generation. Each of these steps take advantage of existing industrial design and test tools. The novel components of this framework that are important for maximizing crosstalk on the critical path have been highlighted.

11.4.1 Parasitic Extraction

Since the pattern generation relies on coupling capacitance for coupling estimation, physical layout is required. For the implementation, the physical parasitic effects of the layout were extracted using Synopsys Star-RCXT [23]. Since the estimate is based on the parasitic coupling capacitance of the layout, another custom tool can be used to extract only the coupling capacitance. However, since the goal is to generate maximum crosstalk on the critical path, extracting the full RC parasitics for timing analysis during this step are also useful.

A 3-D extraction of the layout is obtained and will acquire wire resistances, capacitance to ground, and coupled capacitance. To maintain reasonable runtimes, only parasitics above the minimum coupling threshold are extracted, which can be modified to include greater or fewer capacitances. This initial

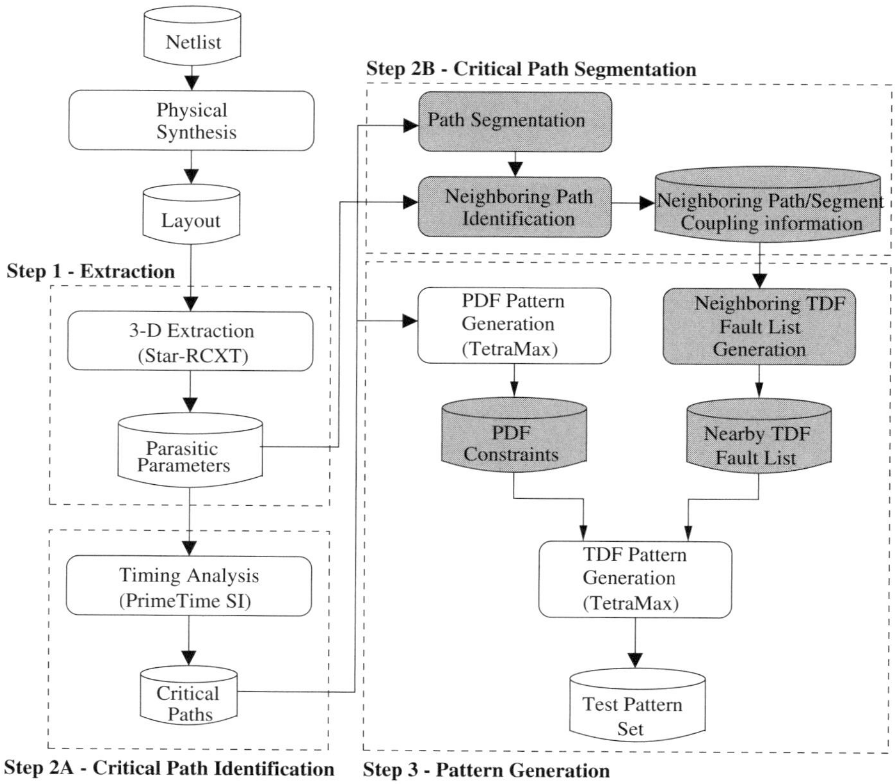

Fig. 11.7. Flow diagram of timing-aware pattern generation to maximize crosstalk in an identified critical path.

filtering criteria limits the number of nets that will be considered as neighboring nets. Depending on the threshold level, the number of neighboring nets potentially could be much higher than necessary by using this form of identification, but the nets that are further away would have significantly smaller coupling capacitance than the closer neighboring nets.

11.4.2 Critical Path Identification and Segmentation

The critical path is identified using PrimeTime SI [23]. PrimeTime SI uses the parasitic information from the extraction tool to determine the critical path. Based on this information, it is assumed the reported path is the critical path that will be used in the remainder of the flow. Future work will incorporate other factors like power supply noise, process variation, and operating conditions into critical path selection. Once identified, the critical path is segmented and coupling information is extracted for neighboring path selection.

Since the tool stores the parasitic coupling information, customized scripts can be used to report this for each segment of the critical path rather than parsing through the parasitics file reported by Star-RCXT. If a segment is coupled to a net in several locations, the PrimeTime SI will sum the coupling capacitances into a single value. Only those segments with a coupling capacitance above a user specified threshold will be identified as a neighboring path.

11.4.3 Test Pattern Generation

The test pattern generation itself can be divided into three (3) components. First, PDF pattern generation to provide constraints during TDF generation. Second, creating the transition delay fault list based on the neighboring nets. Finally, TDF pattern generation.

The PDF pattern generation is rather straight-forward with little modification to default settings. A robust PDF pattern is generated while leaving all don't-care values unfilled. To ensure the pattern is compatible with TDF generation, the pattern is applied using an LOC clocking scheme. From the PDF pattern, any states filled with a care-bit are then extracted and utilized as a constraint during TDF pattern generation.

As addressed in Section 11.3.2, the fault sites for TDF testing are based on the neighboring nets of the critical path, which were identified in Step 2B. The direction of the transition fault is determined by the desired effect on the critical path. When attempting to increase propagation delay, the faults are added in the opposite direction of the critical path segment transition direction. The direction of the transition direction on the critical path segment is specified during critical path identification.

TDF pattern generation is performed with an LOC clocking scheme to remain consistent with the PDF pattern and only needs the constraints extracted from the PDF pattern and the transition delay fault list from the neighboring nets of the critical path in order to generate a pattern that will induce crosstalk onto the critical path. Since each of these items are acquired during pre-processing steps, existing ATPG tools are capable of handling the flow presented. In doing so, a pattern is generated that induces crosstalk such that it will increase the propagation delay through the critical path.

11.5 Experimental Results and Analysis

Rhe pattern generation flow was was implemented on a Linux x86 architecture with a 3.0GHz processor and 1 GB of RAM. The pattern generation procedure was implemented using a combination of C and scripts that used the native

TCL support available in many of the Synopsys tools [23]. Several ISCAS'89 benchmarks were run through physical synthesis using the Cadence Generic Standard Cell Library 3.0 (GSCLib3.0) [24]. Astro [23] was used to perform the placement and routing of the standard cells.

After extracting the parasitic information and identifying the critical paths, the neighboring nets can be specified. In Figure 11.8, a segment of the identified critical path is highlighted in yellow (bold white line in black & white print). The coupled aggressor nets that were identified as neighboring nets are highlighted in red (bold dark lines in black & white print). As can be seen in the figure, the neighboring nets are spaced closely and parallel to the critical path for a majority of the length.

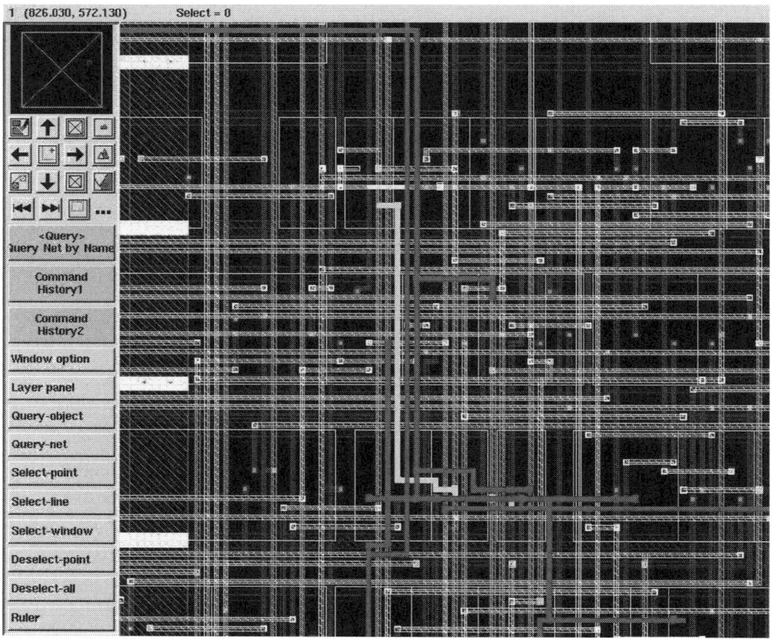

Fig. 11.8. A portion of s38584 showing a segment of the critical path along with the neighboring nets that are coupled with it. The segment is highlighted in yellow (bold white line in black & white print) with the neighboring nets highlighted in red (bold dark lines in black & white print).

PDF Patterns for the critical paths of benchmarks s9234, s15850, s38417, and s38584 were each generated and used as constraints during the final TDF step. To verify the patterns generated by the presented flow, a transistor level simulation was performed using a SPICE model of the layout and NanoSim/VCS [23] to perform the simulations. The results of both the SPICE

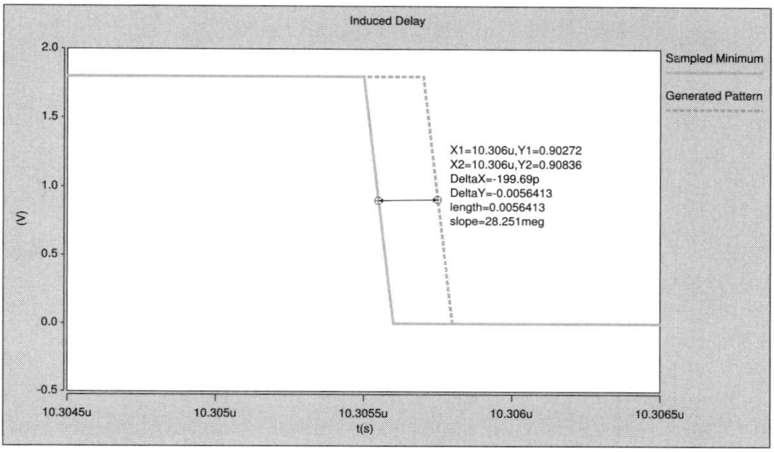

Fig. 11.9. Falling transitions at the endpoint of a critical path for benchmark s38584 for a pattern that was generated using the presented method and a pattern that with minimal delay. The critical path experiences approximately a 200ps increase in delay due to induced crosstalk.

Table 11.1. Percentage increase in propagation delay induced by pattern generated using identified neighboring fault list.

Benchmark	Max % Difference	% Increase Over Min	# of PDF Constraints	Neighb. TDF Fault Sites	Fault Coverage(%)	FaultCoverage w/o Constraints (%)
s9234	7.74	7.04	13	58	25.86	32.76
s15850	2.07	2.05	21	806	28.41	31.76
s38417	10.74	0.40	78	232	9.91	25.43
s38584	7.06	5.42	32	94	30.85	42.55

model simulations and pattern generation flow are shown in Table 11.1. To compare the proposed technique, a random sample of patterns were taken and simulated the results. The values in column 2 of the table shows the percentage difference between the fastest and slowest delays of the random sample. Column 3 compares the delay of our technique with that of the fastest delay measured from the random samples. The delay induced by maximizing transitions on the neighboring nets in s38584 is shown in Figure 11.9. The dashed line represents the falling transition of the generated pattern while the solid line is the propagation delay of the sampled pattern. Column 4 shows the number of constraints (care-bits) generated from the PDF pattern. The fault list size (number of neighbors) for TDF pattern generation is shown in column 5. In columns 6 and 7, the fault coverage for a single generated pattern during TDF generation are shown while being constrained by the PDF care-bits and unconstrained, respectively.

From these results, the proposed flow was able to generate patterns, when minimally constrained by the PDF pattern, that were able to induce crosstalk on the critical path such that it caused the comparable delay to the sampled maximum. As the number of constraints due to the PDF pattern become larger, it becomes more difficult for the ATPG to generate a pattern that can achieve a high fault coverage and create transitions in the correct direction on the neighboring nets. This is particularly true for s38417. When not constrained a single pattern is able to detect 25% of the neighboring faults, however, when constrained, fewer than 10% of the faults are detected and the delay is increased less than a percent. Benchmark s38584 shows an middle ground between the two extreme cases of s15850 and s38417. For s38584, the fault coverage suffers a modest drop due to PDF constraints, which is reflected in the percentage increase over the minimum in column 3.

The order the neighboring faults are detected may also be effecting the fault coverage. If the ATPG is seeking out the hard-to-detect faults first, the ATPG may be defining a larger number of don't-care bits leaving fewer bits to assign to detect other faults. Also, the ATPG may be detecting neighboring fault sites that have nets that are weakly coupled to the critical path segments. Intelligent selection of fault sites may be necessary to reach a higher percentage increase in delay.

As technology nodes continue to shrink, interconnect sidewalls remain to scale disproportionately for the sake of keeping resistance low and length of interconnects continues to become longer, it is expected that the coupling capacitance to increase and the number of neighbors to grow even greater. If these neighboring nets remain outside the cone of the constrained inputs, the method should ease the amount of work the ATPG must do to generate transitions on these neighboring nets. With more neighboring nets switching in the opposite direction, the more delay the critical path will experience.

11.6 Summary

In this chapter, a structural pattern generation flow is presented that uses current path delay and transition delay fault pattern generation algorithms to magnify the effect of crosstalk on critical paths. The physical layout of the design is used to determine which nets have the most impact (i.e. coupling effects) on the critical path and are classified as neighboring nets. These neighboring nets are used as fault sites and the path delay fault pattern used to test the critical path are used during transition delay test pattern generation. Initial results show that this pattern generation flow can increase the propagation delay through a critical path if the path delay fault pattern does not provide too many constraints on the fan-in cones of the neighboring fault sites.

References

1. International Technology Roadmap for Semiconductors (ITRS), 2005
2. S. Yang, C. Papachristou, and M. Tabib-Azar, "Improving Bus Test via IDDT and Boundary scan," in Proc. *Design Automation Conf. (DAC'01)*, pp. 307312, 2001.
3. M. Tehranipour, N. Ahmed and M. Nourani, "Testing SoC Interconnects for Signal Integrity Using Extended JTAG Architecture," *IEEE Transactions on CAD*, vol. 23, issue 5, pp. 800-811, May 2004.
4. M. Tehranipour, N. Ahmed, and M. Nourani, "Multiple Transition Model and Enhanced Boundaryscan Architecture to Test Interconnects for Signal Integrity," in Proc. *Int. Conf. Computer Design (ICCD'03)*, 2003.
5. M. Tehranipour, N. Ahmed and M. Nourani, "Testing SoC Interconnects for Signal Integrity Using Boundary Scan," in proc. *IEEE VLSI Test Symposium (VTS'03)*, Napa, CA, pp. 158-163, 2003.
6. N. S. Nagaraj, P. Balsara, and C. Cantrell, "Crosstalk Noise Verification in Digital Designs with Interconnect Process Variations," in Proc. *Int. Conf. VLSI Design*, pp. 365-370, 2001.
7. M. Cuviello, S. Dey, X. Bai, and Y. Zhao, "Fault Modeling and Simulation for Crosstalk in System-on-Chip Interconnects," in Proc. *Int. Conf. Computer-Aided Design (ICCD'99)*, pp. 297-303, 1999.
8. A. Sinha, S. K. Gupta, and M. A. Breuer, "Validation and Test Issues Related to Noise Induced Byparasitic Inductances of VLSI Interconnects," *IEEE Trans. Adv. Packaging*, pp. 329339, 2002.
9. W. Chen, S. Gupta, and M. Breuer, "Test Generation for Crosstalk-Induced Delay in Integrated Circuits," in Proc. *Int. Test Conf. (ITC'99)*, pp. 191200, 1999.
10. A. Krstic, J. Liou, Y. Jiang, K. T. Cheng, "Delay Testing Considering Crosstalk Induced Effects," in Proc. *Int. Test Conf. (ITC'01)*, pp. 558-567, 2001.
11. S. Naffziger, "Design Methodologies for Interconnects in GHz+ ICs," presented at the *Int. Solid-State Conf.*, 1999.
12. R. Senthinatharr and J. L. Prince, *Simultaneous Switching Noise of CMOS Devices and Systems*, Kluwer Academic Publishers, 1994.
13. Y.-S. Chang, S. K. Gupta and M. A. Breuer, "Analysis of Ground Bounce in Deep Sub-Micron Circuits," in Proc. *IEEE VLSI Test Symposium (VTS'97)*, pp. 110-116, 1997.
14. W. Chen, S. Gupta and M. Breuer, "Analytic Models for Crosstalk Delay and Pulse Analysis Under Non-Ideal Inputs," in Proc. *Int. Test Conf. (ITC'97)*, pp. 809-818, 1997.
15. W. Chen, S. Gupta and M. Breuer, "Test Generation for Crosstalk-Induced Faults: Framework and Compuational Results," in Proc. *Asian Test Conf. (ATS'00)*, pp. 305-310, 2000.
16. A. Sinha, S. Gupta and M Breuer, "An Enchanced Test Generator for Capacitance Induced Crosstalk Delay Faults," in Proc. *Asian Test Conf. (ATS'03)*, pp. 174-177, 2003.
17. J. Wang, Z. Yue, X. Lu, W. Qiu, W. Shi, D. M. H. Walker, "A Vector-based Approach for Power Supply Noise Analysis in Test Compaction," in Proc. *Int. Test Conf. (ITC'05)*, 2005.

18. N. Ahmed, M. Tehranipoor and V. Jayaram, "Supply Voltage Noise Aware ATPG for Transition Delay Faults," to appear in *IEEE VLSI Test Symposium (VTS'07)*.
19. A. Kokrady and C.P. Ravikumar, "Static Verification of Test Vectors for IR-drop Failure," in Proc. *Int. Conf. on Computer-Aided Design (ICCAD'03)*, pp. 760-764, 2003.
20. A. Krstic, Y. Jiang and K. Cheng, "Pattern Generation for Delay Testing and Dynamic Timing Analysis Considering Power-Supply Noise Effects," in *IEEE Transactions on CAD*, vol. 20, No. 3, pp. 416-425, 2001.
21. N. Ahmed, M. Tehranipoor and V. Jayaram, "A Novel Framework for Faster-than-at-Speed Delay Test Considering IR-Drop Effects," in Proc. *Int. Conf. on Computer-Aided Design (ICCAD'06)*, 2006.
22. M. Nourani, M. Tehranipoor and N. Ahmed, "Pattern Generation and Estimation for Power Supply Noise Analysis," in Proc. *IEEE VLSI Test Symposium (VTS'05)*, pp. 439-444, 2005.
23. Synopsys Inc., "User Manuals for SYNOPSYS Toolset Version 2006.09," Synopsys, Inc., 2006.
24. http://crete.cadence.com, $0.18\mu m$ standard cell GSCLib library version 2.0, Cadence, Inc., 2005.

12

Testing SoC Interconnects for Signal Integrity

As the technology shrinks and working frequency is already in multi gigahertz range, design and test of interconnects are no longer trivial issues. Stuck-at fault model can detect bridge and open faults. But, transient, timing, and noise related faults cannot be detected using traditional stuck-at and delay test patterns. New design-for-test (DFT) methods and pattern generation algorithms are required to effectively consider such faults in high-speed designs. The issues related to interconnect design and test will soon become dominant as technology scales going into sub-50nm. Specifically, various issues of signal integrity loss including detection and diagnosis are becoming a great challenge.

This chapter presents an enhanced boundary scan architecture to test high-speed interconnects for signal integrity. This architecture includes: a) a modified driving cell that generates patterns according to multiple transition fault model; and b) an observation cell that monitors signal integrity violations. To fully comply with conventional JTAG, two new instructions are used to control cells and scan activities in the integrity test mode.

12.1 Introduction

12.1.1 Technology Scaling Effects on Signal Integrity

The number of cores in a system-on-chip (SoC) is rapidly growing and therefore, the number of interconnects is intensively increased. With shrinking technology, existence of long interconnects in SoCs and rapid increase in the working frequency (currently in multi gigahertz range) of the SoCs, the signal integrity has become a major concern for design and test engineers. Use of nanometer technology in SoCs magnifies the cross-coupling effects among the interconnects. These effects include coupling capacitance and mutual inductance that affect the integrity of a signal by creating noise and delay. The noise effect can appear as overshoot and ringing. The former is known to shorten transistor lifetime [1] and the latter cause intermittent functional errors. Slowdown and performance degradation are often the result of delay effects. Although various parasitic factors for transistors can be well controlled during fabrication, the parasitic capacitances, inductances and their cross-coupling effects on the interconnects are much more difficult to control. These parasitic factors play a significant role in the proper functionality and performance of high-speed SoCs.

Signal integrity is the ability of a signal to generate correct responses in a circuit. It generally includes all effects that cause design to malfunction due to distortion of the signal waveform. According to this informal definition, a signal with good integrity has: (*i*) voltage values at required levels and (*ii*) level transitions at required times. If signal integrity losses (i.e., noise and delay) on an interconnect are within the defined safe margin, they are acceptable since they do no harm. Otherwise, they may cause an intermittent logic-error, performance degradation, shorter life time and reliability concern [2]. For example, an input signal to a flip-flop with good signal integrity arrives early enough to guarantee the setup and hold time requirements and it does not have spikes causing undesired logic transition (ringing).

The impact of process variations and the way they affect circuit operation are important issues in very deep submicron (VDSM) [3]. Process variations and manufacturing defects both may lead to unacceptable level of noise and delay. The goal of design for VDSM phase is to minimize noise and delay. However, it is impossible to check and fix all possible signal integrity problems during VDSM design by design rule checking (DRC), validation and analysis. Process variations and manufacturing defects may lead to unexpected changes in coupling capacitances and mutual inductances between interconnects. They in turn result in loss of signal integrity (e.g. glitches and excessive delay), which may eventually cause logic error and failure in the chip. The impact of spot defects and process variations on the magnitude of inductance induced noise are reported in [4]. The authors reported that the maximum crosstalk pulse considering process variations is almost twice the value for nominal set of parameters. Since it is impossible to predict the occurrence of defects causing noise and delay, signal integrity testing is essential to ensure error-free operation of the chip and must be addressed in manufacturing testing.

In recent years, various methodologies to test signal noise and skew on interconnects, due to different sources, are reported in literature [5] [6]. Regardless of the method used to detect integrity loss, a mechanism is also needed to coordinate activities in integrity test session. Boundary scan test methodology is believed as one of the best choices as it includes the capability of accessing interconnects. It was initially introduced to facilitate testing of complex printed circuit boards.

IEEE 1149.1 Boundary Scan Test Standard

The IEEE 1149.1 Boundary Scan Test Standard [7], also known as Joint Test Action Group (JTAG), has been widely accepted and practiced in the test community. The standard provides excellent testing features with low complexity but was not intended to address high-speed testing and signal integrity loss. The standard, nevertheless, provides testing of core logic and the interconnects among them. Interconnects can be tested for stuck-at, open and short faults. Unfortunately, the standard boundary-scan architecture has shortcomings for timing-related tests. This drawback is due to the time interval between the update of test stimulus and the response capture, an interval which spans at least 2.5 test clock cycles *2.5 TCKs* [8]. In this chapter, the standard boundary-scan architecture is extended to test interconnects for noise and skew violations. While the focus is on the interconnects, any non-modeled fault (inside or outside cores) that manifest itself as integrity loss on interconnects will be also detected by the proposed method.

12.1.2 Overview

Signal Integrity Modeling and Analysis

Various signal integrity problems have been studied previously for radio frequency circuits and recently for high-speed deep-submicron VLSI chips. Maximum aggressor (MA) fault model [9] is one of the fault models proposed for crosstalk. Various approaches to analyze the crosstalk are described in [10] [11] [12]. Specifically, analysis of interconnect defects coverage of test sets is explained in [10]. The authors address the problem of evaluating the effectiveness of test sets to detect crosstalk defects in interconnections of deep submicron circuits. Interconnect design for GHz+ integrated circuits is discussed in [13]. The author observed that chips failed when a specific test pattern (not included in the MA model) was applied to the interconnects due to overall effect of coupling capacitances and mutual inductances. Similarly, according to [14], the worst-case switching pattern to handle inductive effects for multiple signal lines is not included in the MA fault model.

Several researchers have investigated test pattern generation for crosstalk noise/delay and signal integrity [15] [16]. Authors in [15] and [16] proposed test pattern generation for crosstalk-induced noise and delay, respectively. A test pattern generation algorithm based on considering the effect of inputs and parasitic RLC elements of the interconnect has been proposed in [16].

Test Methodologies

There is a long list of possible design and fabrication solutions to reduce signal integrity problems on the interconnect. None guarantees to resolve the issue perfectly [2]. Several self-test methodologies have been developed to test interconnects for signal integrity in high-speed SoCs. A double sampling data checking (DSDC) technique is used to capture noise-induced logic failures in on-chip buses [17]. At-speed test of crosstalk in chip interconnects [5], testing interconnect crosstalk defects using on-chip processor [6], a BIST (Built-In Self-Test) based architecture to test long interconnects for signal integrity and using boundary scan and I_{DDT} for testing buses [18] are some of the proposed methods. Even short interconnects, especially those located near long interconnects, are also susceptible to integrity problem. Therefore, in near future methodologies both short and long interconnects are required to be tested [19].

The IEEE Std. 1149.1 (*dot1*) structures and methods are intended to test static, i.e, DC-coupled, single ended networks. The IEEE Std. 1149.6 [30] standard provides a solution for testing newer digital network topologies, such as AC-coupled differential interconnects on very high speed digital data paths. The approach presented in this chapter is similar to this standard draft in enhancing the JTAG standard and its instructions for testing high-frequency behaviors. However, there are fundamental differences in that the standard draft does not consider coupling effects among the interconnect lines.

However, there is a controversy as to what patterns trigger the maximal integrity loss. Specifically, in the traditional MA model that takes only coupling C into account, all aggressors make simultaneous transitions in the same direction while the victim line is kept quiescent (for maximal ringing) or makes an opposite transition (for maximal delay). When mutual inductance comes to play, researchers presented other ways (e.g. weighted pseudorandom or deterministic) to generate test patterns to create maximal integrity loss [16] [15].

There are two ways to send test patterns on the interconnects. First, conventional scan method which sends test patterns one by one serially. The second method is compressing test patterns and scanning them in then decompress them on chip. There are thousands of interconnects in large SoCs and using conventional method is very time consuming. Researches showed that compressing the test patterns will considerably reduce the test application time. There are several compression techniques for scan based testing [35], [36], and [37]. Test data compression using don't cares [35], test data compression using Golomb codes for SoCs [36] and finding the minimum compacted test sequence without considering the don't cares for continuous scan [37] are some of the proposed techniques.

Integrity Loss Sensor Cell

Due to more and more concerns about signal integrity loss in gigahertz chips and the fact that their occurrence must be captured on chip, researchers presented various on-chip sensors. Many of such integrity loss sensors (ILS) are amplifier-based circuits capable of detecting violation of voltages and delay thresholds. A BIST structure using D flip-flops has been proposed to detect deviation of the propagation delay of operational amplifiers [20]. During test mode, the Op Amp under test is placed in a voltage follower configuration in order to detect its slew-rate deviation, or in a comparator configuration in order to detect its signal propagation delay deviation. A test methodology targeting bus interconnects defects using I_{DDT} and boundary scan has been presented in [18]. In [18] a built-in sensor is integrated within the system. This sensor is an on-chip current mirror converting the dissipated charges into the associated test time.

Reference [21] presented a more expensive but more accurate circuits to measure jitter and skew in the range of few picoseconds. This circuit, called EDTC, samples signals in non-intrusive way and sends out the test information through its low speed serial information. When the cost is not a concern, the concept of accurate signal monitoring has been followed up by researchers even through idea of on-chip oscilloscope [22]. The authors in [22] presented a sample and hold circuit that probes the voltage directly within the interconnects. While expensive, calibration and waveform measurements and even reconstruction all are possible.

To detect delay violation, an integrity loss sensor (ILS) has been designed in [23] which is flexible and tunable for various delay thresholds and technologies. In [23], the acceptable delay region is defined as the time period from the triggering clock edge during which all output transitions must occur. A double sampling data checking based on-line error detector circuit to test multiple-source noise effects in on-chip buses is proposed in [17].

Modified Boundary Scan and IEEE Standards

Most of the early work in testing interconnects using boundary scan method focused on the delivery of deterministic patterns to the interconnects under test at board level. BIST test pattern generators for board level interconnect testing and delay testing are proposed in [24] and [25], respectively. A modified boundary-scan cell using an additional level sensitive latch (called Early Capture Latch or ECL) was proposed in [25] for delay fault testing. A test methodology targeting defects on bus structures using I_{DDT} and boundary scan has been presented in [18]. The motive was to latch the data at the core input pins as soon as the output cells are updated for delay analysis and to capture the input pin data in the capture state. An additional control circuitry is designed in [26], Early Capture Control Register, to control the relative timing between the update in output cells and the falling edge of Early Capture. The area overhead of the special control circuitry is a drawback of this method.

IEEE P1500 standard defines a core test interface between an embedded core and the system logic to facilitate core test reuse and core interoperability [27]. P1500 proposes standardization of the Core Test Wrapper and the Core Test Language (CTL). The architecture consists of test access mechanism (TAM) for delivering the test patterns and responses in parallel, standard core test wrappers that can isolate the cores and provide different test modes, and a test controller for controlling the wrapper and TAM. Serial test access can be always done by using the Serial Interface Layer (SIL) provided by the P1500 wrapper, which is mandatory. The SIL method is similar to boundary scan in terms of serial data transfer and therefore, conceptually the extended SIL architecture can be used for various test applications at the system level in general and in integrity test, in particular.

IEEE 1149.4 mixed-signal test bus standard [28] was proposed to allow accesses to the analog pins of a mixed-signal device. The analog boundary scan cells or analog boundary modules (ABMs) are placed on the analog inputs and outputs. In the digital test mode, the ABMs behave digitally. In analog mode, ABMs allow stimulus (AC/DC current or voltage) to be applied. In addition to the ability to test interconnects using digital patterns, 1149.4 includes the ability to measure actual passive components, such as resistors and capacitors. Another feature of 1149.4 is that each pin acts as a real-time probe, sometimes referred to as a virtual nail. This standard cannot support the high frequency phenomena such as crosstalk on the interconnects. Reference [29] proposes a method to simplify development of a mixed signal test standard by adding the analog interconnect test to 1149.1.

IEEE 1149.6 provides a solution for testing AC-coupled interconnects between integrated circuits on printed circuit boards and in systems [30]. The standard adds a DC blocking capacitor to each interconnect under test to disallow the DC signals. Thus, 1149.6 cannot test integrity loss due to low-speed but very sharp-edge signals that are known to cause overshoots and noise. The sensors in the proposed architecture in this chapter sit in the observation boundary scan cells to detect such scenarios. Finally, using differential drivers in the modified cells in 1149.6 makes the cells more expensive and less flexible in adopting other type of noise detector/sensors.

Various issues on the extended JTAG architecture to test SoC interconnects for signal integrity are reported in [23]. The maximum aggressor (MA) fault model was used to generate and apply test patterns to the interconnects by the modified boundary scan cells placed at the output of a driving core (at the beginning of interconnects). A second type of modified boundary scan cells positioned at the input of a driven core (at the end of interconnects) collect the integrity loss information. The drawback is that MA fault model does not not take inductance effects into account.

In [23], it was assumed that the test patterns have been generated based on a fault model including inductance. The test patterns are scanned by an external tester into the boundary scan cells and applied to the interconnects. The drawback is that the proposed method is time consuming because of scanning the test patterns through scan cells. In this chapter, a new test pattern generation technique is presented which generates and applies test patterns almost at the speed of test clock (TCK). Also, a new fault model is proposed which covers all possible transitions on the interconnects under test. This model also covers MA and some specific cases presented in [13].

12.1.3 Overview

The main contribution of this chapter is an on-chip mechanism to extend JTAG standard to include testing interconnects for signal integrity. Upon this extension noise and skew violations occurring on the interconnects of high-speed SoCs due to any manufacturing defects can be tested using JTAG boundary scan architecture.

Initially, a new fault model is presented called multiple transition (MT), and its corresponding test pattern generation using enhanced boundary scan architecture. The modified driving-end boundary scan cells (called PGBSC) receives few seeds and generates multiple transition (MT) patterns at speed for stimulating integrity violations. The MT pattern set is a superset of MA set and is much more capable of testing the capacitive and inductive couplings among interconnects. However, MT requires larger area overhead and generate significantly larger number of patterns.

The modified receiving-end boundary scan cells (called OBSC) monitor signals received from the system interconnect. Integrity loss sensor cells (ILS) are incorporated in the boundary scan cell which record the occurrence of signal entering the vulnerable region over a period of operation. Using two new instructions in JTAG architecture, the integrity test information is sent out for final test analysis, reliability judgment and diagnosis.

In the second part of this chapter new cells are designed to implement MA model and generate corresponding patterns. Since all aggressor lines in MA model switch at the same direction, therefore it is easier to implement using minor modifications to boundary scan cells. To reduce test time and test data volume, a new compression method is developed. The regularity existing in the MA model patterns provide the opportunity of reducing the test data volume significantly.

12.2 Testing Interconnects Using Multiple Transition (MT) Fault Model

The MA fault model [9] is a simplified model used by many researchers mostly so far for crosstalk analysis and testing of long interconnects. This model, shown in Figure 12.1, assumes the signal traveling on a victim line V may be affected by signals/transitions on other aggressor line(s) A in its neighborhood. The coupling can be represented by a generic coupling component Z. The result, in general, could be noise (causing ringing and functional error) and delay (causing performance degradation). However, there is a controversy as to what patterns trigger the maximal integrity loss. Specifically, in the traditional MA model that takes only coupling C into account, all aggressors make a same simultaneous transition in the same direction while the victim line is kept quiescent (for maximal ringing) or makes an opposite transition (for maximal delay). When mutual inductance comes into play, some researchers showed that the MA model may not reflect the worst case and presented other ways (pseudorandom, weighted pseudorandom or deterministic) to generate test patterns to create maximal integrity loss [15] [16]. However, implementing MA model on boundary scan cells required smaller modification.

As reported in [13], a chip failed when the nearest aggressor lines change in one direction and the other aggressors in the opposite direction. This and many similar carefully chosen scenarios (e.g. [15] [16]) cannot be covered by the MA fault model. Exhaustive testing covers all situations but is very time consuming because the number of test patterns is huge. Additionally, exhaustive or pseudorandom patterns include some cases that aggressors are in quiescent mode and obviously do not maximally affect the victim line for noise and delay. Therefore, they need not be considered in the model or pattern generators. Based on these observations and empirical evidence reported by researchers, a new fault model and its corresponding test set are defined which covers *all* transitions on victim and *multiple* transition on aggressors.

As a motivating example, Figure 12.2 shows the simulation results of two MT-patterns (*i*) 0110110 → 1001001 and (*ii*) 0110101 → 1001010 and one the MA-pattern i.e. 1110111 → 0001000 applied to a seven interconnect system while the middle wire is the victim and the others are aggressors. Extraction and simulation are done by OEA tool (BUSAN) [38] and TISPICE [39], respectively. The interconnect model is a distributed RLC and coupling capacitances and mutual inductances are considered between the lines using OEA tool for $0.18\mu m$ technology. As shown, the MT-patterns create more delay compared to the MA-pattern, ranging from 35 to 70ps depending on the buffer size. Therefore, the MA-patterns may not be able to generate maximum noise/delay on the victim line especially when inductance is included. In general, this is the main reason why some researchers (e.g. [13]) reported scenarios in which test patterns not covered by the MA model failed a chip.

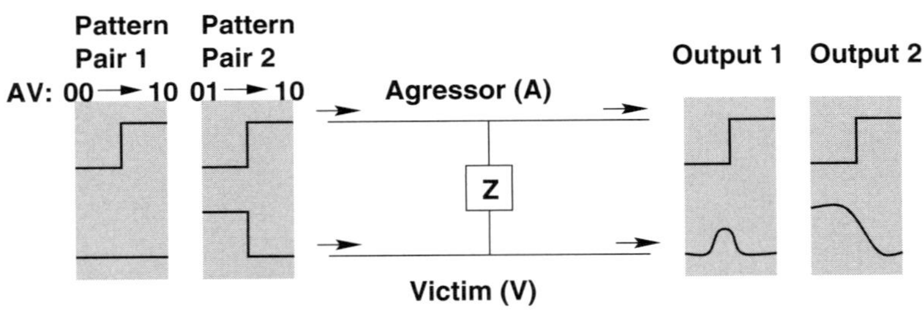

Fig. 12.1. Signal integrity fault model [32].

The main idea behind the multiple transition (MT) fault model is given a single victim and limited number of aggressors, all possible transitions on victim and multiple direction transitions on aggressors are covered. The basis of the MT model is still the effect of coupling components as shown in Figure

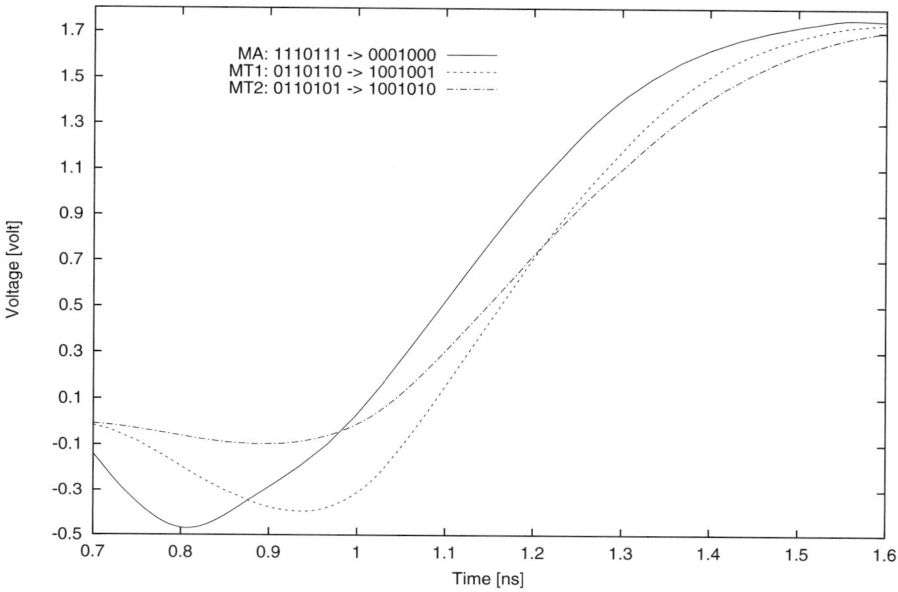

Fig. 12.2. Comparison between the MT and MA models [32].

12.1. In the MT model all possible transitions on the victim and aggressors are applied, while in the MA model only a subset of these transitions are generated. Another difference between the MT and MA is that the aggressors in the MA always change in the same direction (see Figure 12.3). Briefly, the MA-pattern set is a subset of the MT-pattern set. Note carefully that the MT model is not an exhaustive model because it does not cover quiescent cases of aggressor lines for which integrity loss will not be obviously maximal. For example, assume that there are three interconnects and the middle one is victim. Figure 12.4 shows all possible transitions on three line interconnects where the middle one is victim based on the MT fault model. The test patterns for signal integrity are vector-pairs. For example when the victim line is kept quiescent at '0' (column 1), four possible transitions on the aggressors are examined. For example, the first pair is '000' and '101' in which aggressors change from '0' to '1'. The MA-patterns (a subset of MT-patterns) are shaded in Figure 12.4.

Four cases are examined for each victim line when victim line is quiescent at '0', '1' or changes from '0' to '1' or '1' to '0'. As shown in Figure 12.4, the number of required test patterns to cover all possible transitions on the three interconnects is $4*2^{3-1}=16$ when the middle line is in the victim mode. The total number of required test patterns is 3*16=48 when all three lines

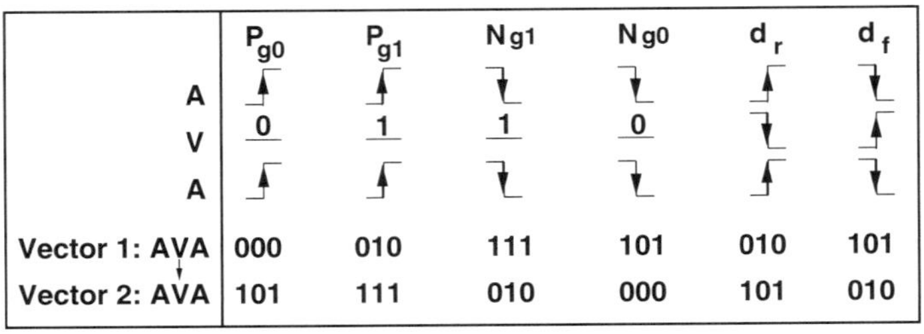

Fig. 12.3. The MA fault model and test patterns [32].

Quiescent at 0 x0x x0x	Transition 0 → 1 x0x x1x	Quiescent at 1 x1x x1x	Transition 1 → 0 x1x x0x
000 101	000 111	010 111	010 101
001 100	001 110	011 110	011 100
100 001	100 011	110 011	110 001
101 000	101 010	111 010	111 000

Fig. 12.4. All transitions on a three interconnect system that the MT and MA (shaded) models generate [32].

are examined in the victim mode. Therefore, in general the number of test patterns for a group of m interconnects is $N_P = m \cdot 4 \cdot 2^{m-1} = m \cdot 2^{m+1}$. When m increases the number of test patterns increases exponentially. Simulations show that in an interconnect system the lines which are far away from the victim cannot affect much on the victim line [13]. Therefore, the number of lines (aggressor) after and before the victim line can be limited. k is defined as the *locality factor* that is empirically determined showing how far the effect of aggressor lines remain significant. Briefly, total number of patterns to be generated will be $m \cdot 2^{m+1}$, where $m = 2k + 1$. By choosing k (e.g. $k < n$ or $k << n$, where n is the total number of interconnects under test) user can do tradeoff between time and accuracy.

Figure 12.5 shows the simulation results, using OEA and TISPICE tools [38] [39], of different number of aggressors in the victim neighborhood while victim line is quiescent at 0 for $V_{dd} = 1.8V$. As shown, when the number of interconnects on either side of the victim increases the noise on the victim increases. The glitch height and period mainly depends on the driver strength. As the driver strength increases, the glitch height increases and the period decreases. The peak noise voltage difference between two locality factors of k=3 and k=4 is $V_{peak}(k = 4) - V_{peak}(k = 3)$=0.048v which for many applications can be assumed negligible. If that is the case, k=3 is the locality factor in the simulation. The percentage difference between glitches of different distributions with respect to k decreases with increase in driver strength [9].

Note that finding such locality factor is technology-dependent (e.g. parasitic RLC values) and application-dependent (e.g. depending on driven core or shielding techniques). However, once a user, based on the application and accurate simulation provides it, the total number of patterns and time to test integrity faults will be significantly reduced. Our observation is that in practice, for current technologies and simulation based on the reduced order models [41], a small k (often less than 4) can produce accurate behavior of the interconnects. In any case by choosing k user can always tradeoff between longer simulation time and accuracy.

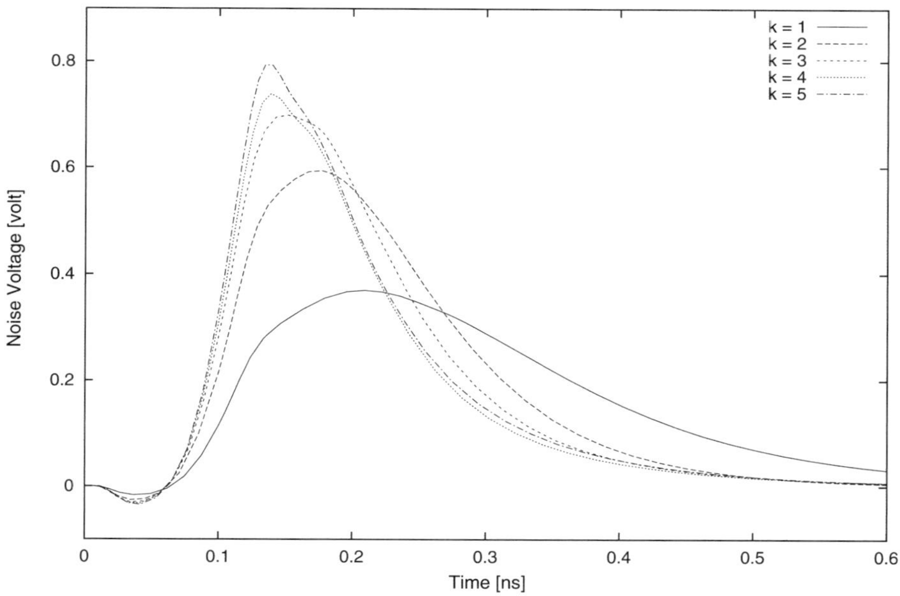

Fig. 12.5. Simulation results for different k [32].

12.2.1 Enhanced Boundary Scan Cells

Boundary scan is a widely used test technique that requires boundary scan cells to be placed between each input or output pin and the internal core logic. The standard provides an efficient mechanism for functional testing of the core logic and the interconnects. Figure 12.6 shows a conventional standard boundary scan cell (*BSC*) with shift and update stages. $Mode = 1$ puts the cell in the test mode. The data is shifted into the shift register (*Shift-DR* state) during scan operation. Test patterns scanned into the boundary scan cells through the scan-in port *TDI* are applied in parallel during the *Update-DR* state (*UpdateDR signal*). Circuit response is captured in parallel by the boundary scan cells connected between internal logic and output pins and is scanned out through the scan-out port *TDO* for observation [28].

Using the JTAG standard (*IEEE 1149.1*), the interconnects can be tested for stuck-at, open and short faults. This is possible by *EXTEST* instruction by which the TAP controller isolates the core logic from the interconnects using the BSCs. However, *EXTEST* was not intended to test interconnects for signal integrity. New cells and instructions are proposed for signal integrity testing. For this purpose, some minor modifications are applied to the standard architecture to target the interconnects for signal integrity.

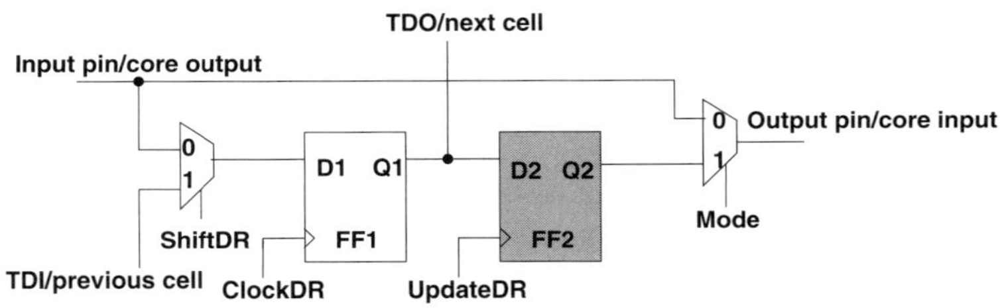

Fig. 12.6. A standard boundary scan cell [32].

Pattern Generation BSC (PGBSC)

Analysis of the MT fault model test vector-pair shows that in some transitions the victim line should remain quiescent, while aggressor lines change. In some other transitions, both the victim and aggressors lines change. Additionally, in all cases the aggressor lines change from one value to another ('0' to '1' or '1' to '0') at every clock, while the victim line value changes every two clocks. This important observation helps us to design a circuit to efficiently generate the MT test patterns. Figure 12.7 shows the test vectors of Figure 12.4 resorted to make the point clear. Each row clearly shows that the victim changes every two clocks and aggressors change every clock.

Each row in Figure 12.7 needs one seed. For example, given seed='000', the test patterns in the first row are generated, which are ('000','101'),

('101','010'), ('010','111') and ('111','000'). For covering all possible transitions as shown in Figure 12.7, four seeds are required. In the above three interconnect system, the seeds are '000', '001', '100' and '101'. The total number of required seeds to cover all lines in victim mode in a three interconnect system is 3*4=12. For a group of m interconnects, where $m = 2k + 1$, the total number of seeds are $N_S = m \cdot 2^{m-1}$=$(2k+1) \cdot 2^{2k}$, in which 2^{m-1} shows all possible combinations of $m - 1$ aggressor lines. Note carefully that as discussed in Section 12.2, since the locality factor k is very small, total number of seeds remains very limited. For example, to stimulate maximum delay in interconnect according to the results in [14], only two seeds are sufficient.

Seed	Quiescent at 0 x0x x0x	Transition 0 → 1 x0x x1x	Quiescent at 1 x1x x1x	Transition 1 → 0 x1x x0x
000	000 101	→ 101 010	→ 010 111	→ 111 000
001	001 100	→ 100 011	→ 011 110	→ 110 001
100	100 001	→ 001 110	→ 110 011	→ 011 100
101	101 000	→ 000 111	→ 111 010	→ 010 101

Fig. 12.7. Resorted test pattern for a three interconnect system [32].

The MT test patterns cover the MA test patterns. In Figure 12.7, the shaded patterns show the MA patterns which are generated based on only two seeds. It shows that after applying the first seed, '000', the generated test patterns cover the P_{g_0}, d_f, and P_{g_1} faults. The generated test patterns after applying the second seed, '101', cover d_r, N_{g_0} and N_{g_1}. Therefore, by such reordering only two seeds are sufficient for covering all 12 test patterns in the MA fault model.

As mentioned before, a pair of test vectors are required to test interconnects for signal integrity. These patterns can be applied to the interconnects using conventional boundary scan architecture. For applying each pair, the first and second patterns are scanned into the conventional BSCs and stored in FF_2 and FF_1, respectively (see Figure 12.6). Then, using *UpdateDR*, they are applied to the interconnects. Scanning and applying patterns in this way is very straightforward but needs a large number of clocks which increases the overall test time. A hardware-based method is proposed for test pattern generation based on the MT fault model. Test pattern generation is performed at the input side of the interconnects, that is the output side of a core which drives the interconnects. The new BSC that generate test patterns is called pattern generation BSC (PGBSC).

Boundary scan cell can be utilized to support the MT model. FF_2 in conventional boundary scan cell (shaded flip-flop in Figure 12.6) can be used to generate test patterns based on a given seed. FF_1 is used to initialize FF_2. A T flip flop divides the clock by half and *UpdateDR* plays the same role of clock in driving FF2 (see Figure 12.8). First, the seed comes from TDI and is sent into the FF_2 through FF_1. Victim/aggressor select signals are then scanned into $FF1$ through TDI to select victim and aggressor lines. Note carefully that only seeds need to be scanned in instead of the exact test patterns. This significantly reduces the number of required clock cycles for applying test patterns to the interconnects in a system-chip.

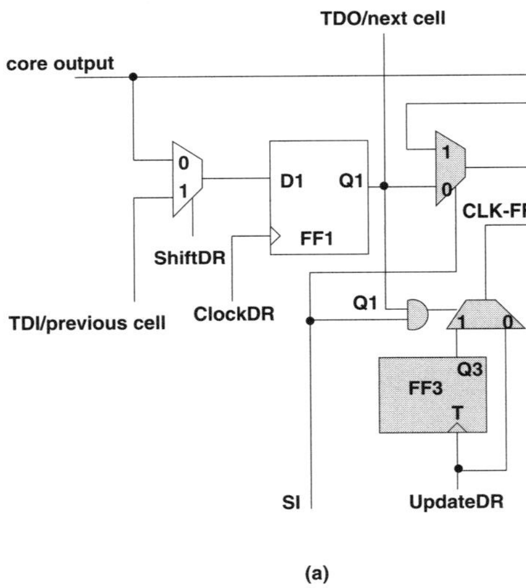

PGBSC Mode	Q1	SI
Victim	1	1
Aggressor	0	1
Non-integrity	x	0

(b)

Fig. 12.8. PGBSC structure [32].

In addition to its normal mode, PGBSC should work in two new operational modes, victim and aggressor in signal integrity test mode. The PGBSC architecture is shown in Figure 12.8(a). Note that the additional components in PGBSC are located in the test path and therefore no additional delay is imposed in the normal mode. Only one extra control signal (SI) is needed in this architecture. The SI signal is generated by a new instruction, to be explained in Section 12.2.2. The PGBSC generates the required test patterns for covering the MT fault model. Figure 12.8(b) shows the operation modes of the PGBSC. Depending on the select line of the MUX attached to FF3, this architecture has three modes:

1. **Victim mode:** Q_3 is selected. *UpdateDR* is divided by two and applied to the FF_2. By every two *UpdateDRs*, the complemented data is generated in $\overline{Q_2}$ and it is transferred to the output pin.

2. **Aggressor mode:** *UpdateDR* is selected, but PGBSC is in signal integrity mode. *UpdateDR* is applied to the FF_2. By each *UpdateDR*, the complemented data is generated in $\overline{Q_2}$ and it is transferred to the output pin through feedback and Q_2.

3. **Non-integrity mode:** *UpdateDR* is selected. It is the normal test mode of the PGBSC (e.g. conventional boundary scan test) and *UpdateDR* is applied to the FF_2.

Figure 12.9 shows the operation of a PGBSC. If PGBSC is in the victim mode, *UpdateDR* is divided by two and generates $CLK\text{-}FF_2$. If the initial value in Q_2 is '0', then $\overline{Q_2}$ is '1' and is applied to D_2 through the feedback. Every two activations of *UpdateDR*, the content of FF_2 is complemented. On the other hand, if PGBSC is in the aggressor mode, $CLK\text{-}FF_2$ has the same frequency as *UpdateDR* and by each *UpdateDR* the content of FF_2 is complemented. As shown in Figure 12.8, $\overline{Q_2}$ is complemented by each $CLK\text{-}FF_2$ while Q_2 is applied to the output (to the interconnect). As a practical point, interconnects under integrity test are physically close and their corresponding PGBSC cells are often close. Therefore, skew on the *UpdateDR* signal used for synchronizing pattern application should not be a problem.

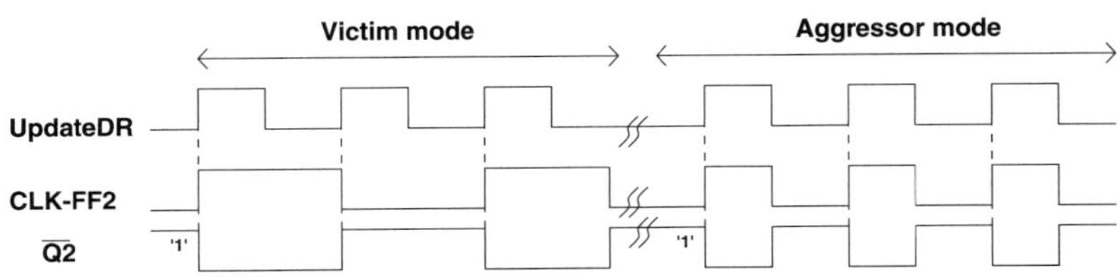

Fig. 12.9. Operation of the PGBSC [32].

Each interconnect can act as a victim or an aggressor. Therefore, in each test session the victim interconnect should be specified. After performing the test process on a victim, it will become an aggressor for other new victims. Briefly, for complete interconnect testing, the victim line rotates. One of the major advantages of limiting the number of aggressor lines is that parallel testing of the interconnects becomes possible. Because, in each step of test at most k agressor lines each are selected as aggressors before and after the victim line. An encoded data is used to specify the victim which is called *victim-select data*.

Table 12.1 shows the victim-select data for a n-bit interconnect system, to be scanned and stored in FF1 when the locality factor is k=2. After specifying the victim, the test vectors are generated by the PGBSC and applied to the interconnects. Then, the new victim line is specified and the process will be repeated for the new victim. As shown in Table 12.1, when a pattern '100100...100' is scanned into n PGBSCs, the first line is victim and the next two lines are aggressors for the first interconnect as well as the fourth one. Therefore, $\lceil n/(k+1)\rceil$=$\lceil n/3\rceil$ victims are tested simultaneously. As shown, with one clock the victims locations rotates ('0100100...10'). Only two rotations are enough for covering the whole interconnects to act as the victim and aggressor.

Table 12.1. Victim-selest data in a n-*bit* interconnect system when $k = 2$.

Victim location	Victim-select data
VAAVAA...VAA	100100...100
AVAAVAA...VA	0100100...10
AAVAAVAA...V	00100100...1

The generic behavior of test pattern generation and applying procedure is shown in Figure 12.10. This behavior will be executed by a combination of automatic test equipment (ATE) and the TAP controller. The first seed is applied to the new BSCs as an initial value into FF_2. The victim and aggressors are selected with victim-select data scanned into FF_1 and then the cells are set in SI mode to start generating test patterns. After generating test vectors and applying them to the interconnects, a new victim is selected and the process will be repeated with the same initial value. Note carefully that at the end of testing one victim (line 8) again the same seed would be in FF_2 (See also Figure 12.7 for an example). Therefore, there is no need to scan in the seed for the next victim. This significantly reduces the overall test time. The same process will be repeated for all seeds.

•Comment on External Test Patterns:

The PGBSC is implemented based on slight modification of the boundary scan cell. The MT test patterns are generated and applied by the PGBSC. While the PGBSC cell supports the MT model, they can be also efficiently used for external (deterministic or weighted pseudorandom) patterns. These patterns are applied by the external tester through TDI input pin. More specifically, the PGBSC cells provide extra control and flexibility to scan in patterns and apply them to the interconnects and even mix the MT model with other fault models. The flexibility of the PGBSC cell can be further explored for various tradeoffs. For example, a simple yet efficient compression

```
for (i= 1 to N_S)
{
    Scan seed i into FF_2
    Activate signal integrity test mode (SI=1)
    Scan the first victim-select data
    For (j=1 to k)   //Total # of shifts for victim-select data
    {
        Apply 4 UpdateDRs.          // Pattern generation
        Shift one '0' into FF1      // Selecting new victim
    }
}
```

Fig. 12.10. Test pattern generation procedure using PGBSC [32].

technique is explained for the enhanced boundary scan architecture using the PGBSC cells to minimize delivery time in [23].

Observation BSC (OBSC)

A new BSC is proposed at the receiving side of the interconnects which can employ any integrity loss sensor (ILS) such as the one presented in [23]. Figure 12.11(a) shows the new BSC named observation BSC (OBSC). As shown, the ILS cell added to the receiving cells, captures signals with noise and delay at the end of an interconnect. When it receives a signal with integrity problem (noise or skew violation) it produces a pulse at its output and the FF is set to '1'. The ILS is activated by the signal cell enable (*CE = '1'*). If *CE = '0'*, the ILS is disabled but the captured data in its flip-flop remain unchanged. The OBSC operates in two modes as summarized in Figure 12.11(b).

1. **ILS mode:** ILS flip-flop is selected. In this case, the captured ILS data is scanned out every *Shift-DR* state through the scan chain for final evaluation.

2. **Non-integrity mode:** In this mode, ILS is isolated and each OBSC acts as a conventional (non-integrity) BSC.

In the SI test mode, as Figure 12.11 shows, ILS FF can be read and scanned out for final evaluation. Before starting the scan out process, the content of the ILS FF need to be transferred to FF1. In this case, *sel* should be zero. Therefore, SI and *ShiftDR* signals should be '1' and '0', respectively. When the scanning out process starts, D_1 is transferred to Q_1 to be used as a TDI for the next cell. After sending the value of ILS FF to the Q_1, the scan chain must be formed. In this case, during the *Shift-DR* state the TDI input must be connected to the FF1. Therefore, the ILS path should be isolated by *sel*='1' (*SI*='1' and *ShiftDR*='1').

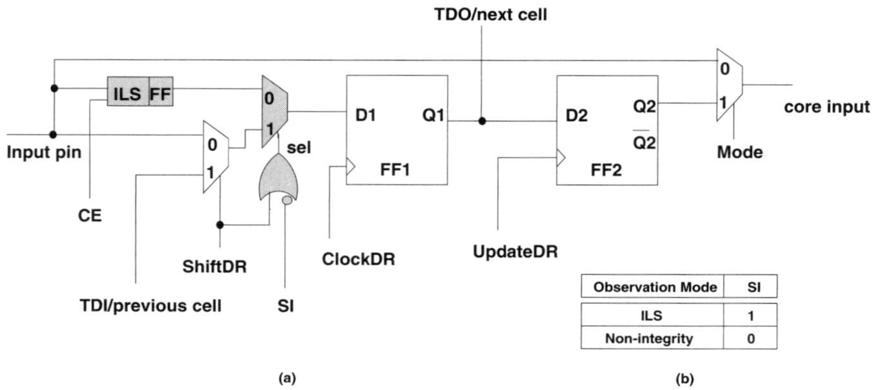

Fig. 12.11. OBSC structure [32].

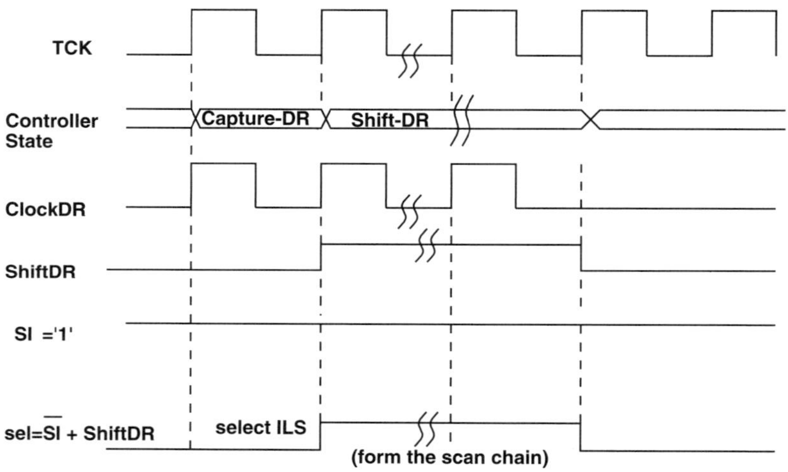

Fig. 12.12. Operation of the OBSC [32].

As shown in Figure 12.11, $\overline{SI}$ and *ShiftDR* are ORed together for selecting the ILS path for transferring the ILS FF to D_1 and scanning out the data in the chain. Figure 12.12 shows the dependency of *sel* to the *SI* and *ShiftDR*. As shown, in *Capture-DR* state, ILS FF is selected and then in *Shift-DR* state scan chain is formed and data is scanned out depending on how many wires are under test. Table 12.2 shows the truth table of signal *sel*. Additional control signals (i.e. *SI* and *CE*) are generated by a new instruction, to be explained in Section 12.2.2. There are three methods of observation:

1. **Method 1:** To capture and scan out the ILS data only once after applying the entire test pattern set and covering all victims.

Table 12.2. Truth Table of signal *sel* [32].

SI	*ShiftDR*	*sel*
1	0	0
1	1	1
0	x	1

2. **Method 2:** To capture and scan out the ILS data N_S times, after the application of a group of patterns (e.g. the patterns generated based on a specified seed) for each victim.

3. **Method 3:** To capture and scan out the ILS data right after applying each test pattern.

The first method has the least test time and a disadvantage of not being able to determine which transitions have caused the fault in an interconnect. The second method provides more information, to determine which set of transitions or faults caused the violation in the interconnects, at the expense of more test time. Finally, the third method is the most informative one, but it is extremely time consuming. In Section 12.2.3 the experimentation and comparison results of these methods will be reported.

12.2.2 Test Architecture

Figure 12.13 shows the overall test architecture with n interconnects between Core i and Core j in a two-core SoC. The five standard JTAG interface (TDI, TCK, TMS, $TRST$ and TDO) are still used without any modification. The Test Access Port (TAP) controller and instruction register provide various control signals to the boundary scan cells (Figure 12.6). The mandatory instruction PRELOAD is initially used to scan data into the boundary scan cells. Two new instructions are defined to be used for signal integrity test, one to activate the PGBSCs to generate test patterns and the other for reading out the test results. As shown in Figure 12.13, the cells at the output pins of *Core i* are changed to the PGBSCs and the cells at the input pins of *Core j* are changed to the OBSCs. The remaining cells are standard BSCs which are present in the scan chain during the signal integrity test mode. In case of bidirectional interconnects, the boundary scan cells used at both ends is a combination of the PGBSC and OBSC to test the interconnects in both directions.

The ILS in the OBSCs does not need any special control and automatically capture occurrence of integrity losses. After all patterns are generated and applied, the signal integrity information stored in the ILS FF is scanned out to determine which interconnect has a problem. In Subsection 12.2.1, various methods have been presented to readout the information. For example,

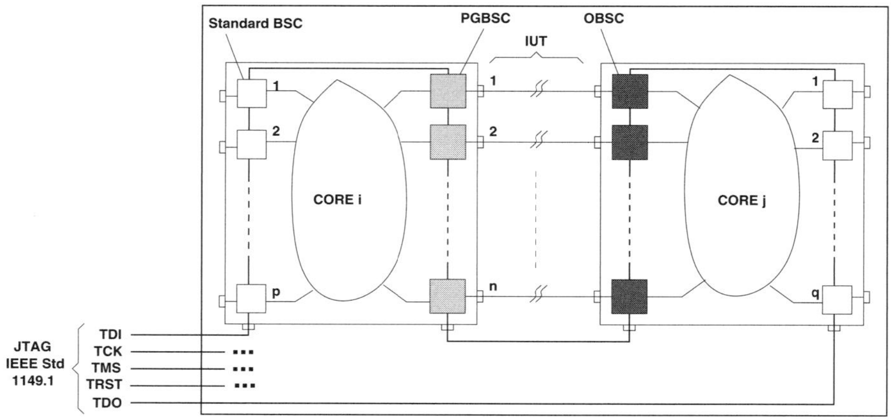

Fig. 12.13. Test Architecture [32].

method 1 can be used to minimize the overall test application time since the information in the cells is scanned out once per interconnect as opposed to once per pattern.

Clock Analysis

In this section, an analysis is presented for the required clocks to apply the test patterns and read out the test results for different methods. Assume n, p and q are the number of interconnects under test, the number of boundary scan cells in the scan chain, before the first PGBSC and after the last OBSC, respectively (see Figure 12.13).

In the conventional BSC, test patterns are scanned in and applied to the interconnects one-by-one. In a n-wire interconnect network, the number of required clocks to scan and apply the MT test patterns by conventional boundary scan architecture through TDI is $N_{clk\_BS} = m \cdot N_P(n+4)+p$, where N_p is the total number of the MT-patterns. When the PGBSC is used, only seeds are scanned into the cells and the other test patterns are generated internally. Having n interconnects, the required clocks in the enhanced boundary scan method is $N_{clk\_EBS} = N_S(2n+8k)+p$, where k is the locality factor defined in Section 12.2. This formula essentially counts clocks when the architecture executes the test generation algorithm shown in Figure 12.10. Tables 12.3 and 12.4 summarize the exact number of the required clocks for test generation/application and readout, respectively. Comparing conventional and enhanced boundary scan, the test application time reduction (TR) will be:

$$TR\% = \frac{N_{clk\_BS} - N_{clk\_EBS}}{N_{clk\_BS}} \cdot 100\%$$

Table 12.3. Pattern generation and application clock analysis [32].

Test Architecture	N_{clk}
Conventional BSC	$m \cdot N_P(n+4)+p$
Enhanced BSC	$N_S(2n+8k)+p$

Table 12.4. Clock analysis of the readout methods [32].

Readout Methods	N_{clk}
Method 1	n+2q
Method 2	$N_S \cdot n$+2q
Method 3	$N_P(n$-1)/2+2q

New Test Instructions

IEEE 1149.1 allows user to define some optional instructions [8]. Two new instructions G-SITEST and O-SITEST are added to the IEEE 1149.1 instruction set for the proposed architecture. The new instructions are only used in signal integrity mode for testing the interconnects.

•G-SITEST Instruction:

This instruction is used for test pattern generation using the enhanced boundary scan architecture. Seeds typically would be loaded onto the boundary scan cells by using the PRELOAD instruction before loading the G-SITEST instruction. Thus, when the G-SITEST instruction takes place in the *Update-IR* controller state, the related signals will be activated. The G-SITEST targets the PGBSCs and enables *SI*=1 during execution of instruction. It also enables the ILS cells (i.e. *CE*=1) to capture the signal integrity information. The victim-select data is then shifted into FF1 of the PGBSCs during the *Shift-DR* state and the patterns for the MT fault model are generated every *Update-DR* state as explained in Section 12.2.1. Three *UpdateDRs* are required to generate three test patterns per victim line for each initial value. Each boundary scan cell at the output of the core whose interconnect is under test will behave according to the G-SITEST instruction. The flow of data through the PGBSC is shown in Figure 12.14. *Core i* executes the PRELOAD instruction and the cores before *Core i* execute the BYPASS instruction to scan in the initial data (seeds) from the system input. The seeds are scanned in (broken lines in Figure 12.14) while other necessary changes to generate patterns are done internally using the additional components (shaded components in Figure 12.14) in the PGBSC cell.

•O-SITEST Instruction:

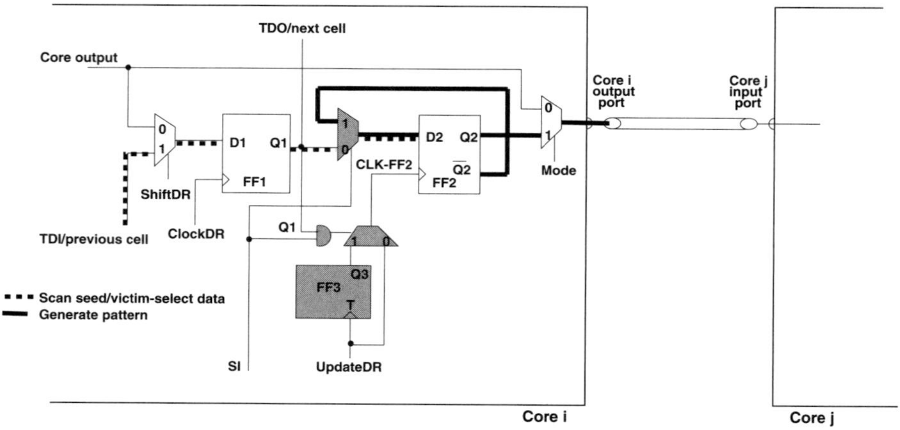

Fig. 12.14. Test data flow while the G-SITEST instruction is executed [32].

This instruction is loaded after execution of the G-SITEST instruction and when the loss information has been stored in the ILS FF. Assuming all test patterns have been applied to the interconnects, the O-SITEST instruction is used to capture and scan out the content of the ILS FF. After the instruction is decoded in the *Update-IR* controller state, the control signals *SI*=1 and *CE*=0 (to deactivate ILS) are generated. With *CE*=0, the ILS will not receive input data anymore as the O-SITEST instruction is executed. The flow of data through the OBSC cell is shown in Figure 12.15. All cores after *Core j* execute the BYPASS instruction to scan out the data to the system output.

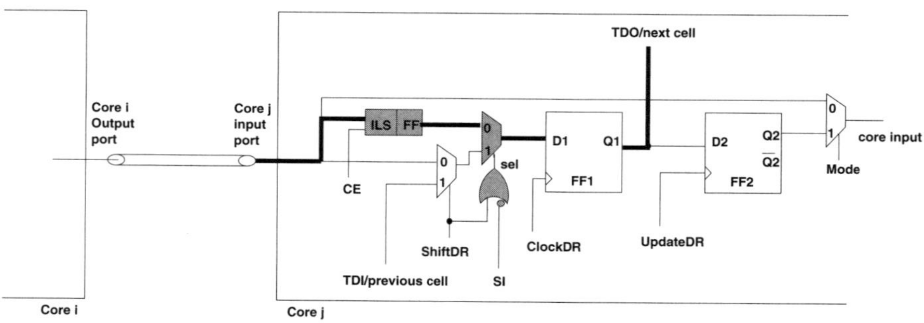

Fig. 12.15. Test data flow while the O-SITEST instruction is selected [32].

Having two instructions in the integrity test mode is necessary in enhanced boundary scan architecture. The G-SITEST enables the PGBSC to generate and apply test patterns and simultaneously, the OBSC captures the signals at the receiving-end of interconnects. After test application is performed, reading out the integrity information is begun by using the O-SITEST instruction. These two different operations cannot be done by only one instruction because FF1 is reused for different purposes. Specifically, the content of the ILS FF which is captured into FF1 and then shifted to the next cells FF1 will be overwritten by the next data captured by ILS FF cells during the *Capture-DR* state in the next test pattern application cycle.

Instruction Register

The instruction register allows an instruction to be shifted in. The instruction can be used to select the test to be performed or the test data register to be accessed or both. The instruction shifted into the instruction register will be latched such that changes of an instruction occur only in the *Update-IR* and *Test-Logic-Reset* controller states. Mandatory (BYPASS, SAMPLE, PRELOAD, and EXTEST) and optional (IDCODE, INTEST, and RUNBIST) instructions have already been defined in the IEEE Standard 1149.1. Here, two new optional instructions are added (G-SITEST and O-SITEST) for use in signal integrity mode. In general, the instructions coding, the size of the instruction register and implementation of the TAP controller may be slightly changed. In general, in the enhanced architecture the TAP states do not change but more control signals are generated in some states when the new instructions are executed.

Test Algorithm

Figure 12.16 summarizes the test process and instructions used in the signal integrity mode. First, the seed is scanned into the boundary scan cells by using the PRELOAD instruction. By loading the *G-SITEST* instruction into the instruction register, the new BSCs which target signal integrity test for the interconnects are set in signal integrity mode. *Update-IR* makes the G-SITEST instruction operational and sets $SI=1$ and $CE=1$. Then, test patterns are generated by the PGBSCs and applied to the interconnects and simultaneously ILS captures the signals at the end of interconnects and detect the violations if any. After test application process, the integrity loss data is stored in the ILS FF and must be read out. This is done using the O-SITEST instruction. The instruction first deactivates the ILS because during the scan out operation some new data will be scanned in and it may be applied to the interconnects in the *Update-DR* controller state before the scan-out phase is complete. This may cause ILS to lose integrity loss data stored in the previous cycle. Finally, the scanning out process is performed as explained in Section 12.2.2.

```
RESET
FOR (j=1 to N_S)
{
    PRELOAD seed j    // Initialize the PGBSCs
    G-SITEST   //Pattern generation and application
    FOR each victim line
        Apply 3 UpdateDRs
}
O-SITEST   //Reading out the integrity loss data
```

Fig. 12.16. Signal integrity test process.

12.2.3 Implementation and Simulation Results

ILS Cell Used in This Work

While any ILS sensor, such as those presented in [20] [18], can be used for the purpose of integrity loss detection, for the purpose of simplicity, cost and experimentation, a new ILS cell has been developed. In what follows, the functionality and the characteristics of this cell are briefly explained.

The new ILS is a delay violation sensor shown in Figure 12.17. The cell consists of two parts, the sensor and the detector. The acceptable delay region (ADR) is defined as the time interval from the triggering clock edge during which all output transitions must occur. The test clock is used to create a window which determines the acceptable skew region. The signal input a is in the acceptable delay period if its transition occurs during the period when b is at logic '0'. Any transition that occurs during the period when b is at logic '1' is passed through the transmission gates to the detector, XNOR gate. The XNOR gate is implemented using dynamic precharged logic. The output c becomes 1 when a signal transition occurs during $b = 1$ and remains unchanged till $b = 0$, the next precharge cycle. The output c is used to trigger a flip flop.

Figure 12.18 shows the timing behavior of the different input signals to the cell. The transmission gates are closed when signal $b = 0$ and their outputs hold to the previous values. The output of the ILS cell is determined by the timing of signals a and X. Three different cases for transition on signals a and X are illustrated as follows,

1. **Case 1:** The transition on signals a and X occurs during $b = 0$, i.e., inside the period t_{Inv1}. The transmission gate outputs change simultaneously and the XNOR gate output c is at logic '0'.

2. **Case 2:** Signal a makes a transition when $b = 0$ and X changes when $b = 1$. The transmission gate outputs have a delay difference of $(t_a + t_{Inv2}) - (t_{nand} + t_{Inv1})$. Signal c changes to '1' when this delay difference is greater or equal to XNOR gate delay. Equation 1 determines the minimum delay of signal a from the clock edge detected by the cell,

i.e, the ADR period.

3. **Case 3:** The transition on signals a and X occurs during $b = 1$, i.e., outside the period t_{Inv1}. The transmission gate outputs have a delay difference equal to t_{Inv2} and the XNOR gate output c changes to logic '1'.

$$ADR = t_{a,min} = t_{XNOR} + t_{Inv1} + t_{nand} - t_{Inv2} \quad (1)$$

Equation 1 has two parameters t_{Inv1} and t_{Inv2} which can be adjusted to tune the cell to detect the required delay. The two parameters are independent of each other. The minimum detectable delay ($t_{a,min}$) can be decreased by increasing t_{Inv2} or decreasing t_{Inv1}. t_{Inv2} can be decreased until the duration is enough to precharge the dynamic logic. Table 12.5 shows the minimum delay detectable by changing the number of cascaded inverters with delay of t_{NOT} each to form Inverter 1 and 2. Note that depending on the application and the threshold for delay and noise, the ILS cells may need to be retuned for different designs. Fortunately, most of sensors [18] [21] are easily tunable.

Figure 12.19 shows the SPICE [39] simulation of the cell, implemented using 0.18 μm technology, for two transitions of signal at input a. The first transition of the signal occurs at 0.2 ns when $b = 0$ and the output remains zero. The second transition occurs at 3.5 ns when $b = 1$ and exceeds the acceptable delay period and the output c goes to 1 till b goes to zero. This delay sensor can detect any transition faults (e.g. due to crosstalk) that result in excessive delay. The pulse produced on c can be fed to a flip-flop to store delay occurrence for further readout/analysis.

Table 12.5. Minimum detectable delay of ILS cell [32].

Inverter1 (t_{Inv1}) [$\times t_{NOT}$]	*Inverter2* (t_{Inv2}) [$\times t_{NOT}$]	*Minimum Detectable Delay* (*ADR*) [ps]
5	1	450
3	3	250
1	5	100

Simulation Results

The enhanced boundary scan cell and architecture is implemented using Synopsys synthesizer [40]. The total area overhead for the conventional BSC cell and the enhanced BSC cell (including ILS) is shown in Table 12.10. The enhanced cells are 28-46% more expensive compared to the conventional one. Considering the overall cost of boundary scan architecture (cells, controller,

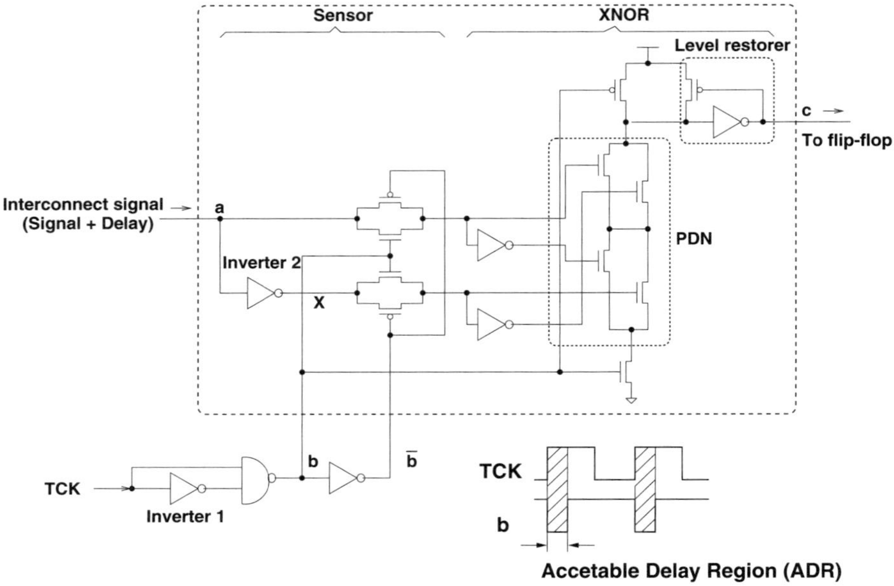

Fig. 12.17. Integrity loss sensor (ILS) cell [32].

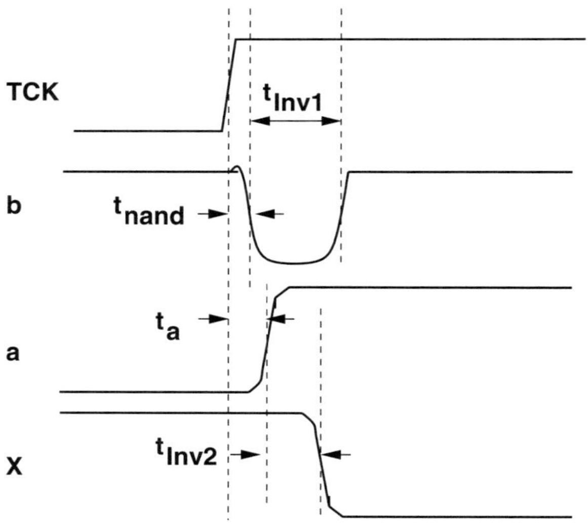

Fig. 12.18. ILS signals and their timing behavior [32].

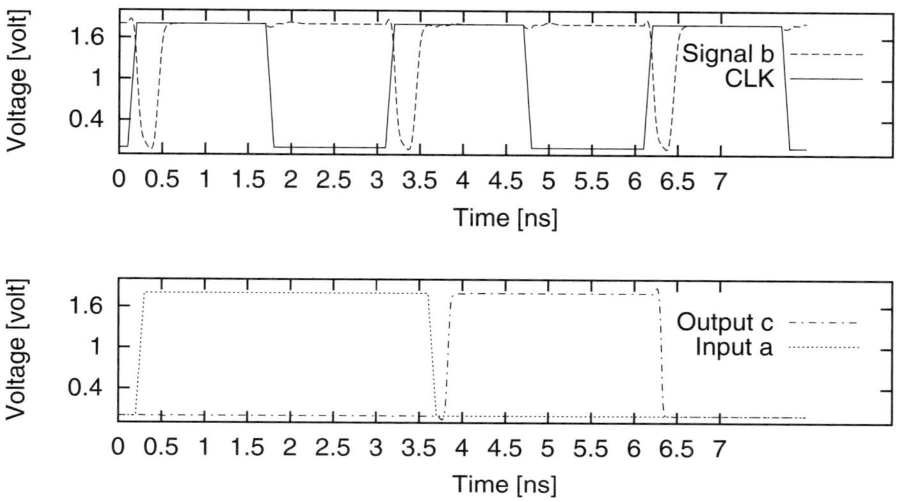

Fig. 12.19. Spice simulation of the ILS cell [32].

etc.) additional overhead of components is still negligible (less than 5% of overall cost). Moreover, to lower the cost user still has option of inserting/changing cells on only interconnects of interest (e.g. those that are susceptible to noise or those that feed glitch/delay sensitive circuits).

Table 12.6. Cost analysis for boundary scan cells.

Test Architecture	*Cost[NAND]*		
	Sending	*Observing*	*Bidirectional*
Conventional Cells	26	26	78
Enhanced Cells	36	38	100
Area Increase %	38.5	46.2	28.2

Table 12.7 shows the comparison between using enhanced boundary scan to generate and apply the MT-patterns and conventional boundary scan to scan in and apply the same set of patterns. The application time reduction ratio is between 86 to 92% for various scenarios. Table 12.11 compares three methods described in Subsection 12.2.1 for different number of interconnects under test (n) and locality factor (k) assuming q=0. As expected the number of clocks required for Methods 1 is significantly lower than Method 3. However, Method 3 provides much more information about type and location of the integrity faults. In method 2, one scan-out operation has been performed per

Table 12.7. MT-pattern application time [32].

Methods	*MT-Pattern Application Time [Cycle] (p=0)*					
	n=8		*n=16*		*n=32*	
	k=2	*k=3*	*k=2*	*k=3*	*k=2*	*k=3*
N_{clk_EBS}	2560	17920	3840	25088	6400	39424
N_{clk_BS}	19200	150528	32000	250880	57600	451584
TR%	86.1%	88.5%	88.3%	90.3%	88.9%	91.8%

victim line. Essentially, method 2 can be used to tradeoff test time versus accuracy.

Table 12.8. Observation test time for three methods [32].

Methods	*Observation Test Time [Cycle] (q=0)*					
	n=8		*n=16*		*n=32*	
	k=2	*k=3*	*k=2*	*k=3*	*k=2*	*k=3*
Method 1	8	8	16	16	32	32
Method 2	640	3584	1280	7168	2560	14336
Method 3	2560	14336	5120	28672	10240	57344

Table 12.9 compares the maximum noise voltage and skew between the MT and the MA fault models for different k's. For k=2, the MT and MA show almost the same maximum noise and delay. The simulation for k=3,4 shows that considerable increase in delay can be stimulated by the MT model compared to the MA model while the peak noise is slightly more.

Table 12.9. MT and MA Comparison [32].

k	*MT*		*MA*	
	$V_{noise}[V]$	*Delay[ps]*	$V_{noise}[V]$	*Delay[ps]*
k=2	0.351	470.5	0.348	468.5
k=3	0.408	472.3	0.404	450.7
k=4	0.420	483.4	0.411	440.9

12.3 Testing Interconnects Using MA Model

MA model assumes all aggressor line transition in the same direction. This model basically is simpler than MT model introduced earlier in this chapter. It also require less modification to the boundary scan cells and the same integrity loss sensor can be used in the observation bounday scan cell. In this section, the focus is on a comparession method to reduce overall test time. There is no modification required for the pattern generation cells to apply the patterns and the observation boundary scan cell (OBSC) described in the previous section can be used to catch signal integrity issues on the interconnects causing delay effects.

12.3.1 Test Data Compression

In conventional boundary scan architecture (BSA), test patterns are scanned in one-by-one and applied to the interconnects. For example in a n-bit interconnect using the maximum aggressor (MA) fault model, 12 test patterns are applied to each victim line and $12n$ clocks are required to apply the test patterns on only one victim line. With rotating victim line among n interconnects, the overall number of clock (test application time) is $12n^2$. Of course, MA is a simplistic model. Using more sophisticated models or having large number of interconnects in an SoC results in huge number of test patterns and thus, the compression becomes a necessity. In this section, a simple yet efficient compression technique is explained for the enhanced boundary scan architecture.

The key points in this approach is twofold. First, the method is a straightforward lossless compression that constructs the compressed bit stream by identifying the maximum similarity between two adjacent patterns and overlapping them. Second, since this compression is neither destructive nor reorders patterns, no additional decompressor hardware is needed. The decompression process is only performed by the automatic test equipment (ATE) by controlling the JTAG TMS control input.

When test patterns are generated, often plenty of don't-cares (up to 90% exist in the test pattern set. This is also true for patterns generated for signal integrity, especially if a locality metric (to limit exploring the pattern space) is considered. In any case, the assumption is that the test set consists of same-length patterns (l) that include don't cares. Figure 12.20 shows the basic idea of compressing two patterns V_i and V_j (of length $l = 16$) by overlapping their bits as much as possible by taking advantage of their don't cares. In this example, the compressed data (V_i, V_j) requires only 21 clocks to scan in as opposed to 16+16=32 clocks if no compression is applied. Note carefully that to decompress a given stream, there needs to be an identifier per pattern (e.g. d_i and d_j in the example) to be able to construct (decompress) the patterns. For purpose of boundary scan test, these identifiers are the number of shifts (i.e. clocks) required before updating the content of BSC cells. Here, the assumption is that the ATE stores the decompression data (d values such that $0 \leq d \leq l$) and during scanning in the bit stream, it activates TMS (Test Mode Select) signal after d number of clocks. TMS signal enables the TAP controller to generate appropriate controls for the signal integrity test

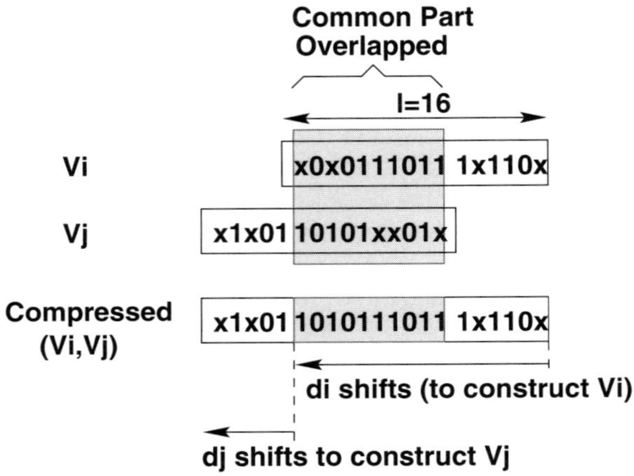

Fig. 12.20. Basic compression of two vectors (V_i, V_j).

(e.g. EX-SITEST). Therefore, there is no additional decompression hardware needed in the architecture.

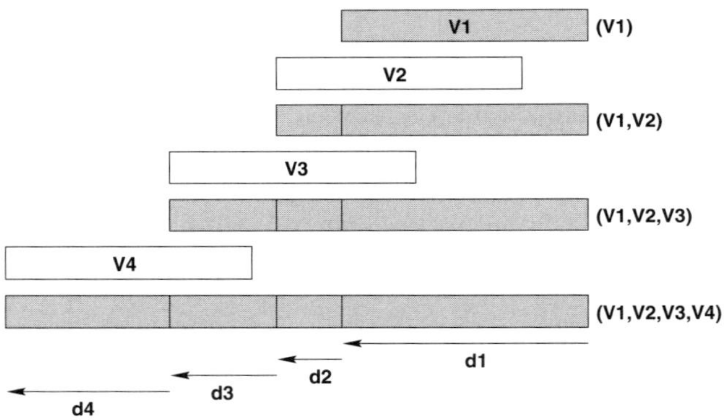

Fig. 12.21. The constructive compression technique.

The compression process, as explained before, continues by compressing one pattern at a time to the bit streams obtained up to that point. Figure 12.21 shows the concept of this constructive process for four patterns $V_1, \cdots, V_4$. The process is repeated until all of the patterns compressed. The decompression data (e.g. $d_1, \cdots, d_4$) are stored in the ATE and will be used to control the TMS signal in the test mode. When there are n patterns $(V_1, \cdots, V_n)$ each of length l, the total length of the compressed data (i.e. total number of clock cycles) needed to deliver them to the interconnects under test will be $\sum_{i=1}^{n} d_i$

which is expected to be much smaller than $l \cdot n$. Thus, for a given set of n patterns the compression rate will be $\frac{l \cdot n - \sum_{i=1}^{n} d_i}{l \cdot n}\%$.

12.3.2 EX-SITEST Instruction and Test Process

A new instruction is proposed to be added, i.e. *EX-SITEST*, to the IEEE 1149.1 instruction set for the new test architecture. This instruction is similar to the EXTEST instruction with an additional control signal, SI activated. In the *Update-IR* state, the instruction is decoded and ($SI = 1$) is generated. The output cells act as standard BSCs and the input cells act as OBSCs. The signal F is captured during the *Capture-DR* state and shifted out every clock cycle during the *Shift-DR* state. In this case, TAP controller sates will not change. Some changes are required in instruction decoding. Figure 12.22 shows the data flow of the EX-SITEST instruction between the cores.

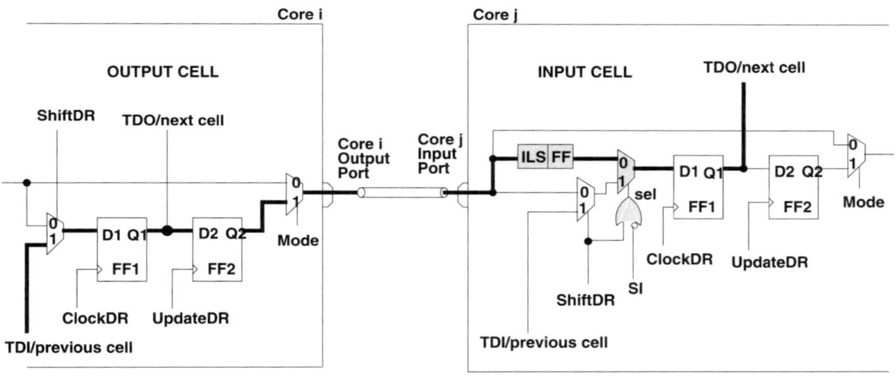

Fig. 12.22. Test data flow for EX-SITEST instruction.

The tap controller IR is loaded with EX-SITEST instruction. Then, all test patterns are applied to the interconnects and simultaneously ILS cells capture the signals at the end of interconnects and detect the violation if any. After test application process, the stored results in the ILS cell FFs must be read. One of the three methods mentioned in Section 12.2.1 can be used for the observation process. For example, using method 3, all the test patterns are applied and then the integrity information is read out once.

12.3.3 Results

The enhanced boundary scan cell and architecture is implemented by Synopsys synthesizer. The total area overhead for conventional BSA cell and

Table 12.10. Cost analysis for boundary scan cells.

Test Architecture	*Cost[NAND]*		
	Sending	*Observing*	*Bidirectional*
Conventional Cell	26	26	78
ILS Cell	26	45	97

enhanced BSA cell (ILS) is shown in Table 12.10. The ILS cell is almost twice expensive compared to the conventional one.

Table 12.11 shows a comparison between three methods described in Subsection 12.2.1. The table shows that the number of clocks required for methods 3 is significantly lower than method 1. However, method 1 provides much information about type and location of the integrity faults. In method 2, one scan-out operation will be performed per victim line. Method 2 can be used to tradeoff test time versus accuracy.

Table 12.11. Observation test time comparison.

Methods	*Total Test Time [Cycle]*		
	n=8	*n=16*	*n=32*
Method 1	768	3077	12288
Method 2	64	256	1024
Method 3	8	16	32

Table 12.12 summarizes the compression rate statistics for the proposed technique. In this table, the compression rate for various interconnects are reported. Without judging the quality of the patterns, for comparison reason three pattern sets have been used based on MA model (12 patterns per victim line), deterministic and pseudorandom. The compression rate (as defined in Section 12.3.1) reflects the application time reduction compared to conventional (uncompressed) method. The compression rate is higher for pseudorandom (with large number of don't cares) and is lower for MA (with no don't cares).

Table 12.12. Compression rate for different test pattern sets.

Application Method	*Compression Rate [%]*		
	n=8	*n=16*	*n=32*
MA	37.5	37.8	38.3
Deterministic	46.7	57.2	59.8
Pseudorandom	58.1	61.2	63.8

12.4 Summary

An enhanced boundary scan architecture is proposed for testing signal integrity in SoC interconnects. The enhanced architecture is utilized to generate and apply the MT- and MA-patterns. The proposed MT fault model is a superset of the MA model and is much more capable of testing the capacitive and inductive couplings among the interconnects. While the MA model is simpler with lower number of patterns and low area overhead on boundary scan cells. The proposed architecture detects integrity loss violations on the interconnects based on the widely-used JTAG boundary scan architecture. To do this, additional detector cell, modified scan cells and minor modifications to the TAP controller to handle two new instructions are needed. The advantage of the proposed architecture is that it provides cost effective solution for thorough testing of interconnects using the popular JTAG standard.

References

1. N. Weste and K. Eshraghian, *Principles of CMOS VLSI Design: A Systems Perspective*, Addisson-Wesley, 1993.
2. L. Green, "Simulation, Modeling and Understanding the Importance of Signal Integrity," *IEEE Circuit and Devices Magazine*, pp. 7-10, Nov. 1999.
3. S. Natarajan, M.A. Breuer, S.K. Gupta, "Process variations and their impact on circuit operation," in Proc. *IEEE Int. Symp. on Defect and Fault Tolerance in VLSI Systems*, pp. 73-81, 1998.
4. A. Sinha, S.K. Gupta and M.A. Breuer, "Validation and Test Issues Related to Noise Induced by Parasitic Inductances of VLSI Interconnects," in Proc. *IEEE Trans. on Advanced Packaging*, pp. 329-339, 2002.
5. X. Bai, S. Dey and J. Rajski, "Self-Test Methodology for At-Speed Test of Crosstalk in Chip Interconnects," in Proc. *Design Automation Conf. (DAC'00)*, pp. 619-624, 2000.
6. L. Chen, X. Bai and S. Dey, "Testing for Interconnect Crosstalk Defects Using On-Chip Embedded Processor Cores," in Proc. *Design Automation Conf. (DAC'01)*, pp. 317-322, 2001.
7. IEEE Standard 1149.1-2001, *"Standard Test Access Port and Boundary-Scan Architecture"*, IEEE Standards Board, 2001.
8. M. Bushnell, V. Agrawal, *Essentials of Electronics Testing*, Kluwer Publishers, 2000.
9. M. Cuviello, S. Dey, X. Bai and Y. Zhao, "Fault Modeling and Simulation for Crosstalk in System-on-Chip Interconnects," in Proc. *Intern. Conf. on Computer Aided Design (ICCAD'99)*, pp. 297-303, 1999.
10. Y. Zhao and S. Dey, "Analysis of Interconnect Crosstalk Defect Coverage of Test Sets," in Proc. *Int. Test Conf. (ITC'00)*, pp. 492-501, 2000.

11. Nagaraj NS, P. Balsara and C. Cantrell, "Crosstalk Noise Verification in Digital Designs with Interconnect Process Variations," in Proc. *Int. Conf. on VLSI Design*, pp. 365-370, 2001.
12. W. Chen, S. Gupta and M. Breuer, "Analytic Models for Crosstalk Delay and Pulse Analysis Under Non-Ideal Inputs," in Proc. *Int. Test Conf. (ITC'97)*, pp. 809-818, 1997.
13. S. Naffziger, "Design Methodologies for Interconnects in GHz+ ICs," Tutorial Lecture in *Int. Solid-State Conf.*, 1999.
14. Y. Cao, et. al, "Effective On-Chip Inductance Modeling for Multiple Signal Lines and Application to Repeater Insertion," *IEEE Transactions on Very Large Scale Integration (VLSI) Systems*, vol. 10, no. 6, pp. 799-805, Dec. 2002.
15. W. Chen, S. Gupta and M. Breuer, "Test Generation for Crosstalk-Induced Delay in Integrated Circuits," in Proc. *Int. Test Conf. (ITC'99)*, pp. 191-200, 1999.
16. W. Chen, S. Gupta and M. Breuer, "Test Generation in VLSI Circuits for Crosstalk Noise," in Proc. *Intern. Test Conf. (ITC'98)*, pp. 641-650, 1998.
17. Y. Zhao, L. Chen and S. Dey, "On-line Testing of Multi-source Noise-induced Errors on the Interconnects and Buses of System-on-Chips," in Proc. *Int. Test Conf. (ITC'02)*, pp. 491-499, 2002.
18. S. Yang, C. Papachristou, and M. Tabib-Azar, "Improving Bus Test Via I_{DDT} and Boundary Scan," in Proc. *Design Automation Conf. (DAC'01)*, pp. 307-312, 2001.
19. International Technology Roadmap for Semiconductors 2001 (http://public.itrs.net).
20. I. Rayane, J. Velasco-Medina and M. Nicolaidis, "A Digital BIST for Operational Amplifiers Embedded in Mixed-Signal Circuits," in *Proc. VLSI Test Symp. (VTS'99)*, pp. 304-310, 1999.
21. S. Tabatabaei and A. Ivanov, "An Embedded Core for Sub-Picosecond Timing Measurements," in Proc. *Int. Test Conf. (ITC'02)*, pp. 129-137, Oct. 2002.
22. F. Caignet, S. Delmas-Bendhia and E. Sicard, "The Challenge of Signal Integrity in Deep-Submicrometer CMOS Technology," in Proc. of the IEEE, vol. 89, no. 4, pp. 556-573, April 2001.
23. M. H. Tehranipour, N. Ahmed and M. Nourani, "Testing SoC Interconnects for Signal Integrity Using Boundary Scan", in Proc. *VLSI Test Symposium (VTS'03)*, pp. 163-168, 2003.
24. C. Chiang and S. K. Gupta, "BIST TPGs for Faults in Board Level Interconnect via Boundary Scan", in Proc. *VLSI Test Symposium (VTS'97)*, pp. 376-382, 1997.
25. K. Lofstrom, "Early Capture for Boundary Scan Timing Measurement", *Proc. ITC,* pp. 417-422, 1996.
26. J. Shin, H. Kim and S. Kang, "At-Speed Boundary-Scan Interconnect Testing in a Board with Multiple System Clocks", in Proc. *Design, Automation and Test in Europe (DATE'99)*, pp. 473-477, 1999.

27. IEEE P1500 standard, *http://grouper.ieee.org/groups/1500/*.
28. IEEE 1149.4 standard, *http://grouper.ieee.org/groups/1149/4/*.
29. L. Whetsel, "Proposal to Simplify Development of a Mixed Signal Test Standard," in Proc. *Int. Test Conf. (ITC'96)*, pp. 400-409, 1996.
30. IEEE 1149.6 Working Group, *http://grouper.ieee.org/groups/1149/6/*, 2002.
31. N. Ahmed, M. H. Tehranipour and M. Nourani, "Extending JTAG for Testing Signal Integrity in SoCs", in Proc. *Design, Automation and Test in Europe (DATE'03)*, pp. 218-223, 2003.
32. M. Tehranipour, N. Ahmed and M. Nourani, "Testing SoC Interconnects for Signal Integrity Using Extended JTAG Architecture," *IEEE Transactions on CAD*, vol. 23, issue 5, pp. 800-811, May 2004.
33. J. Wakerly, *Digital Design, Principles and Practices*, Prentice Hall, 2000.
34. S. Naffziger, "Design Methodologies for Interconnect in GHz+ ICs," Tutorial, ISSCC 1999.
35. S. Kajihara, K. Taniguchi, I. Pomeranz and S. M. Reddy, "Test Compression using Don't care Identification and Statistical Encoding," in Proc. *IEEE Int. Workshop on Electronic Design, Test and Application (DELTA'02)*, pp. 413-416, 2002.
36. A. Chandra and K. Chakrabarty, "Test Data Compression for System-on-Chip Using Golomb Codes," in Proc. *VLSI Test Symp. (VTS'00)*, pp. 113-120, 2000.
37. S. Wang and S. Chiou, "Generating Efficient Tests for Continuous Scan," in Proc. *Design Automation Conf. (DAC'01)*, pp. 162-165, 2001.
38. OEA International, Inc., http://www.oea.com.
39. TI-SPICE3 User's and reference manual, *1994 Texas Instrument Incorporation*, 1994.
40. Synopsys Design Analyzer, "User Manual for SYNOPSYS Toolset Version 2000.05-1," Synopsys, Inc., 2000.
41. M. Celik, L. Pileggi and A. Odabasioglu, *IC Interconnect Analysis*, Kluwer Publishers, 2002.

Index

Printed in the United States of America